In this series original contributions of mathematical research in all fields are contained, such as

— research monographs
— collections of papers to a single topic
— reports on congresses of exceptional interest for mathematical research.

This series is aimed at promoting quick information and communication between mathematicians of the various special branches.

In diese Reihe werden Originalbeiträge zu allen Gebieten der mathematischen Forschung aufgenommen wie

— Forschungsmonographien
— Sammlungen von Arbeiten zu einem speziellen Thema
— Berichte von Tagungen, die für die mathematische Forschung besonders aktuell sind.

Die Reihe soll die schnelle Information und gute Kommunikation zwischen den Mathematikern der verschiedenen Fachgebiete fördern.

Manuscripts
in English and German comprising at least 100 pages and not more than 500 pages can be admitted to this series. With respect to a quick publication the manuscripts are reproduced photomechanically. Authors who are interested in this series please turn directly to the 'Akademie-Verlag'. Here you will get more detailed information about the form of the manuscripts and the modalities of publication.

Manuskripte
in englischer und deutscher Sprache, die mindestens 100 Seiten und nicht mehr als 500 Seiten umfassen, können in diese Reihe aufgenommen werden. Im Interesse einer schnellen Publikation werden die Manuskripte auf fotomechanischem Weg reproduziert. Autoren, die an der Veröffentlichung entsprechender Arbeiten in dieser Reihe interessiert sind, wenden sich bitte direkt an den Akademie-Verlag. Sie erhalten dort genauere Informationen über die Gestaltung der Manuskripte und die Modalitäten der Veröffentlichung.

B. Heinrich

Finite Difference Methods on Irregular Networks

Mathematical Research **Mathematische Forschung**

Wissenschaftliche Beiträge
herausgegeben von der
Akademie der Wissenschaften der DDR
Karl-Weierstraß-Institut für Mathematik

Band 33

Finite Difference Methods on Irregular Networks

bv
B. Heinrich

Finite Difference Methods on Irregular Networks

A Generalized Approach to Second Order Elliptic Problems

by

Bernd Heinrich

Akademie-Verlag Berlin 1987

Autor:

Dr. Bernd Heinrich
Technische Hochschule Karl-Marx-Stadt
Sektion Mathematik
PF 964
Karl-Marx-Stadt
9010

Die Titel dieser Reihe werden vom Originalmanuskript der Autoren reproduziert.

ISSN 0138-3019
ISBN 3-05-500328-4

Erschienen im Akademie-Verlag Berlin,
Leipziger Straße 3-4, DDR 1086 Berlin

Lizenznummer: 202 · 100/410/86
Printed in the German Democratic Republic
Gesamtherstellung: VEB Kongreß- und Werbedruck, 9273 Oberlungwitz
Lektor: Dr. Reinhard Höppner
LSV 1065
Bestellnummer: 763 674 3 (2182/33)
02800

PREFACE

The finite difference and finite element methods are powerful tools for the approximate solution of differential equations governing diverse physical phenomena, and there is extensive literature on these discretization methods. In the last two decades, some extensions of the finite difference method to irregular networks have been described and applied to solving boundary value problems in science and engineering. For instance, "box integration methods" have been widely used in electronics. There are several papers on this topic, but a comprehensive study of these methods does not seem to have been attempted.

The purpose of this book is to provide a systematic treatment of a generalized finite difference method on irregular networks for solving numerically elliptic boundary value problems. Thus, several disadvantages of the classical finite difference method can be removed, irregular networks of triangles known from the finite element method can be applied, and advantageous properties of the finite difference approximations will be obtained.

The book is written for advanced undergraduates and graduates in the area of numerical analysis as well as for mathematically inclined workers in engineering and science. In preparing the material for this book, the author has greatly benefited from discussions and collaboration with many colleagues who are concerned with finite difference or (and) finite element methods.

Thanks are due to many colleagues for their interest and useful discussions on the subject of this monograph. In particular, the author wishes to thank Acad. Prof. A.A. Samarskiĭ and Prof. F. Kuhnert for encouragement and support, Dr. U. Langer for helpful hints and reading the manuscript, Dr. B. Legler for improving the English, and Mrs. I. Scholz for typing the manuscript.

The author welcomes in advance comments and critical remarks.

Karl-Marx-Stadt, April 1986 Bernd Heinrich

CONTENTS

1. INTRODUCTION

1.1. Preliminary remarks. This monograph is devoted to the study of finite difference methods on irregular networks for the numerical solution of boundary value problems (BVPs) $Au = F$ with second order elliptic differential equations $Lu = f$. Problems of this type describe diffusion and heat conduction phenomena, sometimes involving convection, and other field problems which are of interest in physics and technology. Thus, we should mention the torsion of elastic prismatic bars, the transverse deflection of membranes under a lateral load, the distribution of the potential in irrational ideal fluid flow or fluid flow in porous media, neutron diffusion in nuclears reactors, and electromagnetic field distributions as well as electrostatic potential distributions, where the last two field problems can often be encountered in the modelling of electric motors and semiconductor devices, respectively.

The finite difference method (FDM) and the finite element method (FEM) are the commonest methods for solving such elliptic problems, and they are widely used in practice. Today, the basic problems of numerical analysis of the FEM are solved. It is well known that the FEM has some valuable advantages, e.g.

- the flexible approximation of the given domain Ω by irregular networks (triangulations),
- the preservation of symmetry, if symmetric operators are discretized,
- the simple, "nondifferential" approximation of "natural" boundary conditions, i.e. boundary conditions with the conormal derivative,
- the convergence of the approximate FEM-solution to the generalized solution of the corresponding BVP $Au = F$,
- the applicability of Hilbert space methods to the mathematical justification of the FEM.

But there are also aspects of the FDM which seem to be useful in practice, e.g.

- the simple construction (derivation) and handling of finite difference approximations,
- the local and illustrative modelling of the discretization in the neighbourhood of a grid point,
- the universality of the FDM with respect to various types of differential equations.

Nevertheless, there are some disadvantages of the "classical" FDM, which are caused by the approach to discretizing the BVP in its classical formulation. Thus, they are visualized in the rigidity of the networks, in the explicit approximation of natural boundary conditions by difference quotients, in the loss of symmetry (not always) after discretizing symmetric problems, and in the reference to classical solutions $u \in C^k(\bar{\Omega})$ only, often for $k \in \{2,3\}$.

Notwithstanding all that, the FDM is frequently used by engineers and scientists, which fact will become evident from a glance at the literature. This statement is also supported by the trend of using personal computers where discretization methods which are easy to handle seem to be preferred.
For some time, several attempts to improve the classical FDM have been made and described in the literature, as e.g. nonclassical approaches to the FDM via generalized formulations of the BVP Au = F:
- using the variational equation of the BVP, see e.g. ASTRAKHANTSEV [1], FREHSE [1], GIRAULT [1,2], LADYZHENSKAYA [1], SAMARSKIǏ/ANDREEV [1], TEMAM [1,2],
- using the minimum problem associated with the BVP, cf. CONCUS [1], MICHLIN/SMOLIZKI [1], SAMARSKIǏ/ANDREEV [1], SCHWARZ et al. [1], STEPLEMAN [1], VARGA [1].

It seems that advantages and disadvantages of the discretization methods essentially depend on the formulation of the BVP employed for the discretization procedure. For instance, if a weak variational formulation for elliptic problems of second order will be taken, then derivatives of first order only must be approximated, irregular networks can be used, and it is not necessary to approximate explicitly the conormal derivative on the boundary of the given domain. But here, some algorithmic advantages known from the classical FDM cannot be taken for granted any longer.

In the last two decades, in the literature pertaining to science and technology another approach to the FDM has been described, which preserves the algorithmic advantages of local modelling and simple handling of the classical FDM and which eliminates essential disadvantages of the classical FDM. This approach has been given different names, which inevitably reflect the physical background of this method in field problems, as e.g. "box integration method", "box method", "balance method" or "finite control volume method". This method, which we agree to call "balance method" troughout this book, is presented in the literature on engineering as a discretization method which can be physically interpreted. Thus, the balance method is often applied to the numerical solution of field problems in electro- and magnetostatics and related areas, especially in electronics; cf. BANK et al. [1], GREENFIELD/DUTTON [1], HEIMEIER [1], McNEAL [1], POHL [1], REICHERT [1], SELBERHERR [1], SELBERHERR et al. [1], SLOTBOOM [1], WEILAND [1], WINSLOW [1], WOLFF/MÜLLER [1]. For the solution of problems in mechanics by the balance method, see e.g. BOOK [1], GOSMAN et al. [1], REISSMANN [1,2,3,4], REISSMANN/HAUG [1], or see BECK et al. [1] for heat conduction problems. The investigation of the balance method as a generalized FDM and as a subject treated from the mathematical point of view set in only relatively late, see e.g. FRYASINOV/MASLYANKINA [1], FRYASINOV [1,2,3,4], HARTWIG/WEINELT [1,2], HEINRICH [5,6,7,8,9,10], LAZAROV [1,2], MITCHELL/GRIFFITHS [1]. SAMARSKIJ [1], SAMARSKIǏ/ANDREEV [1], WEINELT [1,2],

VARGA [1], and there are a lot of unsolved mathematical problems which merit further study.

<u>1.2. Scope of monograph.</u> In the present monograph, the author intends to make some contributions to the further development of the FDM, especially, to the mathematical justification of the generalized (nonclassical) FDM on the basis of the balance method mentioned in Section 1.1. The presentation will be made on the basis of linear elliptic differential equations of second order and plane domains Ω, where the usual types of boundary conditions, inclusively mixed ones, and curvilinear boundaries $\partial\Omega$ are admitted. The main topics of this monograph are the following:

- the construction and explicit presentation of finite difference schemes (FDSs) for approximating elliptic problems of second order on irregular networks of triangles and (or) rectangles, i.e. on triangulations known from the FEM as well as from the classical FDM,
- the development of some tools of "discrete analysis", e.g., discrete analogues of the Friedrichs-Poincaré-inequality, of the trace theorem, of Green's formula, etc.,
- the study of the properties of the difference operators and the associated matrices, especially with respect to monotonicity (M-matrices), symmetry and positive definiteness,
- the derivation of a priori estimates for the difference operators acting on grid functions, where discrete analogues of the C-, W_2^1- and W_2^{-1}-norms are involved,
- the error estimation and convergence proofs for the FDM, where irregular networks of triangles and (or) rectangles as well as generalized solutions $u \in W_2^k(\Omega)$ $(k \geqslant 2)$ of the BVPs $Au = F$ are admitted.

The FDM considered here is a variant of generalized finite difference approaches and yields the following results, which overcome the essential disadvantages of the classical FDM:

- domains Ω with curvilinear boundary can be approximated by irregular or locally irregular networks consisting of triangles used in the FEM, and finite difference approximations can be derived on such networks,
- the construction of the FDSs $A_h y = F_h$ approximating the BVP $Au = F$ is algorithmically simple and consists in the local modelling of so-called "balance equations" on some "finite regions" $\mathcal{K}(x)$, which are associated with each grid point x and are called "boxes",
- the FDM considered here and called into existence by practising engineers and scientists can be justified mathematically; it is applicable to a wide range of elliptic problems and has an illustrative physical background: the balance of "physical sets" on the boxes $\mathcal{K}(x)$ which are subsets of the given domain Ω,
- the conormal derivative contained in natural boundary conditions can be approximated "nondifferentially" in a simple manner and in the same

way as in the FEM, i.e., it is superfluous to derive finite difference quotients approximating the conormal derivative,

- essential qualitative properties of the operator of the BVP can be preserved, as for instance, symmetry, positive definiteness and the property of being "of monotonic type" (maximum principles, inverse isotone operators),
- the proofs of existence, stability and convergence of solutions of the FDSs $A_h y = F_h$ can be carried out by means of methods in n-dimensional Hilbert spaces and spaces with partial ordering,
- the smoothness assumptions $u \in C^k(\bar{\Omega})$ $(k \geq 2)$ on the solutions u of the BVPs $Au = F$, which are often considered in the FDM, can be weakened to $u \in W_2^k(\Omega)$ $(k \geq 2)$ and error estimates and convergence proofs utilizing the Bramble-Hilbert Lemma can be given.

<u>1.3. Plan of monograph, comments.</u> The subject and the aim of this monograph as well as the results obtained are summarized in Chapter 1. Some problems, which are related with this work and are not included here or treated elsewhere, are also mentioned.

<u>Chapter 2:</u> In Section 2.1., the BVP $Au = F$ and the assumptions on A, u and F, inclusively on the domain Ω and its boundary $\partial\Omega$, will be formulated. The linear differential operator L associated with the BVP $Au = F$ is elliptic, of second order and, temporarily, formally selfadjoint. Mixed boundary conditions and corners on the boundary are always admissible, but with the implicit limitation of these difficulties by smoothness assumptions on the solution u of the type $u \in W_2^2(\Omega)$ in most cases. Irregular networks of triangles and (or) rectangles as well as locally irregular or regular networks, but also the corresponding grids $\bar{\omega}$, are described in Section 2.2. For such "primary networks", two sorts of "secondary networks" (Section 2.3.) will be introduced by means of two types of "boxes" $\mathfrak{X}(x)$: PB- and MD-boxes, which are defined by the <u>p</u>erpendicular <u>b</u>isectors and by the <u>med</u>ians of the triangles (rectangles) respectively. Chapter 2 contains a series of notations, definitions and assumptions, e.g. also grid regularity conditions, which are needed for the description and the correct definition of the primary and secondary networks as well as of the FDSs $A_h y = F_h$ derived subsequently. To grasp the symbolism, the reader should take a good look at the figures contained in Chapter 2.

<u>Chapter 3:</u> In Section 3.1., the balance method is described, and in Sections 3.2. and 3.3., this method will be used to derive FDSs $A_h y = F_h$ approximating the BVP $Au = F$ on irregular networks and for the two types of secondary networks introduced in Chapter 2, i.e. for PB- and MD-boxes. In comparison with earlier papers on the FDM via balance method, several generalizations will be treated, which are important for applications. Thus, mixed boundary conditions, curved boundaries $\partial\Omega$ of the type $\partial\Omega \in C^{0,1} \cap PC^2$ (" $\partial\Omega$ is piecewise continuously

curved"), domains Ω which need not be convex, and weaker smoothness assumptions on the parameters as well as on the solution u of the BVP $Au = F$ are admitted. McNEAL [1] and WINSLOW [1] seem to be the first to have tackled the PB- and MD-approaches, respectively. Other papers on the derivation or application of FDSs $A_h y = F_h$ via balance approach have been cited in Section 1.1. It is worth noting that secondary networks of the type (PB) and (MD) are also utilized in the FEM, e.g. for constructing finite element schemes with lumping approximation in the secondary part of the elliptic operator, cf. IKEDA [1], THOMASSET [1].

Chapter 4: This chapter deals with the discussion of several analytical and matrix properties of the difference operators A_h on irregular networks, with A_h from Chapter 3. In Section 4.2., the essential topic is the study of the monotonicity of the difference operators A_h and of the matrices associated with A_h. This property will be guaranteed by a monotonicity condition $M(A_h, \bar{\omega})$, cf. (4.2.1), which is equivalent to the so-called "irreducibly weakly diagonally dominant positive type operators" A_h. The condition $M(A_h, \bar{\omega})$ has often been used in the literature, with various denotations, see e.g. BRAMBLE/HUBBARD [1], CIARLET [1], COLLATZ [1,2], HEINRICH [2,3,4], IKEDA [1], LORENZ [1], MITCHELL/ GRIFFITHS [1], SAMARSKIJ [1], TÖRNIG [1] and VARGA [1]. By means of restrictions imposed on the angles of the triangles, which have not yet been considered in the literature on the FDM, the monotonicity condition $M(A_h, \bar{\omega})$ can be satisfied. Our angle conditions are mainly restrictions to pairs of angles $\theta_{\pm}$, e.g. $\theta_- + \theta_+ \leq \pi - \theta_0$ ($\theta_0 > 0$ or $\theta_0 = 0$), and these are weaker than those required in several papers on the FEM, cf. CIARLET [2], CIARLET/RAVIART [1], HÖHN/TÖRNIG [1] and IKEDA [1], where triangulations of weakly or strictly acute type are demanded, i.e. all angles satisfy the relations $\theta \leq \pi/2$ or $\theta < \pi/2$, respectively. Angle conditions similar to ours are mentioned by REID [1] and RUAS SANTOS [1]. But here we shall also consider operators div (k grad u) with $k \neq$ const. Another point of interest studied in Section 4.2. is the comparison of FDSs $A_h y = F_h$ obtained via method (PB) and (MD), which gives an insight into common and different properties of the PB- and MD-schemes $A_h y = F_h$.

In Section 4.3., discrete analogues of scalar products in L_2- and W_2^1-spaces as well as discrete counterparts of the L_2-, W_2^1-, W_2^{-1}- and C-norms for grid functions defined on irregular networks will be introduced. Then, equivalence statements on various types of norms and a discrete analogue of the well-known trace theorem will be presented. Section 4.4. contains the derivation of discrete analogues of Green's formula and of inequalities of Friedrichs-Poincaré-type for grid functions on irregular networks of triangles and (or) rectangles. For the special case that the domain Ω is a rectangle and is covered by a network of rectangles, inequalities of Friedrichs-Poincaré-type can also be found in SAMARSKIĬ/ ANDREEV [1]. Discrete forms of Friedrichs' inequality in connection with the FEM and for networks of triangles are discussed in ŽENIŠEK [1]. The inequalities of Friedrichs-Poincaré-type derived in this monograph are convenient tools for proving the positive definiteness of A_h, i.e. $\|y\|_o^2 \leq C_o(A_h y,y)$ with $C_o = O(1)$, and various a priori estimates. The results are such that the symmetry and the positive definiteness of the difference operator A_h can always be conserved, if the operator of the BVP Au = F is endowed with such properties.

In Section 4.5., using discrete C-, L_2-, W_2^1- and W_2^{-1}-norms of the grid functions, i.e. $\|\cdot\|_C$, $\|\cdot\|_o$, $\|\cdot\|_1$ and $\|\cdot\|_{-1}$, as well as discrete L_2-scalar products (.,.), and applying the results of Section 4.4., we shall get, for instance, the estimates $\|y\|_1^2 \leq C_1(A_h y,y)$ and $\|y\|_1 \leq C_1\|A_h y\|_{-1}$ or $\|y\|_1 \leq C_1\|A_h y\|_-$. These inequalities confirm the discrete "V-ellipticity" and the W_2^{-1}-W_2^1-stability of the difference operator A_h, respectively. The norm $\|\cdot\|_-$ is an upper bound of $\|\cdot\|_{-1}$ and can be easily calculated. Inequalities of the type $\|y\|_C \leq C_2|\ln h|^{1/2}\|y\|_1$, which express the "weak imbedding" of the discrete space $W_2^1(\bar{\omega})$ into the space $C(\bar{\omega})$, will also be obtained. The proof of these inequalities is based in part on OGANESYAN/RUKHOVETS [2] or on OGANESYAN et al. [1, part 1]. Furthermore, a priori estimates of the type $\|y\|_C \leq C_3\|A_h y\|_C$ will be derived, where the monotonicity condition $M(A_h,\bar{\omega})$ from Section 4.2. and so-called "majorant functions" are employed. Thus, some customary techniques of proving the C-C-stability of A_h are extended to locally irregular networks and, in some cases, to irregular networks. The a priori estimates established in Section 4.5. will also be used for proving error estimates and convergence results, cf. Chapter 5.

Chapter 5: This chapter deals with the splitting, classification and estimation of errors appearing when applying the FDM via balance method on irregular networks approximating domains Ω, with curved boundaries $\partial\Omega \in C^{o,1} \cap PC^2$. The errors are caused by the cubature, quadrature and finite difference formulas as well as by the curved boundary $\partial\Omega$. Since the error also depends on the type of the secondary network, the error splitting must be considered separately for PB- and MD-schemes. In Section 5.1., various splittings and a priori estimates of the error z:= y - u will be derived, where y and u denote the exact solutions of the FDS $A_h y = F_h$ and the BVP Au = F, respectively. Convergence results for solutions $u \in C^l(\bar{\Omega})$ $(l > 2)$ will be given and, moreover, a new technique of error estimation for the FDM on irregular networks and generalized solutions u satisfying the smoothness assumption $u \in W_2^l(\Omega)(l > 2)$ will be outlined.

Section 5.2. and 5.3. are devoted to the estimation of the discrete L_2-norm of the error functional $\varkappa$, which is, properly speaking, a weak norm of the error of the principal part of A_h. Section 5.4. yields estimates of the errors of the secondary part of the difference operator A_h and of the right-hand side F_h. In contrast to the bulk of the literature on the FDM, here solutions $u \in W_2^2(\Omega)$, domains Ω with boundaries $\partial\Omega \in C^{o,1} \cap PC^2$, various types of boundary conditions and, at the same time, networks which are irregular in the whole domain Ω are admitted in the present study. Here, imbedding and trace theorems, affine transformations of domains and functionals as well as the Bramble-Hilbert Lemma, cf. BRAMBLE/HILBERT [1], are involved. The Bramble-Hilbert Lemma on the estimation of linear functionals on Sobolev spaces will be utilized for $W_2^2(\Omega)$-functions in a similar sense as Taylor's expansion of the error for $C^2(\bar{\Omega})$-functions.

In Section 5.5., the error estimates derived in Sections 5.2., 5.3. and 5.4. are collected and exploited to prove estimates of the type $\|y-u\|_1$ $M_o h^l$ and $\|y-u\|_C \leq M_1 |\ln h|^{1/2} h^l$, for $l \in \{1, \frac{3}{2}, 2\}$, which imply the convergence $y \to u$ as $h \to 0$ with respect to grid norms $\|.\|_1$ and $\|.\|_C$ being discrete analogues of the W_2^1- and C-norm, respectively. The exponents $l = 1$, $l = \frac{3}{2}$ and $l = 2$ are associated with irregular, locally irregular and regular networks, respectively, and the estimates indicate the rate of convergence $y \to u$ as $h \to 0$. Additionally, some smoothness assumptions must be fulfilled, e.g. $u \in W_2^{1+j}(\Omega)$ ($j=l=1$ or $l \leq j=2$). The results obtained for PB- and MD-schemes do not coincide, but they are similar in many cases. For special BVPs $Au = F$, viz. the Dirichlet problem on rectangles Ω, which are covered with a network of rectangles, similar results were described by WEINELT [1] and LAZAROV [1].

<u>Chapter 6:</u> The aim of this chapter consists in the extension of the FDM via balance approach to BVPs $A^b u = F$ with nonsymmetric operators, which contain a so-called convection term with derivatives of first order i.e. $b_i \frac{\partial u}{\partial x_i}$ ($i=1,2$). Thus, defining some upwind difference quotients on irregular networks of triangles and (or) rectangles as well as various difference quotients which are more or less central, several nonsymmetric difference operators A_h^b and FDSs $A_h^b y = F_h$ approximating $A^b u = F$ will be proposed. Assuming the existence of so-called upwind triangles, the difference operators A_h^b can be constructed such that the monotonicity condition $M(A_h^b, \bar{\omega})$, cf. (4.2.1), is fulfilled. Exploiting this property of A_h^b, a priori estimates of the type $\|y\|_C \leq C_4 \|A_h^b y\|_C$, with the discrete C-norm $\|.\|_C$, can be derived. For $|b_i| \leq b_o$ ($i=1,2$) and sufficiently small b_o, the positive definiteness of A_h^b can be proved and leads to the estimate $\|y\|_1^2 \leq C_5 (A_h^b y, y)$. This is the starting point of deriving some a priori estimates with the discrete W_2^1- and W_2^{-1}-norms. The error of the FDS $A_h^b y = F_h$ can be split into the part originating from the symmetric operator A_h and the remaining part caused by the approximation of the convection term. By means of the estimation techniques developed in Chapter 5 and using some extensions, the error can be estimated for solutions $u \in C^k(\bar{\Omega})$ as well as for $u \in W_2^k(\Omega)$, $k > 2$. If y and u denote the solutions of the FDS $A_h^b y = F_h$ and the BVP $A^b u = F$, respectively, then the convergence $y \to u$ as $h \to 0$ with respect to the discrete C-

and W_2^1-norms can be proved. Thus, we shall get $\|y-u\|_C \le M_2 h$ for $u \in C^3(\bar{\Omega})$ and locally irregular networks, without restrictions on b_o (b_o fixed). Moreover, the estimates $\|y-u\|_1 \le M_3 h^l$ and $\|y-u\|_C \le$ $\le M_3 |\ln h|^{\frac{1}{2}} h^l$ will be obtained, for $u \in W_2^{1+j}(\Omega)$, $j \in \{1,2\}$, sufficiently small b_o, and with $l \in \{1, \frac{3}{2}, 2\}$, where l depends on the regularity of the parameters of the BVP $A^b u = F$ and of the network.

Chapter 7: This chapter contains concluding remarks on the FDM via balance approach on irregular networks. Various other problems and papers are quoted, which are related to the topics of the present monograph. Thus, boundary conditions containing a skew derivative $\frac{\partial u}{\partial \mu}$ and the treatment of the corresponding BVPs $A^\mu u = F$ by the FDM on irregular networks are discussed, cf. HEINRICH [7,10]. Then, weaker smoothness assumptions on the coefficients of the elliptic differential operator as well as on the solution u are considered. The latter problem, especially $u \in W_2^1(\Omega)$ for standard domains Ω, is extensively studied in the monograph of LAZAROV/MAKAROV/SAMARSKIĬ [1]. Citing or reviewing several papers, we confirm the applicability of the FDM on irregular networks to variational inequalities (HARTWIG/WEINELT [2]), differential equations with nonlinear elliptic, fourth-order or parabolic operators, and on other primary (secondary) networks as well as for domains $\Omega \subset R^N$, with arbitrary $N \geq 1$. Some remarks about the comparison of several finite difference and element schemes will be made. Finally, computational aspects of the FDM on irregular networks and some implementations will be mentioned.

Appendices: At the end of this monograph, six appendices are added, where mathematical tools widely used in this monograph are compiled. The greatest part of the assertions and relations is taken from analysis and prepared in such a way that their application is straightforward. The assertions and relations encountered in the appendices involve basic notations of Sobolev spaces, the estimation of linear functionals on Sobolev spaces, the extension of functions, some relations of geometry connected with discretization methods, imbedding and trace theorems, and finally, affine transformations of coordinates, derivatives, integrals, difference quotients etc.

Lastly, when reading this monograph, the following facts should be borne in mind. Several abbreviations such as FDM, BVP, FDS etc. are listed on p. 205. For easier access to the figures scattered in this book, there is a register (p. 205) indicating the pages where the figures can be found. The appendices are cited under their abbreviations DI, ES, EX, GE, IM and TR, which are introduced on p. 8 (contents).

2. BOUNDARY VALUE PROBLEMS AND IRREGULAR NETWORKS

<u>2.1. A class of elliptic problems.</u> We consider the following plane BVP with a formally selfadjoint elliptic differential operator L of second order and mixed boundary conditions:

$$(Lu)(x) := -\sum_{i,j=1}^{2} \frac{\partial}{\partial x_i} \left(k_{ij}(x) \frac{\partial u}{\partial x_j}\right) + c(x)u = f(x), \quad x \in \Omega \subset R^2, \qquad (2.1.1a)$$

$$(lu)(x) := \begin{cases} u = g(x), \quad x \in \Gamma_1, & (2.1.1b) \\ \sum_{i,j=1}^{2} k_{ij}(x) \frac{\partial u}{\partial x_j} n_i = g(x), \quad x \in \Gamma_2, & (2.1.1c) \\ \sum_{i,j=1}^{2} k_{ij}(x) \frac{\partial u}{\partial x_j} n_i + \alpha(x)u = g(x), \quad x \in \Gamma_3, & (2.1.1d) \end{cases}$$

with $\bigcup_{i=1}^{3} \Gamma_i = \Gamma := \partial\Omega$. For the sake of brevity, the symbol x of the point $x = (x_1, x_2)^T \in R^2$ is often omitted, and, instead of (2.1.1), we briefly write

$$Au = F \quad (x \in \bar{\Omega} = \Omega \cup \Gamma), \; A = (L,l)^T, \; F = (f,g)^T. \qquad (2.1.2)$$

Furthermore, the following notations will be employed:

$K := (k_{ij})_{2x2}$, the matrix of the coefficients k_{ij},

$L_o u := -\sum_{i,j=1}^{2} \frac{\partial}{\partial x_i} (k_{ij} \frac{\partial u}{\partial x_j})$, the principal part of Lu,

$(K\nabla u, \nabla v) := \sum_{i,j=1}^{2} k_{ij} \frac{\partial u}{\partial x_j} \frac{\partial v}{\partial x_i}$, the bilinear differential expression which corresponds to L_o, (2.1.3)

$(K\nabla u, n) = \sum_{i,j=1}^{2} k_{ij} \frac{\partial u}{\partial x_j} n_i$, the conormal derivative of u corresponding to L_o, with the normal $n = (n_1, n_2)^T$,

$\nabla u := (\frac{\partial u}{\partial x_1}, \frac{\partial u}{\partial x_2})^T$, the gradient of u,

and (.,.) is here the usual scalar product in R^2. Analytical considerations concerning BVP (2.1.1) or at least special cases of (2.1.1) can be found in books or papers by AZIZ [1], COSTABEL/STEPHAN [1], GAJEWSKI et al. [1], GILBARG/TRUDINGER [1], GRISVARD [1], KONDRAT'EV/OLEINIK [1], LADYZHENSKAYA/URAL'TSEVA [1], LADYSHENSKAYA [1], NEČAS [1], ODEN/REDDY [1], WIGLEY [1], WLOKA [1].

Now we introduce several assumptions on the domain Ω and its boundary Γ.

<u>Assumption 2.1. - $V(\bar{\Omega})$.</u> Let Ω be a two-dimensional bounded domain (an open and connected subset of R^2) with a piecewise smooth boundary $\Gamma := \partial\Omega$, where

$$\Gamma \in C^{0,1} \cap PC^k, \quad k \in \{2,3\}, \qquad (2.1.4)$$

i.e. Γ is Lipschitz-continuous and at least twice piecewise continuously differentiable; for $C^{0,1}$-boundaries, see e.g. ADAMS [1], GAJEWSKI et al. [1], KUFNER et al. [1], NEČAS [1], WLOKA [1]. The value $k = 3$ is sometimes required in accordance with the assumptions made in several papers devoted to analytical results concerning singularities caused by corners on Γ and mixed boundary conditions, e. g. in OGANESYAN/RUKHOVETS [1,2], WIGLEY [1]. The boundary Γ or subsets Γ_T of Γ can be represented by

$\Gamma = \bigcup_{i=1}^{3} \Gamma_i$, Γ_i: boundary part with boundary conditions of i-th kind, where $\Gamma_i \in C^{0,1} \cap PC^k$, $k \in \{2,3\}$,

$\Gamma_T = \bigcup_{j=1}^{m_T} \Gamma_T^j$, i.e. Γ_T consists of a finite number of simple arcs Γ_T^j of class C^k, $k \in \{2,3\}$, and $\Gamma_T^j (j=1,2,\dots,m_T)$ is at least continuously curved. ≡ (2.1.5)

We often take $\Gamma_T \in \{\Gamma, \Gamma_1, \Gamma_2, \Gamma_3, \Gamma_{23}\}$, $\Gamma_{23} := \Gamma_2 \cup \Gamma_3$, or Γ_T may be another subset of Γ. The arcs Γ_T^j, $j \in \{1,2,\dots,m_T\}$, are assumed to be continuous images of the segment $[0,1]$, i.e., the sets Γ_T^j and Γ_T are closed. The interior of these and other closed sets M is denoted by int M or $\overset{\circ}{M}$. The sets $\overset{\circ}{\Gamma}_i$ from (2.1.5) are supposed to be pairwise disjoint, i.e. $\overset{\circ}{\Gamma}_i \cap \overset{\circ}{\Gamma}_j = \emptyset$ $(i \neq j)$, and, the same should be valid for the subsets $\overset{\circ}{\Gamma}{}_T^j$, with Γ_T^j from (2.1.5). There is at most a finite number of corner points $P_k \in \Gamma$ with interior angles β_k between the two tangents, where $0 < \beta_k < 2\pi$, see Fig. 1a.

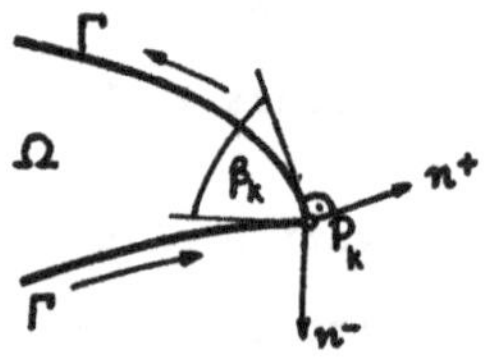

Fig. 1a

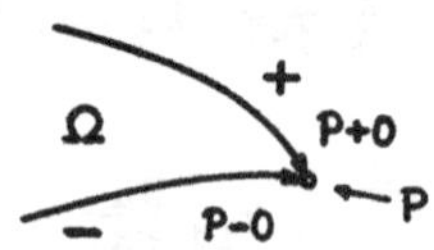

Fig. 1b

The outward normal to Γ is denoted by $n := (n_1, n_2)^T$, and, at the corner points $P_k \in \Gamma$, the left- and right-hand limits $n^{\pm}(P_k) := n(P_k \pm 0)$ are considered. The boundary of Ω and other domains will always be traversed counterclockwise, cf. Fig. 1a. If the parameters (coefficients, right-hand side) of the boundary condition as α or g have a jump at $x \in \Gamma$, e.g. at the collision points $P_c \in \Gamma_{i_1} \cap \Gamma_{i_2}$ $(i_1 \neq i_2;\ i_1, i_2 \in \{1,2,3\})$ or at the corner points $P_k \in \Gamma$, then the functions α, g are considered at such points P in the sense of left- and right-hand limits, i.e. $\alpha^{\pm}(P) := \alpha(P \pm 0)$, $g^{\pm}(P) := g(P \pm 0)$, cf. Fig. 1b.

The coefficients of the operator A and the right-hand side F of problem (2.1), (2.2) should fulfil some smoothness assumptions and others needed for constructing finite difference analogues to (2.1.1) and the derivation of error estimates depending on the degree of smoothness of these parameters. Clearly, such assumptions are also important for the existence, uniqueness and regularity theory of solutions of problem (2.1.1). Concerning this topic, we only refer to the literature (e.g. GILBARG/TRUDINGER [1], GRISVARD [1], KONDRAT'EV/OLEINIK [1], LADYZHENSKAYA/URAL'TSEVA [1], NEČAS [1], WIGLEY [1]).

The notation C^s or PC^s (s: integer) is used for s times continuously or piecewise continuously differentiable functions (and boundaries), respectively. For instance, "$\varphi \in PC^s(\Gamma_T)$", $\Gamma_T \subset \Gamma$, means that φ is s times piecewise continuously differentiable on Γ_T in the sense of $\varphi \in C^s(\Gamma_T^j)$ for the smooth arcs $\Gamma_T^j \in C^k$ ($j=1,2,\dots,m_T$, $0 \le s \le k$), with $\Gamma_T = \bigcup_{j=1}^{m_T} \Gamma_T^j$. Hence, φ is bounded since Γ_T and Γ_T^j from (2.1.5) are assumed to be closed. We shall deal with Sobolev spaces $W_p^s(\Omega)$ (s: integer), where $p = 2$ will be preferred in this paper. The spaces W_2^s can also be defined for one-dimensional manifolds which are sufficiently smooth. For Sobolev spaces, see ADAMS [1], AZIZ [1], GAJEWSKI et al. [1], KUFNER et al. [1], NEČAS [1], WLOKA [1]. For instance, we shall use $L_2(\mathring{\Gamma}_T)$ and $W_2^s(\mathring{\Gamma}_T^j)$, for $\Gamma_T \subset \Gamma$ and $\Gamma_T^j \in C^k$, $0 \le s \le k$, and, for brevity, the symbol "$\circ$" indicating the interior of a set will be often omitted, i.e., we shall write $L_2(\Gamma_T)$, $W_2^s(\Gamma_T^j)$ etc. Moreover, the symbolism "$\varphi \in PW_2^s(\Gamma_T)$", $\Gamma_T \subset \Gamma$, means that $\varphi \in W_2^s(\Gamma_T^j)$ holds for smooth arcs $\Gamma_T^j \in C^2$ ($s \le 2$; $j=1,2,\dots,m_T$), with $\Gamma_T = \bigcup_{j=1}^{m_T} \Gamma_T^j$. Assumptions on the coefficients of A and on the right-hand side F are summarized in

<u>Assumption 2.2. - V(A,F).</u> Let k_{ij}, c, f, α and g from (2.1.1) fulfil the following smoothness assumptions, where the kind of smoothness and the range of the integer s, $s \in \{0,1,2\}$, will be exactly stated later when needed:

$k_{ij} \in C^s(\bar{\Omega})$ $(i,j=1,2)$ for $s \in \{1,2\}$;

$c \in C^s(\bar{\Omega})$ for $s \in \{0,1,2\}$ or $c \in W_2^2(\Omega)$ in some cases;

$f \in C^s(\bar{\Omega})$ or $f \in PC^s(\bar{\Omega})$ or $f \in W_2^s(\Omega)$, $s \in \{0,1,2\}$;

$\alpha \in PC^s(\Gamma_3)$ for $s \in \{0,1,2\}$ or $\alpha \in PW_2^s(\Gamma_3)$ for $s \in \{1,2\}$; (2.1.6)

$g \in C(\Gamma_1)$ at least, $g \in PC^s(\Gamma)$ for $s \in \{0,1,2\}$, $s = s(\Gamma^j)$ $(\Gamma = \bigcup_{j=1}^{m} \Gamma^j)$ in some cases, or $g \in PW_2^s(\Gamma)$ for $s \in \{1,2\}$.

In some papers, e.g. in OGANESYAN/RUKHOVETS [1], $s = 3$ is demanded in $PC^s(\Gamma)$, which is assumed to be fulfilled if the results of these papers are quoted. Furthermore, let the coefficients k_{ij}, c and α satisfy the following conditions: There are positive constants k_0, k_1 $(0 < k_0 \le k_1)$ such that the inequalities

$$k_0(\xi,\xi) \le ((K(x)\xi,\xi) \le k_1(\xi,\xi) \text{ hold for all } \xi \in R^2 \text{ and } x \in \bar{\Omega}, \quad (2.1.7a)$$

with $K = K(x)$ from (2.1.3), $k_{12} = k_{21}$, i.e. L is uniformly elliptic, furthermore,

$$c(x) \ge 0 \text{ for } x \in \Omega, \quad \alpha(x) \ge \alpha_0 > 0 \text{ for } x \in \Gamma_3. \quad (2.1.7b)$$

Sometimes α is used on $\Gamma_{23} := \Gamma_2 \cup \Gamma_3$ with the convention $\alpha|_{\Gamma_2} = 0$. Moreover, throughout this paper we suppose that the sets Γ_1, Γ_3 and $\Omega_c := \{x \in \Omega: c(x) > 0\}$ are not empty simultaneously, i.e.

$$\text{mes } \Gamma_1 + \text{mes } \Gamma_3 + \text{mes } \Omega_c > 0. \equiv \qquad (2.1.8)$$

Finally, we assume that there is a solution u of problem (2.1.1), where u fulfils appropriate smoothness assumptions summarized below in V(u). The uniqueness of a solution u is clearly a consequence of (2.1.8), see e.g. GILBARG/TRUDINGER [1, p. 198], but the degree of smoothness depends essentially on the smoothness of the parameters of (2.1.1), on the angles β of the corners on Γ and on the fact that mixed boundary conditions occur, see AZIZ [1], AZZAM/KREYSZIG [1], COSTABEL/STEPHAN [1], DOBROWOLSKI [1], GRISVARD [1], KONDRAT'EV/OLEINIK [1], WIGLEY [1]. In general, the "shift theorem" which states that $f \in W_2^k(\Omega)$ and sufficiently smooth parameters in A, F imply $u \in W_2^{k+2}(\Omega)$ does not hold and even $f \in W_2^2(\Omega)$ or $f \in C^2(\bar{\Omega})$ can lead to $u \in W_2^1(\Omega)$ only. Moreover, the smoothness assumptions on the parameters made in (2.1.6) are not sufficient, in general, for the smoothness postulated for the solution u in V(u), cf. e.g. $g \in C(\Gamma_1)$. But it is well known from the literature that the solution u from (2.1.1) can be split into an expansion u_s of a finite number of singular functions $u_s^{(i)}$ and a regular part u_r, i.e. $u = u_s + u_r$. For important special cases of (2.1.1) with mixed boundary conditions and (or) corners on Γ, it is proved in the literature that for $f \in L_2(\Omega)$ we have $u_r \in W_2^2(\Omega)$, $u_s \in W_2^1(\Omega)$, $u_s \in C(\bar{\Omega}) \cap C^k(\Omega)$ (k=2 at least), the identity Lu = f holds a.e. (almost everywhere) on Ω and the traces of u_s and ∇u_s on Γ_T^j have certain asymptotic properties, see COSTABEL/STEPHAN [1,2], DOBROWOLSKI [1], GRISVARD [1], OGANESYAN/RUKHOVETS [1,2], OGANESYAN et al. [1], SCHATZ/WAHLBIN [1], WIGLEY [1]. In these and in many other papers, it is demonstrated that the singular part u_s is regular, if it is considered sufficiently far away from points on Γ (corner and collision points) which can cause a singularity, i.e., the "ill character" of u_s is of local nature only. Even $u_s \equiv 0$ and $u \equiv u_r \in W_2^k(\Omega)$, $k \in \{2,3\}$, can occur if the parameters are sufficiently smooth and the angles β of the corners sufficiently small. Sometimes the convexity of Ω is sufficient to state this for k = 2. Results concerning classical solutions were discussed, for instance, by AZZAM/KREYSZIG [1], WIGLEY [1]. It should be noted that discontinuous coefficients k_{ij} in (2.1.1) can also imply considerable singularities, cf. DOBROWOLSKI [1]; but this is not considered here. According to these foregoing remarks on the smoothness of the solution u, it seems to be reasonable to make assumption V(u).

<u>Assumption 2.3. - V(u)</u>

1. Let u be the weak solution of (2.1.1) with the representation $u = u_r + u_s$ and the following smoothness assumptions:

a) $u \in W_2^k(\Omega)$ with $k \in \{2,3\}$, i.e. $u \equiv u_r$ and $u_s \equiv 0$; (2.1.9)

hence, u is a regular solution;

or more weakly:

b) $u \in W_2^1(\Omega)$, with $u_r \in W_2^2(\Omega)$ and $u_s \in W_2^1(\Omega) \cap C(\bar{\Omega})$, (2.1.10)

$Lu_s \in L_2(\Omega)$, $u_s \in W_2^2(\Omega \setminus \mathcal{K})$, $\mathcal{K} = \bigcup_{j=1}^{J} \mathcal{K}_j$, where $\mathcal{K}_j$ is a sufficiently small circle-neighbourhood of a singular point $P_j \in \Gamma$, and $u \in W_2^1(\mathcal{K}_j)$. For the conormal derivative $(K\nabla u,n)$ from (2.1.3), $(K\nabla u,n) \in L_p(\Gamma_{23}^j)$ is assumed, with $p = 2$ for $\Gamma_{23}^j \cap \Gamma_1 = \emptyset$ (i.e. Γ_{23} without contact to Γ_1) and with some fixed $p > 1$ for $\Gamma_{23}^j \cap \Gamma_1 \neq \emptyset$ (Γ_{23}^j in contact with Γ_1), Γ_{23}^j from (2.1.5). Furthermore, formula (2.1.16) is assumed to be fulfilled for $v \equiv 1$.

2. Let the solution u of (2.1.1) fulfil the differential equation $Lu = f$ and the boundary condition $lu = g$, with parameters from (2.1.6), in the following sense:

$$Lu := L_0 u + cu = f \text{ a.e. in } \Omega, \tag{2.1.11a}$$

$$lu := \begin{cases} u = g \text{ pointwise on } \Gamma_1, & (2.1.11b) \\ (K\nabla u,n) + \alpha u = g \text{ a.e. on } \Gamma_{23}. \equiv & (2.1.11c) \end{cases}$$

It is obvious that the solution u satisfying (2.1.9) or (2.1.10) can be handled as a continuous function (after changing the values of u on a set of measure zero). This is a consequence of Sobolev's imbedding theorem cited in Appendix IM. Moreover, the assumptions on the trace $(K\nabla u,n)|_{\Gamma_{23}}$ made in (2.1.10) actually mean that u is, in the neighbourhood of Γ_{23}, somewhat smoother than $u \in W_2^1(\Omega)$, since from $u \in W_2^1(\Omega)$ we can conclude $(K\nabla u,n) \in W_2^{-1/2}(\Gamma_T^j)$ only. But Sobolev spaces $W_2^k(\Omega)$ with noninteger k will not be considered here. For instance, in COSTABEL/STEPHAN [1], GRISVARD [1] and NEČAS [1] smoothness results of the type $u \in W_2^{1+\varepsilon}(\Omega)$, $0 < \varepsilon < 1$, are mentioned. Otherwise, analytical results qualifying (2.1.10) as an acceptable assumption in the case of boundary singularities (corners, mixed boundary conditions) can be found in the literature, e.g. in COSTABEL/STEPHAN [1]. Finally, let us mention some variational formulations of the BVP (2.1.1), which are given in the literature mostly after homogenization of the boundary conditions, at least concerning the Dirichlet condition. We assume that $g|_{\Gamma_1} \equiv 0$ is given or that $g|_{\Gamma_1} (\neq 0)$ admits a sufficiently smooth extension to Ω such that after homogenization of the Dirichlet condition the data f and $g|_{\Gamma_{23}}$ have the same character as supposed in V(A,F), cf. (2.1.6). Hence, we are now in position to give the following variational formulation, see e.g. AZIZ [1], CIARLET [2], GILBARG/TRUDINGER [1], ODEN/REDDY [1]. Find

$$u \in V : \quad a(u,v) = f(v) \quad \text{for all } v \in V, \tag{2.1.12a}$$

with

$$V := \overset{\circ}{W}{}^1_{2,\Gamma_1}(\Omega) := \{v \in W^1_2(\Omega) : v|_{\Gamma_1} = 0\}, \tag{2.1.12b}$$

$$a(u,v) := \int_\Omega \{(K\nabla u, \nabla v) + cuv\}dx + \int_{\Gamma_{23}} \alpha uv\, ds,$$

$$f(v) := \int_\Omega fv\, dx + \int_{\Gamma_{23}} gv\, ds,\ K \text{ from } (2.1.3),\ \alpha|_{\Gamma_2} = 0. \tag{2.1.12c}$$

Additionally, (2.1.12a) can be equivalently formulated as a minimum problem. Find

$$u \in V:\ J(u) = \min_{v \in V} J(v),$$

with

$$J(v) := \frac{1}{2}\int_\Omega \{(K\nabla v, \nabla v) + cv^2 - 2\, fv\}dx + \frac{1}{2}\int_{\Gamma_{23}} \{\alpha v^2 - 2gv\}\, ds,\ \alpha|_{\Gamma_2} = 0. \tag{2.1.13}$$

Now, the part 2 of assumption V(u) should be illustrated. Assuming the existence of a solution $u \in V$ and, furthermore, the global regularity in the sense of (2.1.9), i.e. $u \in W^2_2(\Omega)$, we can conclude $Lu \in L_2(\Omega)$ and $lu \in L_2(\Gamma)$, where the latter relation will be clear by $u, \frac{\partial u}{\partial x_i} \in L_2(\Gamma)$, $n_i \in L_\infty(\Gamma_{23})$ and the properties of the coefficients k_{ij}, α in (2.1.6), cf. also Appendix IM (trace theorems). Using Green's formula

$$\int_\Omega (L_0 u)v\, dx = \int_\Omega (K\nabla u, \nabla v)dx - \int_{\partial\Omega} (K\nabla u, n)v\, ds, \tag{2.1.14}$$

with $u \in W^2_2(\Omega)$, $v \in W^1_2(\Omega)$ and L_0, $(K\nabla u, .)$ from (2.1.3), we obtain

$$\int_\Omega \{L_0 u + cu\}\, v\, dx + \int_{\Gamma_{23}} \{(K\nabla u, n) + \alpha u\}\, v\, ds = \int_\Omega fv\, dx + \int_{\Gamma_{23}} gv\, ds \tag{2.1.15}$$

as a consequence of (2.1.12a) and $u \in V \cap W^2_2(\Omega)$. Since (2.1.15) holds for all $v \in C^\infty_0(\Omega)$ (the functions from $C^\infty(\Omega)$ with $v|_\Gamma = 0$) and $C^\infty_0(\Omega)$ is dense in $L_p(\Omega)$ ($1 \le p\ \infty$), the identity $Lu = f$ (2.1.11a) is verified. Inserting $Lu = f$ in (2.1.15) and using some trace and density theorems with respect to Γ, the identity $lu := (K\nabla u, n) + \alpha u = g$ in (2.1.11c) can also be shown. The trace $u|_{\Gamma_1}$ has at least the character $u|_{\Gamma_1^j} \in W^{3/2}_2(\Gamma_1^j)$, $u|_{\Gamma_1} \in L_2(\Gamma_1)$, and due to Sobolev's imbedding theorem, we have $u = 0$ in the sense of $C(\Gamma_1)$, which confirms (2.1.11b) at least for homogeneous Dirichlet data. If u satisfies the weaker assumption (2.1.10) in V(u), then $Lu = f$ in (2.1.11a) is also a consequence of (2.1.12a). To verify (2.1.11c), a generalization of Green's formula (2.1.14) is needed:

$$\int_\Omega (L_0 u)v\, dx = \int_\Omega (K\nabla u, \nabla v)dx - \langle \gamma_1 u, \gamma_0 v\rangle_{\partial\Omega} \tag{2.1.16}$$

for $u \in H^1_2(\Omega) := \{\varphi \in W^1_2(\Omega) : L_0\varphi \in L_2(\Omega)\}$, $v \in W^1_2(\Omega)$,

where $\langle\cdot,\cdot\rangle_{\partial\Omega}$, γ_1 denote the duality pairing of $W^{-1/2}_2(\Gamma) \times W^{1/2}_2(\Gamma)$ and the trace operator (of order 1), respectively. Thus, $\langle \gamma_1 u, \gamma_0 u\rangle_{\partial\Omega}$ is the generalization of $\int_{\partial\Omega}(K\nabla u, n)v\, ds$, if u, v are taken from $H^1_2(\Omega)$, $W^1_2(\Omega)$, respectively. Such generalized Green's formulas can be found e.g. in

COSTABEL/STEPHAN [1, p.6] for $L_0 = -\Delta$ and in COSTABEL [1, p. 58] for analytical coefficients k_{ij} in L_0.

2.2. Irregular networks

2.2.1. Networks of triangles and rectangles. The so-called balance method described in Chapter 3 allows us to construct finite difference schemes (FDS) on different types of networks formed by triangles, convex quadrangles, polar meshes or other curvilinear meshes. In this monograph, networks of triangles, rectangles and the combination of these elements are in the centre of interest and will be described next. Here, a "grid" $\bar{\omega}$ is the set of all nodes of a given network. We shall restrict ourselves to grids having the property $\bar{\omega} \subset \bar{\Omega}$, with $\bar{\omega} = \omega + \gamma, \omega \subset \Omega$ and $\gamma \subset \Gamma$, i.e., the "grid boundary" γ is a subset of Γ. We note that such networks do not yield conforming finite elements in the sense of CIARLET [2]. This is due to the definition of Ω_h, which leads, in general, to $V_h \not\subset V$ (V_h: finite element space), if Ω is not convex.

Assumption 2.4. - $V(\gamma, \Gamma_h, \Omega_h)$. Let γ be a set of grid points which are located on the boundary Γ of the domain Ω, i.e. $\gamma \subset \Gamma$, such that at least all corner points $P_k \in \Gamma_{23} \setminus \Gamma_1$, collision points $P_c \in \Gamma_2 \cap \Gamma_3$ and corner points $P_k = P_c \in \Gamma_1 \cap \Gamma_{23}$ belong to γ. However, for the sake of simplicity we assume right now and throughout this paper that all corner points $P_k \in \Gamma$ and collision points $P_c \in \Gamma_i \cap \Gamma_j$ ($i \neq j$; $i,j = 1,2,3$) are contained in γ. Additionally, the points where the boundary functions α and g have a jump, cf. (2.1.6), are also included in the grid boundary γ. Let h be the real mesh parameter tending to zero, where $0 < h \leq h_0$, h_0 is fixed, and let $\hbar(x,\xi)$ be the straight mesh segment $[x,\xi]$ which connects two adjacent neighbouring grid points $x,\xi \in \gamma \subset \Gamma$. Furthermore, let Γ_h denote the closed polygonal line (or several of them in the case of multiply connected domains) which joins the grid points of γ one after another, such that Γ_h is a polygonal approximation of Γ, with $\Gamma_h = \bigcup_{x,\xi \in \gamma} \hbar(x,\xi)$. Due to $V(\bar{\Omega})$, we are able to put (without loss of generality of the essential concepts and results in this monograph) the grid points $x,\xi \in \gamma$ on Γ such that the mesh segments $\hbar(x,\xi)$ fulfil either

$$\text{a)} \quad \operatorname{int} \hbar(x,\xi) \cap \bar{\Omega} = \emptyset \quad \text{or} \qquad \text{b)} \quad \hbar(x,\xi) \subset \bar{\Omega}, \tag{2.2.1}$$

if $h \leq h_0$ holds and h_0 is sufficiently small. Let Ω_h be the polygonal domain with the boundary $\partial\Omega_h$, where $\partial\Omega_h$ is defined by $\partial\Omega_h := \Gamma_h$, and $\bar{\Omega}_h = \Omega_h \cup \partial\Omega_h$ is an approximation of $\bar{\Omega} = \Omega \cup \Gamma$. ■

Assumption (2.2.1) sometimes simplifies the explanations and means that $\operatorname{int} \hbar(x,\xi)$ is located outside $\bar{\Omega}$ or inside $\bar{\Omega}$, while the boundary $\partial\hbar(x,\xi)$ consisting of the points $x,\xi \in \gamma$ is always a subset of Γ. See Figs. 2a, b.

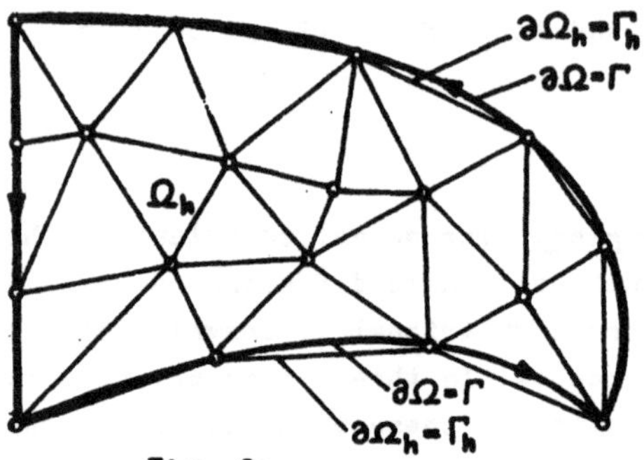

Fig. 2a

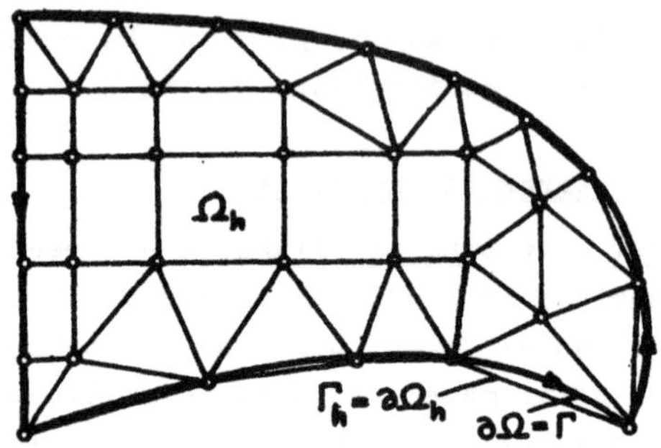

Fig. 2b

Now, we cover the closed original domain Ω and its approximation $\bar{\Omega}_h$ by finite elements e in the sense of a triangulation $\mathcal{T}$ as is well known from the FEM, cf. CIARLET [2, pp. 38, 51] and Figs. 2a, b. We shall restrict ourselves to elements $e \in \{\Delta, \square, ⊿\}$, where Δ, $\square$ and ⊿ denote a straight triangle, a rectangle and a curved triangle, respectively. The set e is assumed to be closed, i.e. $e = \bar{e}$. For the description of such triangulations, finite elements and several grids, see Definition 2.5.

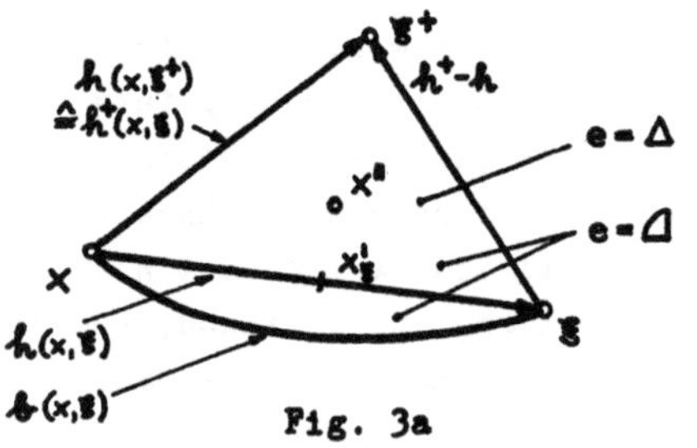

Fig. 3a

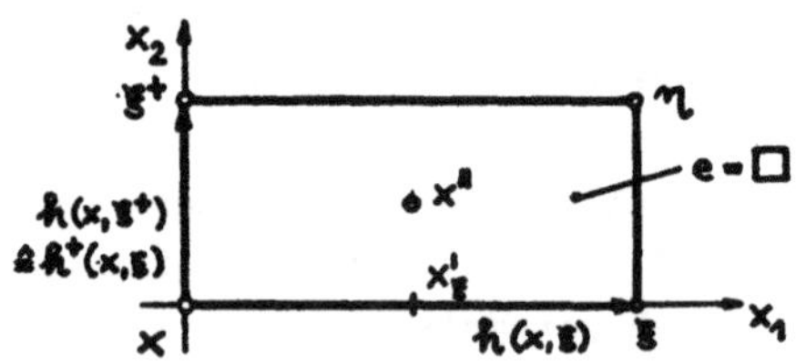

Fig. 3b

<u>Definition 2.5.</u> The following symbols will be introduced.

a) $h(x, \xi)$, the (closed) side of a straight element $e \in \{\Delta, \square\}$, where $h(x, \xi)$ connects the vertices x, ξ of e; at the same time $h(x, \xi)$ is a mesh segment, cf. Figs. 3a, b;

b) $x' := x'_\xi := \frac{1}{2}(x+\xi)$, the midpoint of the mesh segment $h(x, \xi)$;

c) x'', the centre of gravity of the straight element e, e.g. $x'' := \frac{1}{3}(x+\xi+\xi^+)$ for the straight triangle $\Delta(x, \xi, \xi^+)$ with the vertices x, ξ and ξ^+, cf. Figs. 3a, b; (2.2.2)

d) $\bar{\omega}$, the "closed grid", i.e., the set of all grid points which are here identical with the nodes of the network built-up by the triangulation $\mathcal{T}$;

e) $\bar{\omega}'$, the set of the midpoints $x' := x'_\xi$ of all mesh segments $h(x, \xi)$;

f) $\bar{\omega}''$, the set of the centres of gravity x" of all straight elements e used for the triangulation of $\bar{\Omega}_h$,
$\bar{\omega}''_{\Delta}$, $\bar{\omega}''_{\Box}$, the sets of centres of gravity of all triangles Δ and rectangles $\Box$, respectively, which are subsets of $\bar{\omega}''$;

g) e(x"), the finite element e with the index x", cf. Figs. 3a, b;

h) $\triangle$(x"), the curved triangle which has the same vertices x, ξ, ξ^+ as Δ(x"), cf. Figs. 3a; (2.2.2)

i) $S'(x) := \{\xi \in \bar{\omega}: h(x, \xi) \text{ is mesh segment}\}$, the set of the adjacent grid neighbouring points $\xi \in \bar{\omega}$ associated with the grid point x, excluding x itself; the set of the midpoints x'_ξ of $h(x, \xi)$, $\xi \in S'(x)$, is uniquely assigned to S'(x), see e.g. Figs. 5a, b, e and 6a, b;

j) $S(x) := S'(x) + \{x\}$, the "full difference star" of the grid point x, i.e. the "periphery" S'(x) together with the central point x itself, see e.g. Figs. 5a, b, e and 6a, b;

k) $S''(x) := \{x'' \in \bar{\omega}'': x \in e(x'')\}$, the set of indices x" (centres of gravity, if $e \in \{\Delta, \Box\}$) of all elements e adjacent to the grid point x, see e.g. Figs. 5d, e and 6a, b. ≡

We now describe the triangulations $\mathcal{T}$ used for $\bar{\Omega}$ and for the family of polygonal approximations $\bar{\Omega}_h$.

<u>Assumption 2.6. - V($\mathcal{T}_{\triangle}$), V($\mathcal{T}_{\Delta}$).</u> Assume that the domains $\bar{\Omega}$ and $\bar{\Omega}_h$ are covered by the triangulation $\mathcal{T}_{\triangle}$ and $\mathcal{T}_{\Delta}$, respectively, such that

$$\mathcal{T}_{\triangle}: \ \bar{\Omega} = \bigcup_{x'' \in \bar{\omega}''} e(x''), \text{ with } e \in \{\Delta, \Box, \triangle\} \text{ and for } h \leq h_0, \tag{2.2.3a}$$

$$\mathcal{T}_{\Delta}: \ \bar{\Omega}_h = \bigcup_{x'' \in \bar{\omega}''} e(x''), \text{ with } e \in \{\Delta, \Box\} \quad \text{ and for } h \leq h_0, \tag{2.2.3b}$$

where

$$\mathring{e}(x'') \cap \mathring{e}(\bar{x}'') = \emptyset \text{ for } x'' \neq \bar{x}'', \text{with } x'', \bar{x}'' \in \bar{\omega}''(e = \bar{e}), \tag{2.2.3c}$$

and h denotes the mesh-size parameter. The triangulation $\mathcal{T}_{\Delta}$ consists of straight elements Δ , $\Box$ only, and, for the sake of simplicity, the rectangles $\Box$ are used under the assumption that the sides of $\Box$ are parallel to one of the coordinate axes and no rectangle $\Box$ touches curved parts of the boundary Γ. The triangulation $\mathcal{T}_{\triangle}$ consists of the same triangles Δ and rectangles $\Box$, but with the exception that, at curved parts of the boundary Γ, the straight triangles Δ are substituted by curved triangles $\triangle$, where the curved edges of $\triangle$ are formed by the boundary Γ only, see e.g. Figs. 3a, 5b. The intersection of any two elements e is disjoint or consists of a common side, a vertex or e itself, cf. Figs. 2a, b. ≡

Obviously, the only difference between (2.2.3a) and (2.2.3b) consists in the use of triangles $\triangle$(x") and Δ(x") for the approximation of Γ and Γ_h respectively. If Δ (x") from $\mathcal{T}_{\Delta}$ has at least two vertices on Γ_h (consequently on Γ, γ, too), then, in order to get $\mathcal{T}_{\triangle}$ in (2.2.3a), $\Delta(x'')$ will be replaced by the corresponding triangle $\triangle(x'')$, which has curved edges on Γ. The hypotheses imposed upon $\mathcal{T}_{\Delta}$ and $\mathcal{T}_{\triangle}$ are in effect the same as in CIARLET [2, pp. 38, 51] enumerated by $\mathcal{T}_{h1}, \mathcal{T}_{h2}, \ldots, \mathcal{T}_{h5}$. The asymptotic properties of the family of triangulations $\{\mathcal{T}_{\Delta}\}_{h\le h_o}$, $\{\mathcal{T}_{\triangle}\}_{h\le h_o}$ will be given later by VO(PB, $\bar\omega$), VO(MD, $\bar\omega$), etc. The triangulation $\mathcal{T}_{\Delta}$ from (2.2.3b) contains the special cases

$$\bar\Omega_h = \bigcup_{x''\in\bar\omega''}\Delta(x''),\ h\le h_o,\ \bar\omega'' = \bar\omega''_{\Delta},\ \text{i.e. triangles only,} \tag{2.2.4}$$

$$\bar\Omega_h = \bigcup_{x''\in\bar\omega''}\Box(x''),\ h\le h_o,\ \bar\omega'' = \bar\omega''_{\Box},\ \text{i.e. rectangles only.} \tag{2.2.5}$$

For triangulations $\mathcal{T}_{\Delta}$ from (2.2.3b), (2.2.5) involving rectangles $\Box$, the sets S'(x) and S(x) from (2.2.2) must be replaced by $S'_e(x)$ and $S_e(x)$, respectively. The set $S_e(x)$ comprises the vertices of all rectangles $\Box$ and triangles Δ which have x as a common vertex. Some of the neighbouring points ξ of x, namely $\xi\in S'_e(x)\setminus S'(x)$, have no mesh segment $h(x,\xi)$ because the diagonals of the rectangles are not defined to be a mesh segment. For instance, if four rectangles meet at x, then $S_e(x)$ and S(x) denote the 9- and 5-point difference star, respectively, cf. Figs. 4b or 5d. For practical purposes, it is convenient to assume that the curved part $\tilde\Gamma$ of the boundary Γ is approximated by triangles only and the remaining, piecewise straight part $\Gamma\setminus\tilde\Gamma$ by rectangles or triangles. This assumption made in V($\mathcal{T}_{\triangle}$), V($\mathcal{T}_{\Delta}$) and used throughout this monograph is not essential for maintaining the principal assertions, but it enables us to shorten our discussion and to focus on networks of triangles which are of great interest in connection with curved boundaries. Domains composed of rectangles will be covered preferably by rectangular networks.

<u>Definition 2.7.</u> The set of all nodes of the network forms the grid $\bar\omega$, cf. Definition 2.5., where the following subsets will now be specified: 1)

$$\gamma := \bar\omega\cap\Gamma \quad \text{(this coincides with Assumption 2.4.),}$$

$$\gamma_1 := \gamma\cap\Gamma_1,\quad \gamma_{23} := \gamma\cap\{(\Gamma_2\cup\Gamma_3)\setminus\Gamma_1\},\quad \gamma_3 := \gamma_{23}\cap\Gamma_3, \tag{2.2.6}$$

$$\gamma_2 := \gamma_{23}\setminus\gamma_3,\quad \omega := \bar\omega\cap\Omega,\ \overset{*}{\omega} := \{x\in\omega:\ S'(x)\cap\gamma\ne\emptyset\},\ \overset{\circ}{\omega} := \omega\setminus\overset{*}{\omega}. \ \blacksquare$$

That is, we have

$$\bar\omega = \omega+\gamma,\ \omega = \overset{\circ}{\omega}+\overset{*}{\omega},\quad \gamma = \gamma_1 + \gamma_{23} = \gamma_1+\gamma_2+\gamma_3, \tag{2.2.7}$$

where "+" means the same as "$\cup$", but "+" will be used for sets of grid points only. Obviously, all points of the "grid boundary" γ are located on Γ and Γ_h, and, all corner points $P_c\in\Gamma_{23}\setminus\Gamma_1$ belong to γ_{23}. Now grid functions will be considered on $\bar\omega$ as well as on $\bar\omega'$ and $\bar\omega''$, cf. (2.2.2). We assume that each of these sets is uniquely numbered, where $|.|$ denotes the number of the grid points.

1) The symbols γ_1 from (2.2.6) and γ_o from (2.2.11) denote subsets of $\bar\omega$; they should not be confused with the trace operators γ_1, γ_o from (2.1.16).

Definition 2.8. **The following sets will be introduced:**

$$\left.\begin{array}{l} \text{a) } D:=\{y:\bar{\omega}\to R^n\},\ n:=|\bar{\omega}|,\ D \text{ is the set of all grid functions } y \\ \quad \text{defined on } \bar{\omega},\ y \text{ denotes a vector of } R^n \text{ at the same time;} \\ \text{b) } D_0:=\{y\in D: y|_{\gamma_1}=0\},\ \text{the set of all grid functions vanishing on} \\ \quad \gamma_1,\ D_0\subset D; \\ \text{c) } D':=\{y:\bar{\omega}'\to R^{n'}\},\ n':=|\bar{\omega}'|,\ \text{the grid functions} \\ \quad \text{defined on } \bar{\omega}' \text{ (midpoints of mesh segments } \hbar); \\ \text{d) } D'':=\{y:\bar{\omega}''\to R^{n''}\},\ n'':=|\bar{\omega}''|,\ \text{the grid functions on } \bar{\omega}'' \\ \quad \text{(centres of gravity);} \\ \text{e) } \tilde{D}:=\{y:\omega+\gamma_{23}\to R^{\tilde{n}}\},\ \tilde{n}:=|\omega+\gamma_{23}|,\ \text{the grid functions on } \omega+\gamma_{23}; \\ \text{f) } \tilde{D}':=\{y:\omega'+\gamma_{23}'\to R^{\tilde{n}'}\},\ \tilde{n}':=|\omega'+\gamma'_{23}|,\ \text{the grid functions} \\ \quad \text{on } \omega'+\gamma_{23}',\ \text{with } \omega',\gamma_{23}' \text{ from (2.3.14).} \equiv \end{array}\right\} \quad (2.2.8)$$

2.2.2. Locally irregular networks. In order to obtain a simple approximation of the given domain $\bar{\Omega}$, the so-called "locally irregular" networks or grids are often used in applications and theoretical investigations. In the case of bounded plane domains, these are networks or grids with a number of $O(h^{-2})$ equilateral triangles or uniform rectangles and only $O(h^{-1})$ nonuniform ones. To explain this more precisely, we introduce some notions.

Definition 2.9

a) A grid point $x\in\bar{\omega}$ and a difference star $S(x)$ ($S_e(x)$) are called regular, if six equilateral triangles Δ or four rectangles $\square$ of the same shape meet at the grid point x, or, if these four rectangles are subdivided into triangles by diagonals with the same direction, or, for grid points $x\in\gamma_1$, if $S(x)=\{x\}$ holds. See Figs. 4a, b, c for the three typical cases. Otherwise, x and $S(x)$ ($x\in\bar{\omega}\setminus\gamma_1$) are called irregular.

b) A network or the grid $\bar{\omega}$ assigned to the network is called regular, if the triangulation $\mathcal{T}_\Delta$ from (2.2.3b) is built up of equilateral triangles or of rectangles of the same shape or of triangles which arise after subdivision of these rectangles, cf. a). Otherwise, the network or the grid $\bar{\omega}$ is called irregular.

c) A network or the grid $\bar{\omega}$ assigned to the network is called locally irregular, if it is regular with the only exception that the triangles Δ or rectangles $\square$ can be arbitrary in some strips of width $O(h)$. The number of these strips lying in $\bar{\Omega}_h$ is fixed. $\equiv$

Obviously, if a grid $\bar{\omega}$ is regular, then all interior grid points $x\in\mathring{\omega}$ are regular. Grid points $x\in\gamma_{23}$ are irregular, since they are located on the boundary. Locally irregular grids $\bar{\omega}$ can be generated by cutting off or deforming a given infinite regular network under the influence of the boundary Γ or of an interface Γ_0 within Ω, i.e., the regular network will be changed within a strip of width $O(h)$, see e.g. MITCHELL/

GRIFFITHS [1], SAMARSKIJ[1], HEINRICH/FÖRSTER [1]. Another type of locally irregular networks arises, if Ω is divided into n subdomains Ω_s, each Ω_s is covered by a regular network and the mesh width changes near or on the boundaries of the subdomains Ω_s. The advantages of such networks consist in the easy automatic generation on a computer, the topological equivalence to a regular network in most cases, the higher accuracy for sufficiently smooth solutions, the applicability of efficient solvers and in a relatively small data storage for describing the geometry of the network. Otherwise, we always have more grid points than in the case of irregular networks which take into account complicated geometry of Ω and properties of the solution u (e.g. singularities).

Fig. 4a

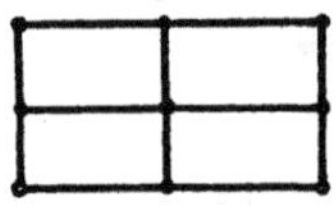
Fig. 4b

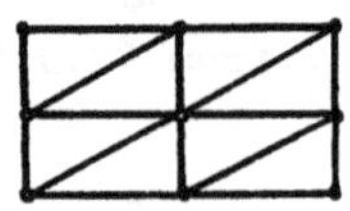
Fig. 4c

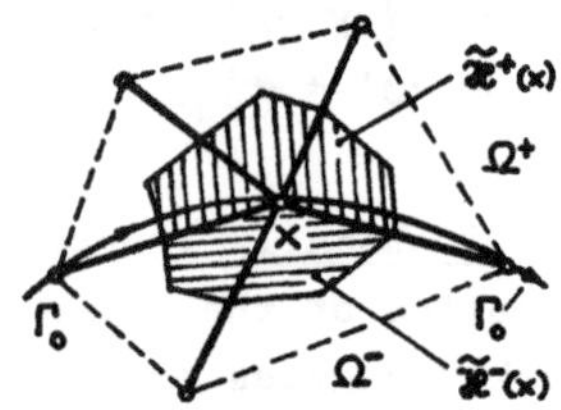
Fig. 4g

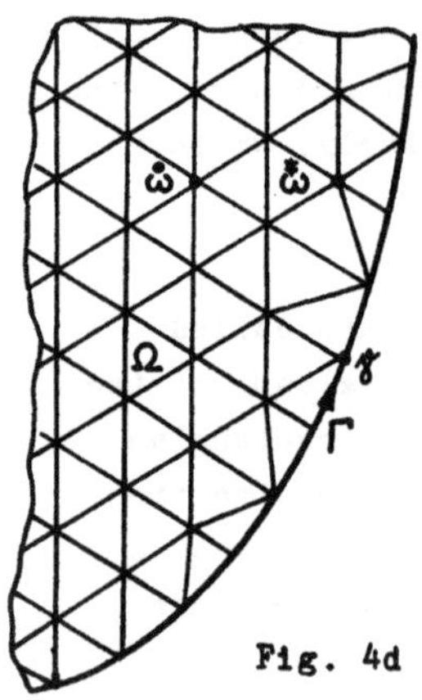
Fig. 4d

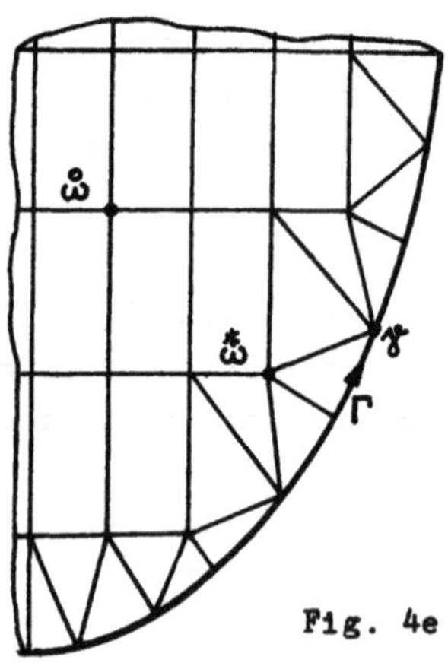
Fig. 4e

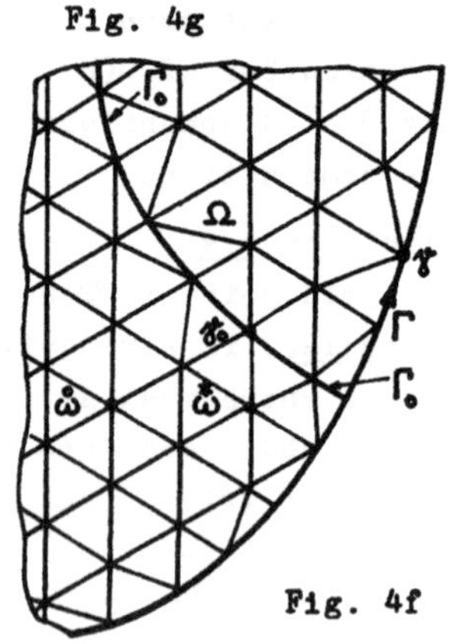
Fig. 4f

In the following, we shall define some locally irregular networks used in this monograph.

<u>Definition 2.10.</u> Locally irregular network of equilateral triangles with Γ-adaption: Let $\overline{\omega}$ be a grid with the properties

a) $\overline{\omega} = \omega + \gamma$ with the subsets $\mathring{\omega}, \overset{*}{\omega}, \gamma$ from (2.2.6), but for the (2.2.9)

b) special case of $\mathring{\omega}$: S(x) is a regular 7-point difference star for each $x \in \mathring{\omega}$, i.e. six equilateral triangles form the meshes for S(x), cf. Fig. 4a. ≡

Hence, we have irregular difference stars S(x) only for grid points $x \in \overset{*}{\omega}$ near the boundary Γ and boundary points $x \in \gamma_{23}$, see Fig. 4d.

<u>Definition 2.11.</u> Locally irregular network with rectangles and triangle adaption to Γ: Let $\overline{\omega}$ be a grid with the properties

a) $\bar{\omega}=\omega+\gamma$ with the subsets $\overset{\circ}{\omega}, \overset{*}{\omega}, \gamma$ from (2.2.6), but for the following (2.2.10)

b) special case of $\overset{\circ}{\omega}$: $\overset{\circ}{\omega}$ consists of the nodes of a regular network of rectangles with mesh lines parallel to the coordinate axes, cf. Fig. 4e. ≡

Therefore, the strictly interior grid points $x\in\overset{\circ}{\omega}$ have a 5- or 9-point regular difference star $S(x)$ or $S_e(x)$, respectively. Irregular grid points originate at most from $\overset{*}{\omega}+\gamma_{23}$, since the elements assigned to $x\in\overset{*}{\omega}+\gamma_{23}$ are arbitrary rectangles and (or) triangles.

Definition 2.12. Locally irregular network of right triangles with Γ-adaption: Let $\bar{\omega}$ be the grid (2.2.10), but with diagonals of the same direction in each rectangle. ≡

Definition 2.13. Locally irregular networks of equilateral triangles with Γ-Γ_0-adaption: If there is an interface Γ_0 caused by discontinuities of the right-hand side f from (2.1.1a) (or of a coefficient in L), with $\Gamma_0\in C^{0,1}\cap PC^2$, then we make an adaption of the network of triangles with respect to $\hat{\Gamma}=\Gamma\cup\Gamma_0$ in the way described for Γ in HEINRICH/FÖRSTER [1], cf. Fig. 4f. Then we have

a) $\bar{\omega}=\omega+\gamma$ with ω and γ as in (2.2.6), but for the following

b) special case of $\overset{*}{\omega},\overset{\circ}{\omega}$: $\overset{*}{\omega}:=\{x\in\omega: S(x)\cap(\gamma_0+\gamma)\neq\emptyset\}$, (2.2.11)
with $\gamma_0:=\omega\cap\Gamma_0$, $\overset{\circ}{\omega}:=\omega\setminus\overset{*}{\omega}$ (cf. footnote on p. 26). ≡

In this case, regular difference stars $S(x)$ arise at the strictly interior grid points $x\in\overset{\circ}{\omega}$, which do not lie on Γ or Γ_0, and we have irregular difference stars $S(x)$ at the grid points $x\in\overset{*}{\omega}+\gamma_{23}$ at most. The set $\overset{*}{\omega}$ comprises the grid points close to γ, γ_0 and on γ_0, cf. Fig. 4f.

2.3. Secondary networks and boxes

2.3.1. General remarks. The so-called "balance method" or "box integration method" (briefly: box method) discussed here is described in the literatur as an appropriate approach to the FDM, especially, if the network is irregular, see e.g. BANK et al. [1], FRYASINOV [3,4], GOSMAN et. al.[1], HEIMEIER [1], HEINRICH [5,6], McNEAL [1], MITCHELL/GRIFFITHS [1], REICHERT [1], REISSMANN [1,2,3], RICE/BOISVERT [1], SAMARSKIĬ/ANDREEV [1], SELBERHERR [1], VARGA [1], WEILAND [1], WINSLOW [1], WOLFF/MÜLLER [1]. This method is used mainly for the approximate solution of field problems and preferably by engineers and physicists due to the fact that the so-called balance equation (2.3.2) has a real physical background. For instance, if we consider Poisson's equation $-\Delta u = f$ in the domain Ω and for $u\in W_2^2(\Omega)$, $f\in L_2(\Omega)$, then we have as a consequence of Gauss' formula, cf. Appendix DI,

$$\int_{\mathcal{X}}\Delta u\,dx=\int_{\partial\mathcal{X}}\frac{\partial u}{\partial n}\,ds \text{ for } \mathcal{X}\subset\bar{\Omega},\ \partial\mathcal{X}\in C^{0,1}\cap PC^2, \tag{2.3.1}$$

which implies

$$-\int_{\partial\mathcal{X}}\frac{\partial u}{\partial n}\,ds=\int_{\mathcal{X}}f\,dx. \tag{2.3.2}$$

Here, $\mathcal{K}$ is a closed "finite region", i.e. $\mathcal{K}=\bar{\mathcal{K}}$, $\mathcal{K}\subset\bar{\Omega}$; $\mathcal{K}$ is related to a finite element. The boundary $\partial\mathcal{K}$ is of the same type as $\partial\Omega$ described in V($\bar{\Omega}$). The relation (2.3.2) expresses the balance of the source term $\int f dx$ and the flux $\int \frac{\partial u}{\partial n} ds$ crossing the boundary $\partial\mathcal{K}$, where u is the diffusing quantity of interest. If $\mathcal{K}$ is a "small" subset of $\bar{\Omega}$ in the sense that $\mathcal{K}$ has approximately the size of a mesh, then the relation (2.3.2) allows us to construct a finite difference equation approximating locally Poisson's equation $-\Delta u = f$. In the following, we use this box method to derive FDSs for more general elliptic differential equations of second order and mixed boundary conditions. At the first stage, it will be necessary to define the finite regions $\mathcal{K}$, the so-called boxes. We present two methods of defining the boxes, namely, boxes via the perpendicular bisectors (PB) and via the medians (MD) of the triangles and rectangles. The author believes that the former was originally studied in McNEAL [1] and the latter in WINSLOW [1], but without strict justification and largely from a physical point of view. We now introduce two types of secondary triangulations which correspond to $\mathcal{T}_{\square}$ and $\mathcal{T}_{\Delta}$.

Assumption 2.14. - V($\mathcal{T}_{\mathcal{K}}$), V($\mathcal{T}_{\tilde{\mathcal{K}}}$). The primary networks $\mathcal{T}_{\square}$ and $\mathcal{T}_{\Delta}$ for $\bar{\Omega}$ and $\bar{\Omega}_h$, cf. (2.2.3), are associated with so-called secondary networks or secondary triangulations $\mathcal{T}_{\mathcal{K}}$ and $\mathcal{T}_{\tilde{\mathcal{K}}}$ by defining boxes $\mathcal{K}(x)$ and $\tilde{\mathcal{K}}(x)$, respectively, for each grid point $x\in\bar{\omega}$. The secondary triangulations are provided with the following properties:

$$\mathcal{T}_{\mathcal{K}}: \bar{\Omega} = \bigcup_{x\in\bar{\omega}} \mathcal{K}(x),\ \dot{\mathcal{K}}(x)\cap\dot{\mathcal{K}}(\bar{x}) = \emptyset \quad \text{for } x,\ \bar{x}\in\bar{\omega},\ x\neq\bar{x}, \text{ and } h\leq h_0, \tag{2.3.3a}$$

$$\mathcal{T}_{\tilde{\mathcal{K}}}: \bar{\Omega}_h = \bigcup_{x\in\bar{\omega}} \tilde{\mathcal{K}}(x),\ \dot{\tilde{\mathcal{K}}}(x)\cap\dot{\tilde{\mathcal{K}}}(\bar{x}) = \emptyset \text{ for } x,\ \bar{x}\in\bar{\omega},\ x\neq\bar{x}, \text{ and } h\leq h_0. \tag{2.3.3b}$$

The boundaries $\partial\mathcal{K}(x)$ of the boxes $\mathcal{K}(x)$ are polygonal for $x\in\omega$ and partially curved for $x\in\gamma$. But the boundary $\partial\tilde{\mathcal{K}}(x)$ of $\tilde{\mathcal{K}}(x)$ is always polygonal, $\tilde{\mathcal{K}}(x)$ and $\mathcal{K}(x)$ coincide for $x\in\omega$ and, for $x\in\gamma$, $\tilde{\mathcal{K}}(x)$ is a polygonal approximation of $\mathcal{K}(x)$:

$$\tilde{\mathcal{K}}(x) = \mathcal{K}(x) \text{ for } x\in\omega,\quad \tilde{\mathcal{K}}(x)\approx\mathcal{K}(x) \text{ for } x\in\gamma. \tag{2.3.3c}$$

Since two types of secondary networks are taken into account, viz. (PB) and (MD), the boxes are distinguished by the subscripts p and m, i.e. we write $\mathcal{K}_p$, $\tilde{\mathcal{K}}_p$ and $\mathcal{K}_m$, $\tilde{\mathcal{K}}_m$, respectively. ≡

Clearly, $\mathcal{T}_{\mathcal{K}}$ and $\mathcal{T}_{\tilde{\mathcal{K}}}$ define respectively subdivisions of $\bar{\Omega}$ and $\bar{\Omega}_h$, but now with respect to the grid $\bar{\omega}$. At the same time, they correspond to $\mathcal{T}_{\square}$ and $\mathcal{T}_{\Delta}$, where the partition of $\bar{\Omega}$ and $\bar{\Omega}_h$ is made by finite elements e. Thus, $\mathcal{T}_{\square}$ and $\mathcal{T}_{\mathcal{K}}$, or $\mathcal{T}_{\Delta}$ and $\mathcal{T}_{\tilde{\mathcal{K}}}$ constitute dual systems of networks.

2.3.2. Secondary networks via the perpendicular bisectors (PB)

Let us now define more precisely the secondary networks of the type (PB) and the boxes $\mathcal{K}_p$, $\tilde{\mathcal{K}}_p$ which were already mentioned in Assumption 2.14.

Definition 2.15. Each grid point $x\in\bar{\omega}$ is associated with boxes $\mathcal{K}_p(x)$ and $\tilde{\mathcal{K}}_p(x)$ which can be described as follows (the subscript p is temporarily omitted): The points of intersection of the perpendicular bisectors of the triangle and rectangle sides, i.e., the circumcentres of Δ and $\square$, around each grid point x will be determined. In the case $x \in \overline{\omega}$, with $\mathfrak{k} \in \omega$, two such circumcentres correspond to each side $\mathfrak{h}(x,\mathfrak{k})$ and, in special cases, they can coincide. These two points will be connected by a perpendicular bisector segment $\mathfrak{S}(x,\mathfrak{k})$, which is oriented counter-clockwise about the point x, see Figs. 5a, b. In the case $x \in \gamma$, with $\mathfrak{k} \in \gamma$, the segment between the point of intersection of the perpendicular bisectors and the "perforation point" of the bisector on Γ will be taken and designated by $\mathfrak{S}(x,\mathfrak{k})$. Moreover, $\mathfrak{S}(x)$ denotes the arc lying on Γ between the two perforation points $x_\Gamma^{\pm} \in \Gamma$. For details, see Fig. 5b (triangles) and Fig. 5e (rectangles). Now the boxes $\mathfrak{B}(x)$ are defined as the region enclosed in the boundary $\partial\mathfrak{B}(x)$ described by

$$\partial\mathfrak{B}(x) := \bigcup_{\mathfrak{k} \in S'(x)} \mathfrak{S}(x,\mathfrak{k}) \cup \mathfrak{S}(x) \text{ for } x \in \overline{\omega},$$

$$\mathfrak{S}(x) := \begin{cases} \emptyset & \\ \mathfrak{S}^-(x) \cup \mathfrak{S}^+(x) & \end{cases} \text{for} \begin{matrix} x \in \omega, \\ x \in \gamma, \end{matrix} \qquad \mathfrak{S}^{\pm}(x) := \widehat{xx_\Gamma^{\pm}}. \tag{2.3.4}$$

The box $\tilde{\mathfrak{B}}(x)$ is given by $\tilde{\mathfrak{B}}(x) := \mathfrak{B}(x)$ for $x \in \omega$ and, for $x \in \gamma$, $\tilde{\mathfrak{B}}(x)$ is a polygonal approximation of $\mathfrak{B}(x)$, where the edges $\mathfrak{S}(x,\mathfrak{k}), \mathfrak{k} \in \gamma$, are modified and $\mathfrak{S}(x)$ is replaced by a polygonal approximation $\tilde{\mathfrak{S}}(x)$. The modifications $\tilde{\mathfrak{S}}(x,\mathfrak{k})(x,\mathfrak{k} \in \gamma)$ originate from $\mathfrak{S}(x,\mathfrak{k})$ by prolongation (shortening) of $\mathfrak{S}(x,\mathfrak{k})$ by the straight segment $\delta(x,\mathfrak{k}) := \overline{x'x_\Gamma}$ located between $x'_{\mathfrak{k}} \in \mathfrak{h}(x,\mathfrak{k})$ and $x_\Gamma \in \Gamma$ in the case a) (case b)) of (2.2.1), see e.g. Fig. 5c. The box $\tilde{\mathfrak{B}}(x)$ is defined by the region enclosed in $\partial\tilde{\mathfrak{B}}(x)$, where

$$\partial\tilde{\mathfrak{B}}(x) := \bigcup_{\mathfrak{k} \in S'(x)} \tilde{\mathfrak{S}}(x,\mathfrak{k}) \cup \tilde{\mathfrak{S}}(x) \text{ for } x \in \gamma, \text{ with } \tilde{\mathfrak{S}}(x,\mathfrak{k}) := \mathfrak{S}(x,\mathfrak{k}) \text{ for } \mathfrak{k} \in \omega,$$

$$\tilde{\mathfrak{S}}(x) := \tilde{\mathfrak{S}}^-(x) \cup \tilde{\mathfrak{S}}^+(x), \; \tilde{\mathfrak{S}}^{\pm}(x) := \overline{xx'_{\mathfrak{k}_\Gamma^{\pm}}}, \quad \mathfrak{k}_\Gamma^{\pm} \in \gamma. \equiv \tag{2.3.5}$$

In Figs. 5b, c, a special situation is shown, where the boundary Γ is even concave (convex) over the segment $\mathfrak{h}(x,\mathfrak{k})$ and, therefore, case a) (case b)) is given. The symbols $\tilde{\mathfrak{S}}^{\pm}(x)$ in (2.3.5) describe the segments between the points x and $x'_{\mathfrak{k}_\Gamma^{\pm}} \in \mathfrak{h}(x,\mathfrak{k}_\Gamma^{\pm})$. We remark that all boundaries $\partial\mathfrak{B}(x)$, $x \in \omega$, and $\partial\tilde{\mathfrak{B}}(x)$, $x \in \gamma$, are purely polygonal, and $\partial\mathfrak{B}(x)$, $x \in \gamma$, may be partially curved due to the arcs $\mathfrak{S}^{\pm}(x)$. We sometimes further split the edges $\mathfrak{S}(x,\mathfrak{k})$, viz.

$$\mathfrak{S}(x,\mathfrak{k}) := \mathfrak{S}^-(x,\mathfrak{k}) \cup \mathfrak{S}^+(x,\mathfrak{k}), \tag{2.3.6}$$

where $\mathfrak{S}^{\pm}(x,\mathfrak{k})$ denotes that part of $\mathfrak{S}(x,\mathfrak{k})$ which is lying "before" (+) and "behind" (-) the point $x'_{\mathfrak{k}} \in \mathfrak{h}(x,\mathfrak{k})$, cf. Fig. 5a. For simplicity of some formulas, we often take the symbol $\mathfrak{S}^o(x,\mathfrak{k})$ instead of $\tilde{\mathfrak{S}}(x,\mathfrak{k})$, i.e.

$$\mathfrak{S}^o(x,\mathfrak{k}) := \tilde{\mathfrak{S}}(x,\mathfrak{k}) := \mathfrak{S}(x,\mathfrak{k}) \text{ for } x \in \overline{\omega}, \mathfrak{k} \in \omega, \quad \mathfrak{S}^o(x,\mathfrak{k}) := \tilde{\mathfrak{S}}(x,\mathfrak{k})$$
$$\text{for } x \in \gamma, \mathfrak{k} \in \gamma \text{ according to Definition 2.15.} \tag{2.3.7}$$

In general, the construction of the boxes $\mathfrak{B}(x)$ and $\tilde{\mathfrak{B}}(x)$ via method (PB) can be performed correctly for such networks only which satisfy some restrictions, e.g., if the following "grid regularity condition"

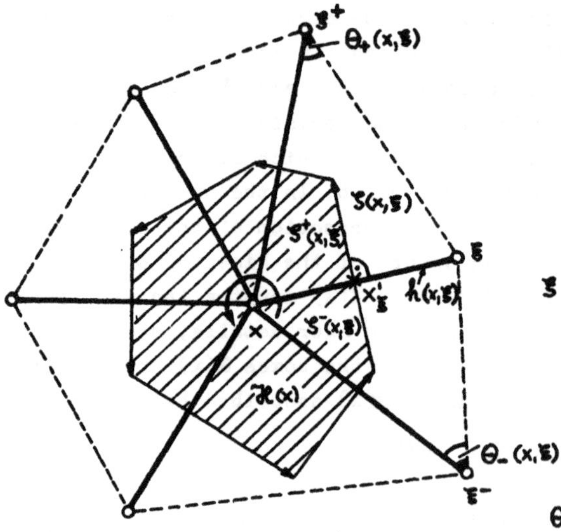

Fig. 5a

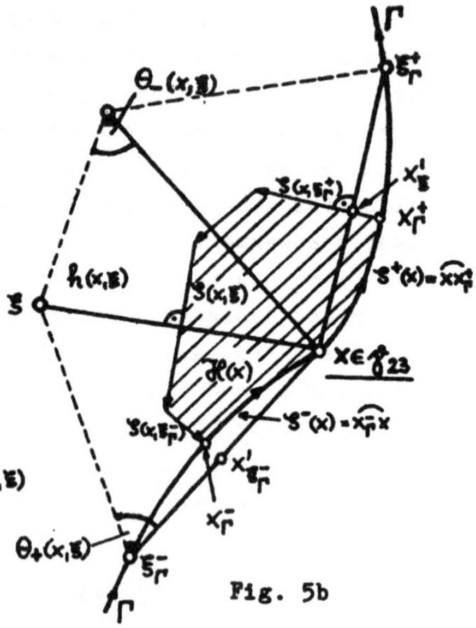

Fig. 5b

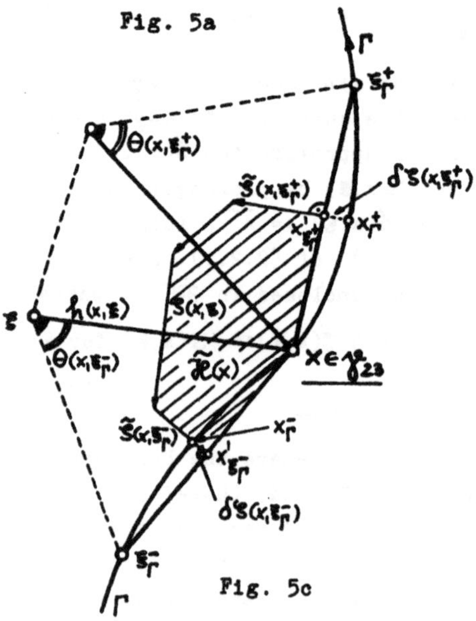

Fig. 5c

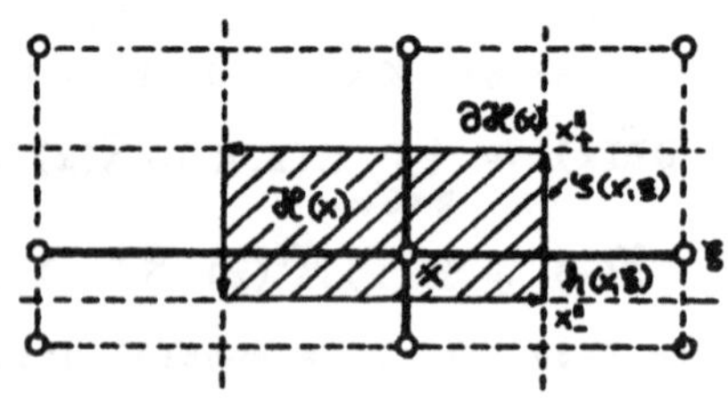

Fig. 5d

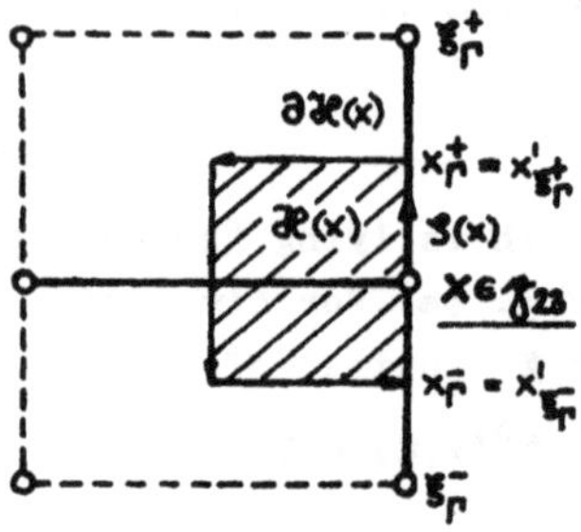

Fig. 5e

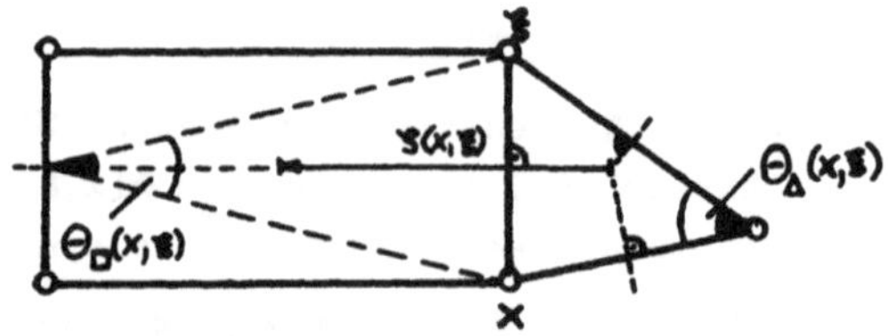

Fig. 5f

$VO(PB,\bar{\omega})$ is fulfilled and, if $h \leq h_o$ holds, where h_o is sufficiently small.

<u>Condition 2.16. - $VO(PB,\bar{\omega})$.</u> The mesh segments $h(x,\xi)$ and the interior angles θ of the triangles Δ satisfy the following conditions T0, W0, W1 W2, W3 (if rectangles occur) for any $h, 0 < h \leq h_o$, where the constants θ_o, ε_o $(0 < \theta_o < \pi/2,\ 0 < \varepsilon_o \leq 1)$ do not depend on h:

T0: $0 < \varepsilon_o h \leq \text{mes}\, h(x,\xi) \leq h/\varepsilon_o$ for all mesh segments $h(x,\xi)$, with $x,\xi \in \bar{\omega}$; (2.3.8a)

W0: $0 < \theta_o \leq \theta \leq \pi - \theta_o$ for the angles θ of all triangles Δ; (2.3.8b)

W1: $\theta_- + \theta_+ \leq \pi$ for a pair of angles $\theta_\pm := \theta_\pm(x,\xi)$, (2.3.8c)
where $\theta_\pm(x,\xi)$ are the angles at the vertices opposite to the side $h(x,\xi)$, which is the common edge of the two triangles $\Delta^\pm$, cf. Figs. 5a, b, and this holds for each mesh segment $h(x,\xi)$ with two adjacent triangles $\Delta^\pm$;

W2: $\theta \leq \pi/2$ for the angle $\theta := \theta(x,\xi)$ at the vertex opposite (2.3.8d)
to the side $h(x,\xi)$ in the case $x,\xi \in \gamma$, and this holds at least for each $h(x,\xi) \subset \Gamma_h$, where Γ_h is the boundary of Ω_h, cf. Fig. 5c; moreover, instead of $\theta \leq \pi/2$ the strengthened inequality
$\theta \leq \pi/2 - \theta_o$ is fulfilled, if int $h(x,\xi) \cap \bar{\Omega} = \emptyset$ (case a) (2.3.8e)
of (2.2.1)) holds or if the case $j = 1$ of the FDS $A_h^j y = F_h^j$ is considered, cf. Fig. 5c (the side $h(x,\xi_\Gamma^-)$ with the angle $\theta(x,\xi_\Gamma^-)$); for $A_h^1 y = F_h^1$ see Theorem 3.1.

W3: $\theta_\square + \theta_\Delta \leq \pi$ for the angles $\theta_\square(x,\xi), \theta_\Delta(x,\xi)$ defined in (2.3.8f)
Fig. 5f, if a rectangle $\square$ and a triangle Δ touch each other along the side $h(x,\xi)$. ■

Clearly, if $VO(PB,\bar{\omega})$ is employed in such cases where the secondary network (PB) is of interest only and FDSs $A_h^j y = F_h^j$ are not taken into account, then the restriction concerning $j = 1$ in (2.3.8e) is meaningless and can be omitted. The conditions W0, T0 are basic assumptions on the asymptotic geometrical behaviour of the network, and W1, W2 concern the angles opposite to the sides $h(x,\xi)$ for the cases int $h(x,\xi) \subset \Omega_h$ and $h(x,\xi) \subset \Gamma_h$, respectively. Sometimes, beside condition W1, W2 from $VO(PB,\bar{\omega})$, we consider the following strengthened angle restrictions

$$\begin{aligned} &W1' : \theta_- + \theta_+ < \pi, \qquad W1'': \theta_- + \theta_+ \leq \pi - \theta_o,\ \theta_o > 0, \\ &W2' : \theta < \pi/2 \qquad, \quad W2'': \theta \leq \pi/2 - \theta_o,\ \theta_o > 0. \end{aligned} \tag{2.3.9}$$

We denote by $VO''(PB,\bar{\omega})$ that grid regularity condition which arises from $VO(PB,\bar{\omega})$ after substituting the angle inequalities W1, W2 by W1", W2" from (2.3.9). We also consider the more general condition $VO''(PB,\omega^1)$.

<u>Condition 2.17.-$VO''(PB,\omega^1)$.</u> Condition 2.16.-$VO(PB,\bar{\omega})$ and W1",W2" instead of W1,W2 for the triangles Δ associated with irregular grid points. If all grid points $x \in \bar{\omega}$ are taken into account, then $VO''(PB,\bar{\omega})$ will be written. ■

The meaning of the inequalities restricting the angles of the triangles follows from Remark 2.18.

Remark 2.18. For the segment $\mathfrak{S}(x,\xi)$ joining the circumcentres of the triangles Δ^- and Δ^+ having $\mathfrak{h}(x,\xi)$ as a common side, the following assertions a) and b) hold under the assumptions W1, W1', W1" from (2.3.8c), (2.3.9):

a) The segment $\mathfrak{S}(x,\xi)$ is correctly defined, i.e., the circumcentres $p^{\pm}$ of the triangles Δ^- and Δ^+, with the common side $\mathfrak{h}(x,\xi)$, follow each other counterclockwise about the point x in the same sense as Δ^+ follows Δ^-.

b) The following inequalities are satisfied (correspondence of the rows):

$$s(x,\xi) := \operatorname{mes} \mathfrak{S}(x,\xi) \begin{cases} = 0 \\ > 0 \\ \geq \varepsilon h > 0 \end{cases} \quad \text{for} \quad \theta_- + \theta_+ \begin{cases} = \pi, \\ < \pi, \\ \leq \pi - \theta_0. \end{cases} \qquad (2.3.10)$$

c) For the "shared" perpendicular bisectors $\mathfrak{S}^{\pm}(x,\xi)$ of the triangles $\Delta(x,\xi,\xi^{\pm})$ with angles $\theta_{\pm}(x,\xi)$ satisfying W2, W2' and W2" from (2.3.8d), (2.3.9), the following relations are fulfilled (correspondence of the rows):

$$s^{\pm}(x,\xi) := \operatorname{mes} \mathfrak{S}^{\pm}(x,\xi) \begin{cases} = 0 \\ > 0 \\ \geq \varepsilon h > 0 \end{cases} \quad \text{for } \theta \begin{cases} = \pi/2, \\ < \pi/2, \\ \leq \pi/2 - \theta_0. \end{cases} \qquad (2.3.11)$$

In (2.3.10) and (2.3.11), ε is a positive constant depending on θ_0, ε_0 from (2.3.8a, b), but ε is independent of h, $h \leq h_0$. ≡

The proof of these assertions rests on well-known basic relations of geometry, cf. Appendix GE. The grid regularity condition VO(PB,$\bar{\omega}$) ensures, as a consequence of Remark 2.18., that distinct boxes $\mathfrak{W}(x)$ do not overlap. In particular, (2.3.8d) and (2.3.8e) guarantee that the coefficients $s^j(x,\xi)$, cf. (3.2.18c), of the difference operator A_h^j from (3.2.18) are at least nonnegative even in the following more complicated cases: $\operatorname{int} \mathfrak{h}(x,\xi) \cap \bar{\Omega} = \emptyset$ for $x,\xi \in \mathcal{J}$, or the difference operator $A_h^j := A_h^1$ (j=1), with higher accuracy at $x\in\mathcal{J}_{23}$, is taken into account. If there are no further restrictions on ε_0 and θ_0 from (2.3.8), then, even under the condition VO(PB,$\bar{\omega}$), the following cases can arise, e.g.: The perpendicular bisector segment $\mathfrak{S}(x,\xi)$ adjacent to $\mathfrak{h}(x,\xi)$ does not intersect the side $\mathfrak{h}(x,\xi)$, but $\mathfrak{S}(x,\xi)$ is a subset of $\Delta^- \cup \Delta^+$, or, the segment $\mathfrak{S}(x,\xi)$ has a nonempty intersection with $\mathfrak{h}(x,\xi)$, but $\mathfrak{S}(x,\xi)$ is not a subset of $\Delta^- \cup \Delta^+$ In order to avoid such complicated cases in some considerations, we shall introduce a strengthened grid regularity condition V1.

Condition 2.19. - V1(PB,$\bar{\omega}$). The condition VO(PB,$\bar{\omega}$) and, additionally, the inequality

$\theta \leq \pi/2$ for each angle θ of Δ and all triangles Δ (2.3.12)

are assumed to be satisfied. ≡

If we suppose the strict inequalities W2' or W2" from (2.3.9) instead of (2.3.12), then we shall write V1' or V1", respectively. In the literature on FEM, see e.g. CIARLET [2], IKEDA [1], networks of triangles with angle restrictions of the type $\theta \leq \pi/2$, $\theta < \pi/2$ or $\theta \leq \pi/2 - \theta_0$ are sometimes said to be of the weakly acute, acute or strictly acute type, respectively.

<u>Remark 2.20</u>. Under the conditions V1, V1' and V1"(PB,$\overline{\omega}$), the following relations hold for the circumcentre $p(\Delta)$ of the triangle Δ and for the portion $\mathcal{S}^{\mathscr{h}}(\Delta)$ of the perpendicular bisector assigned to each side $\mathscr{h}$ of Δ

a) $p(\Delta) \in \Delta$ and $\mathcal{S}^{\mathscr{h}}(\Delta) \subset \Delta$ for the three portions $\mathcal{S}^{\mathscr{h}}(\Delta)$,

b) mes $\mathcal{S}^{\mathscr{h}}(\Delta)$ satisfies inequalities of the type (2.3.11). $\equiv$

The meaning of inequality (2.3.8f) in the case of coupled triangles and rectangles will become apparent by

<u>Remark 2.21.</u> For the segment $\mathcal{S}(x,\xi)$ joining the circumcentres of $\Box$ and Δ having the common side $\mathscr{h}(x,\xi) := \Box \cap \Delta$, inequality (2.3.8f), $\theta_\Box + \theta_\Delta \leq \pi$, suffices to prove mes $\mathcal{S}(x,\xi) \geq \varepsilon h > 0$, with ε independent of h, $h \leq h_0$. $\equiv$

We now introduce notations involving the measures (written in typescript letters) of some geometric sets (expressed by the corresponding script letters) which were mentioned above. Recalling (2.3.3) - (2.3.7) and Definition 2.15. concerning the sets $\mathcal{H}$, $\widetilde{\mathcal{H}}$, $\mathcal{S}$, $\widetilde{\mathcal{S}}$ $(=\mathcal{S}^0)$, $\mathcal{S}^\pm$ and $\widetilde{\mathcal{S}}^\pm$, we denote the measures of these sets by

$$\left.\begin{aligned} &H(x) := \text{mes}\,\mathcal{H}(x),\ h(x,\xi) := \text{mes}\,\mathscr{h}(x,\xi),\ s(x,\xi) := \text{mes}\,\mathcal{S}(x,\xi), \text{ and} \\ &\quad \tilde{s}(x,\xi)(=s^0(x,\xi)),\ s^\pm(x,\xi),\ \tilde{s}^\pm(x,\xi),\ s(x),\ \tilde{s}(x),\ s^\pm(x),\ \tilde{s}^\pm(x) \\ &\quad \widetilde{H}(x) \text{ etc.. quite similarly; additionally,} \\ &h(x) := \tfrac{1}{2}\left\{h(x,\xi_\Gamma^-) + h(x,\xi_\Gamma^+)\right\} \quad \text{for } x, \xi_\Gamma^\pm \in \gamma, \end{aligned}\right\} \quad (2.3.13)$$

and the index (x,ξ) will be often replaced by x'_ξ or x', cf. (2.3.14).

Obviously, $\tilde{s}(x) = h(x)$ holds because of $\tilde{s}^\pm(x) := h(x,\xi_\Gamma^\pm)/2$, where $x, \xi_\Gamma^\pm \in \gamma$.

Beside the primary grid $\overline{\omega}$, the corresponding grid $\overline{\omega}' = \omega' + \gamma'$ of all "flux points" x' will be considered, where $x' := x'_\xi = \xi'_x$ is defined as the midpoint of the mesh segment $\mathscr{h}(x,\xi)$ being a side of a triangle or a rectangle. The sets $\overline{\omega}'$, ω' and γ', together with some other subsets, are introduced in

<u>Definition 2.22.</u> Let $\overline{\omega}'$, ω', γ', γ'_{23} and γ'_2 be the following sets of midpoints of the mesh segments $\mathscr{h}(x,\xi)$:

$$\begin{aligned} &\overline{\omega}' := \{x'_\xi : x \in \overline{\omega}, \xi \in \overline{\omega}\},\ \omega' := \{x'_\xi \in \overline{\omega}' : x \in \omega\}, \\ &\gamma' := \{x'_\xi \in \overline{\omega}' : x \in \gamma, \xi \in \gamma\},\ \gamma'_{23} := \{x'_\xi \in \gamma' : x \in \gamma_{23}\}, \\ &\gamma'_2 := \{x'_\xi \in \gamma'_{23} : x \in \gamma_2\}. \equiv \end{aligned} \quad (2.3.14)$$

Here, ω' and γ' denote respectively the set of interior flux points and boundary flux points. In principle, the symbols x'_ξ and x' are used with the same meaning, but the long form $x'_{\mathfrak{s}}$ of x' is preferred in such cases where the position of $x,\mathfrak{s}$ in $\overline{\omega}$ or the orientation ($\mathfrak{s} \in S'(x)$) is of interest. Therefore, we have equality by definition for some symbols, e.g., $h(x') := h(x'_{\mathfrak{s}}) := h(x, \mathfrak{s})$, $s(x') := s(x'_{\mathfrak{s}}) := s(x, \mathfrak{s})$, etc.

2.3.3. Secondary networks via the medians (MD). In Section 2.3.1., we used the abbreviations (MD) and $\mathfrak{X}_m$, $\tilde{\mathfrak{X}}_m$ in order to indicate that the medians are taken for the secondary subdivision of $\overline{\Omega}$ and $\overline{\Omega}_h$. In this section, we shall turn to the definition of these boxes $\mathfrak{X}_m$, $\tilde{\mathfrak{X}}_m$, where the same symbols are used for analogous notions introduced in Section 2.3.2. Confusion of these symbols will not occur, since the methods (PB) and (MD) are either treated separately in some sections or the subscripts p and m indicate the kind of boxes taken into account.

Definition 2.23. Each grid point $x \in \overline{\omega}$ is associated with boxes $\mathfrak{X}_m(x)$ and $\tilde{\mathfrak{X}}_m(x)$ which can be described as follows (the subscript m is omitted momentarily): The centres of gravity $x'' \in S''(x)$ and the midpoints $x'_{\mathfrak{s}} \in S'(x)$ of the sides of the triangles Δ and rectangles $\square$ adjacent to x will be connected one after another and counterclockwise by straight segments ("partial medians"), such that a polygonal line $\partial\mathfrak{X}(x)$ is formed around $x \in \overline{\omega}$, which is closed for $x \in \omega$; see Fig. 6a. In the case $x, \mathfrak{s} \in \gamma$ with $\operatorname{int} h(x, \mathfrak{s}) \cap \overline{\Omega} = \emptyset$ (case a) in (2.2.1)), the partial median over the triangle side $h(x, \mathfrak{s})$ will be taken only up to the perforation point x_Γ on Γ. Otherwise, for $h(x,\mathfrak{s}) \subset \overline{\Omega}$ (case b)) in (2.2.1)), the end point $x'_{\mathfrak{s}}$ of the partial median will be connected with the midpoint x_Γ of the arc $\widehat{x\mathfrak{s}}$ lying on Γ by a straight segment. The points $x_\Gamma^{\pm}$ mentioned above and located respectively before ("+") and behind ("-") the point $x \in \gamma$ will be fitted by an arc $\mathfrak{S}(x)$ along Γ. Now, $\partial\mathfrak{X}(x)$ has been completely described as a closed curve assigned to $x \in \gamma$, see Fig. 6b, and $\partial\mathfrak{X}(x)$ is partially curved. The curve $\partial\mathfrak{X}(x)$ defines the box $\mathfrak{X}(x)$ as the area enclosed in $\partial\mathfrak{X}(x)$, together with the boundary itself. At $x \in \omega$, we choose $\tilde{\mathfrak{X}}(x) := \mathfrak{X}(x)$, and at the grid point $x \in \gamma$, each box $\mathfrak{X}(x)$ is associated with a polygonal approximation $\tilde{\mathfrak{X}}(x)$. The boundary $\partial\tilde{\mathfrak{X}}(x)$ of $\tilde{\mathfrak{X}}(x)$ arises from $\partial\mathfrak{X}(x)$ by taking the partial medians $\tilde{\mathfrak{S}}(x,x'')$ in the straight triangle $\Delta(x,x'')$ instead of $\mathfrak{S}(x,x'')$ from the curved triangle $\Delta(x,x'')$ near Γ (i.e., take $\tilde{\mathfrak{S}}(x,\mathfrak{s})$ over $h(x,\mathfrak{s})$, with $x,\mathfrak{s} \in \gamma$, instead of $\mathfrak{S}(x,\mathfrak{s})$); cf. (2.3.15), (2.3.16) for $\mathfrak{S}$ $\tilde{\mathfrak{S}}$. Finally, the curve $\mathfrak{S}(x) := \widehat{x_\Gamma^- x} \cup \widehat{x x_\Gamma^+}$ on Γ will be approximated by the polygonal $\tilde{\mathfrak{S}}(x)$ connecting the points $x'_{\mathfrak{s}_\Gamma^-}$, x and $x'_{\mathfrak{s}_\Gamma^+}$; see Fig. 6c. ≡

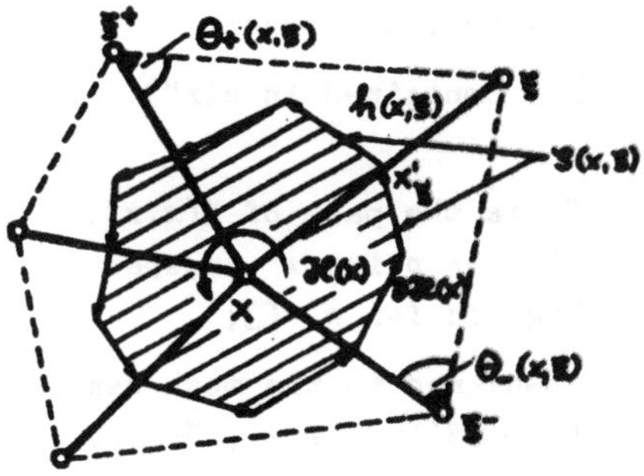

Fig. 6a

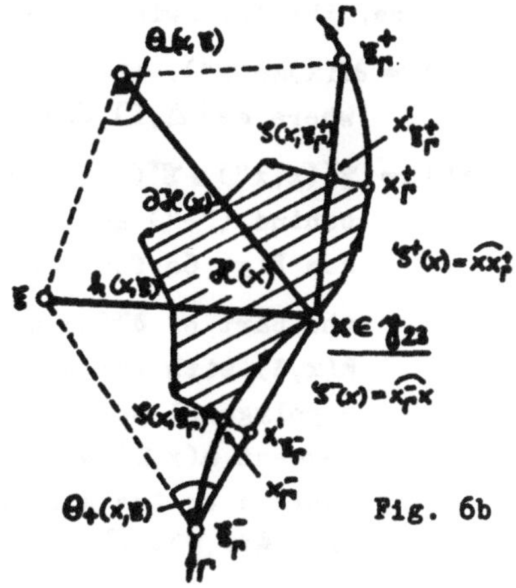

Fig. 6b

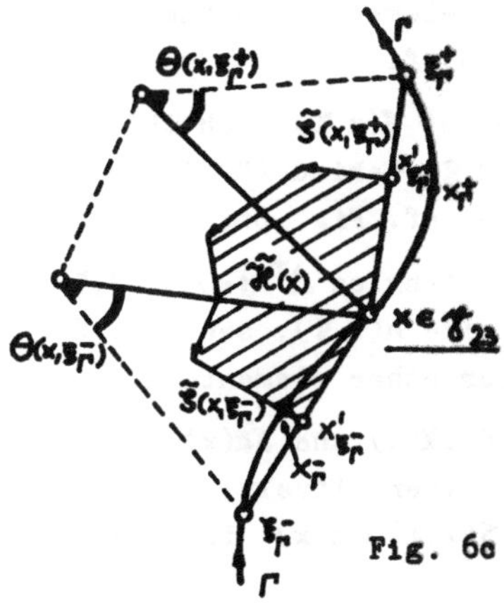

Fig. 6c

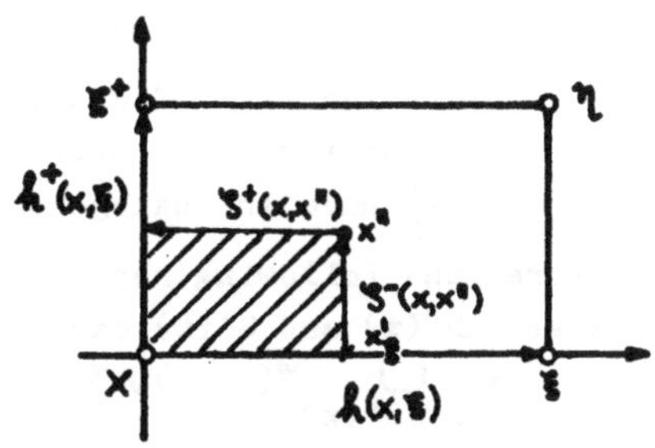

Fig. 6d

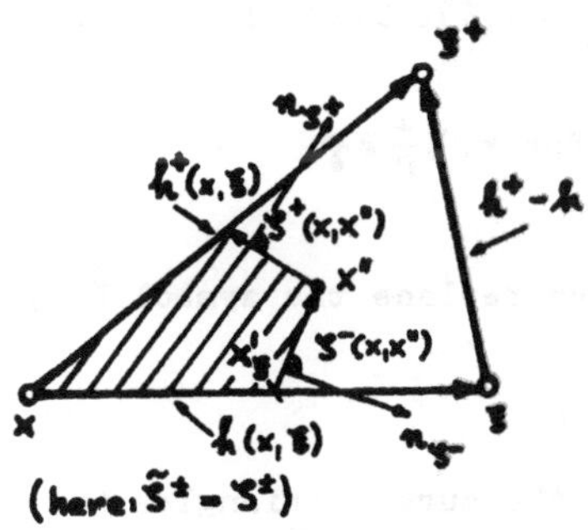

Fig. 6e

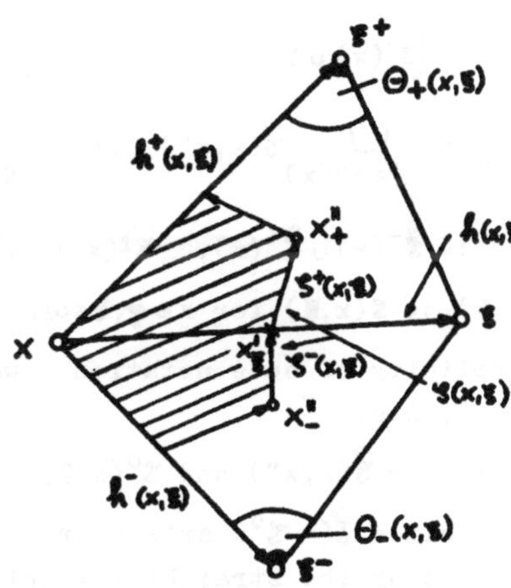

Fig. 6f

Furthermore, the following notations will be introduced:

$$
\left.
\begin{array}{lll}
\mathfrak{S}(x,x'') := \partial\mathfrak{K}(x)\cap e(x''), & \text{the part of } \partial\mathfrak{K}(x) \text{ contained in } e(x''), \\
 & \text{where } e\in\{\Delta,\Box,\triangle\}; \\
\mathfrak{S}(x,x'') := \mathfrak{S}^-(x,x'')\cup\mathfrak{S}^+(x,x''), & \mathfrak{S}^{\pm}(x,x'') \text{ is the part of } \mathfrak{S}(x,x'') \\
 & \text{behind ("-") and before ("+") the point } x'', \text{ respectively;} \\
 & \text{see Fig. 6e for } e=\Delta \text{ and Fig. 6d for } e=\Box; \\
\mathfrak{S}(x,\mathfrak{x}): & \text{the part of } \partial\mathfrak{K}(x) \text{ which is assigned to the mesh segment} \\
 & \mathfrak{h}(x,\mathfrak{x}) \text{ with the running point } \mathfrak{x}\in S'(x), \text{ i.e., } \mathfrak{S}(x,\mathfrak{x}) \text{ is} \\
 & \text{a polygonal portion of } \partial\mathfrak{K}(x) \text{ behind and before the} \\
 & \text{segment } \mathfrak{h}(x,\mathfrak{x}) \text{ which is the common side of two ele-} \\
 & \text{ments } e \text{ or the side of a single element } e \text{ in the} \\
 & \text{case } \mathfrak{h}(x,\mathfrak{x})\subset\Gamma_h, \text{ cf. Figs. 6a, b, f.} \\
\mathfrak{S}(x,\mathfrak{x}) = \mathfrak{S}^-(x,\mathfrak{x})\cup\mathfrak{S}^+(x,\mathfrak{x}), & \text{the splitting of } \mathfrak{S}(x,\mathfrak{x}) \text{ such that} \\
 & \mathfrak{S}^{\pm}(x,\mathfrak{x}) \text{ denote the parts of } \mathfrak{S}(x,\mathfrak{x}) \text{ behind ("-") and} \\
 & \text{before ("+") the segment } \mathfrak{h}(x,\mathfrak{x}), \text{ cf. Fig. 6f;} \\
x''_{\pm}: & \text{the centres of gravity of two triangles } \Delta^{\pm} := \Delta(x,\mathfrak{x},\mathfrak{x}^{\pm}) \\
 & \text{touching each other along the side } \mathfrak{h}(x,\mathfrak{x}), \text{ see Fig. 6f;} \\
 & x''_{\pm} \text{ have an analogous meaning for other elements } e.
\end{array}
\right\}
\tag{2.3.15}
$$

Therefore, the following representations of $\partial\mathfrak{K}(x)$ and $\partial\tilde{\mathfrak{K}}(x)$ using the index sets $S''(x)$ and $S'(x)$ can be employed alternatively:

$$
\left.
\begin{array}{l}
\partial\mathfrak{K}(x) := \bigcup\limits_{x''\in S''(x)} \mathfrak{S}(x,x'')\cup\mathfrak{S}(x) = \bigcup\limits_{\mathfrak{x}\in S'(x)} \mathfrak{S}(x,\mathfrak{x})\cup\mathfrak{S}(x) \quad \text{for } x\in\bar{\omega}, \\
\mathfrak{S}(x) := \begin{cases} \emptyset & x\in\omega, \\ & \text{for} \\ \mathfrak{S}^-(x)\cup\mathfrak{S}^+(x) & x\in\gamma, \quad \mathfrak{S}^{\pm}(x) := \widehat{xx^{\pm}_{\Gamma}}, \end{cases} \\
\partial\tilde{\mathfrak{K}}(x) := \bigcup\limits_{x''\in S''(x)} \tilde{\mathfrak{S}}(x,x'')\cup\tilde{\mathfrak{S}}(x) = \bigcup\limits_{\mathfrak{x}\in S'(x)} \tilde{\mathfrak{S}}(x,\mathfrak{x})\cup\tilde{\mathfrak{S}}(x) \quad \text{for } x\in\gamma,
\end{array}
\right\}
\tag{2.3.16}
$$

with

$$\tilde{\mathfrak{S}}(x) := \tilde{\mathfrak{S}}^-(x)\cup\tilde{\mathfrak{S}}^+(x), \quad \tilde{\mathfrak{S}}^{\pm}(x) := \overline{xx^{\pm}_{\mathfrak{x}_{\Gamma}}}, \text{ for } x, \mathfrak{x}^{\pm}_{\Gamma}\in\gamma,$$

$$\tilde{\mathfrak{S}}(x,\mathfrak{x}) := \mathfrak{S}(x,\mathfrak{x}) \text{ for } x\in\gamma, \mathfrak{x}\in\omega.$$

For simplicity of some notations, we often replace the symbol $\tilde{\mathfrak{S}}$ by $\mathfrak{S}^0$ for all $\tilde{\mathfrak{S}}$, i.e.

$$\mathfrak{S}^0(x,x'') := \tilde{\mathfrak{S}}(x,x'') \text{ and } \mathfrak{S}^0(x,\mathfrak{x}) := \tilde{\mathfrak{S}}(x,\mathfrak{x}). \tag{2.3.17}$$

Clearly, since $\tilde{\mathfrak{S}}(=\mathfrak{S}^0)$ arises from $\mathfrak{S}$, if the curved triangles $\triangle$ are substituted by the straight triangles Δ, we have $\mathfrak{S}\neq\tilde{\mathfrak{S}}\,(=\mathfrak{S}^0)$ at curved edges of triangles $\triangle$ at most, which are located near the boundary Γ only. In analogy to (2.3.13), we employ abbreviations for the measures of some sets mentioned previously. Recalling (2.3.15) - (2.3.17) and Definition 2.23. concerning the sets $\mathfrak{K}$, $\tilde{\mathfrak{K}}$, $\mathfrak{S}$, $\tilde{\mathfrak{S}}\,(=\mathfrak{S}^0)$, $\mathfrak{S}^{\pm}$ and $\tilde{\mathfrak{S}}^{\pm}$, we denote the measures of these sets by

$$\left.\begin{array}{l} H(x) := \text{mes}\, \mathcal{H}(x),\ h(x,\xi) := \text{mes}\, \mathfrak{h}(x,\xi),\ s(x,\xi) := \text{mes}\, \mathfrak{S}(x,\xi), \\ s(x,x'') := \text{mes}\, \mathfrak{S}(x,x''), \text{ and } \tilde{s}(x,\xi)(= s^{o}(x,\xi)),\ \tilde{s}(x,x'')(= s^{o}(x,x'')), \\ s^{\pm}(x,\xi),\ s^{\pm}(x,x''),\ s(x),\ \tilde{s}(x),\ \tilde{s}^{\pm}(x),\ \tilde{H}(x) \text{ etc. quite similarly;} \\ \text{additionally} \\ h(x) := \frac{1}{2}\{h(x,\xi_{\Gamma}^{-}) + h(x,\xi_{\Gamma}^{+})\} \quad \text{for } x,\xi_{\Gamma}^{\pm} \in \gamma, \end{array}\right\} \quad (2.3.18)$$

and the index (x,ξ) will be often replaced by x'_{ξ} or x'. Although some symbols of (2.3.13) and (2.3.18) coincide formally, there are differences in the special interpretation and, in general, the actual coincidence is only given for he mesh segment $\mathfrak{h}(x,\xi)$ and its measure $h(x,\xi)$ as well as for the boundary sets $\tilde{\mathfrak{S}}^{\pm}(x)$, $\tilde{\mathfrak{S}}(x)$ on Γ_h and their measures $\tilde{s}^{\pm}(x)$, $\tilde{s}(x)$, and for $h(x)$, $x \in \gamma$. The mesh segments $\mathfrak{h}(x,\xi)$ and certain other geometric sets, e.g. $\partial\mathcal{H}$ and $\mathfrak{S}$, are provided with orientation; e.g., the symbol $\mathfrak{h}(x,\xi)$ is used to denote the geometric set $\mathfrak{h}(x,\xi) := [x,\xi]$ and, at the same time, the vector $\mathfrak{h}(x,\xi) := \overrightarrow{x\xi}$, with the shortened designation $\mathfrak{h}$ and the vector representation $\mathfrak{h} = (h_1,h_2)^T$, with $h_1,h_2 \in R^1$. Instead of $\mathfrak{h}(x,\xi)$ $h(x,\xi)$, $\mathfrak{S}(x,\xi)$, etc., the brief designations $\mathfrak{h}(x'_{\xi})$, $h(x'_{\xi})$, $\mathfrak{S}(x'_{\xi})$, etc. are often employed, occasionally also $\mathfrak{h}(x')$, $h(x')$, $\mathfrak{S}(x')$, etc. In addition to $\mathfrak{h} := \mathfrak{h}(x,\xi)$, the following related notations are introduced:

$$\mathfrak{h}^{\pm} := \mathfrak{h}^{\pm}(x,\xi) := \mathfrak{h}(x,\xi^{\pm}) := \overrightarrow{x\xi^{\pm}},\quad \mathfrak{h}^{\pm} = (h_1^{\pm},h_2^{\pm})^T, \qquad (2.3.19)$$

moreover, the symbols $\mathfrak{h}^{o}$, $\mathfrak{h}^{\oplus}$ and $\mathfrak{h}^{\ominus}$ for the vectors $\mathfrak{h}$, $\mathfrak{h}^{+}$ and $\mathfrak{h}^{-}$ rotated clockwise by an angle $\pi/2$, i.e.

$$\mathfrak{h}^{o} := (h_2,-h_1)^T,\ \mathfrak{h}^{\oplus} := (h_2^{+},-h_1^{+})^T,\ \mathfrak{h}^{\ominus} := (h_2^{-},-h_1^{-})^T. \qquad (2.3.20)$$

Evidently, for arbitrary vectors $a,b \in R^2$, with $a = (a_1,a_2)^T$, $a^{o} = (a_2,-a_1)^T$ and the scalar product $(.,.)$ of R^2, the following relations hold:

$$\begin{array}{l}(a^{o},b^{o}) = (a,b),\ (a,b^{o}) = -(a^{o},b),\ (a\pm b)^{o} = a^{o} \pm b^{o}, \\ |a^{o}| = |a| = \sqrt{(a,a)},\ (a,b^{o}) = |a||b| \sin\{\sphericalangle(a,b)\}.\end{array} \qquad (2.3.21)$$

For further considerations, the vectors $\mathfrak{h}(x,\xi)$, $\mathfrak{h}^{+}(x,\xi)$, $\mathfrak{S}^{-}(x,x'')$, $\mathfrak{S}^{+}(x,x'')$ in the triangles $\Delta(x,\xi,\xi^{+})$ and the normal $n_{\mathfrak{S}\pm}$ to $\mathfrak{S}^{\pm}$ are needed, cf. Fig. 6e and (2.3.19), (2.3.20), which are connected by the following relations:

$$\mathfrak{S}^{\ominus}+\mathfrak{S}^{\oplus} = \frac{\mathfrak{h}^{\oplus} - \mathfrak{h}^{o}}{2},\ \mathfrak{S}^{\ominus} = \frac{2\mathfrak{h}^{\oplus} - \mathfrak{h}^{o}}{6},\ \mathfrak{S}^{\oplus} = \frac{\mathfrak{h}^{\oplus} - 2\mathfrak{h}^{o}}{6},\ n_{\mathfrak{S}\pm} = \frac{\mathfrak{S}^{\oplus}}{\text{mes}\,\mathfrak{S}^{\pm}}. \qquad (2.3.22)$$

The construction of the secondary network via method (MD) is feasible for any network (2.2.3b) and $h \le h_o$, h_o sufficiently small. The following grid regularity condition VO(MD,$\bar{\omega}$) is assumed to be fulfilled, which describes the asymptotic behaviour of the network. Here, all forms of nondegenerating triangles and (or) rectangles will be admitted.

<u>Condition 2.24. - VO(MD,$\bar{\omega}$).</u> The mesh segments $\mathfrak{h}(x,\xi)$ and the interior angles θ of the triangles Δ satisfy the conditions TO, WO for any h:

$$\begin{array}{l}\text{TO: } 0 < \varepsilon_o h \le h(x,\xi) \le h/\varepsilon_o \text{ for all } h(x,\xi), \text{ with } x,\xi \in \bar{\omega}, \\ \text{WO: } 0 < \theta_o \le \theta \le \pi - \theta_o \text{ for the angles } \theta \text{ of all triangles } \Delta,\end{array} \qquad (2.3.23)$$

and the constants θ_o, ε_o $(0 < \theta_o < \frac{\pi}{2},\ 0 < \varepsilon_o \le 1)$ do not depend on h, $h \le h_o$. ■

3. CONSTRUCTION OF FINITE DIFFERENCE APPROXIMATIONS

3.1. The principle of approximation. The FDM considered in the following is to be understood as an "exterior approximation" of the BVP $Au = F$ in the space R^n, $n := |\bar{\omega}|$, where n tends to infinity as $h \to 0$. The FDSs $A_h y = F_h$, with $A_h \in L(R^n, R^n)$ (family of spaces of linear operators from R^n into R^n) and $F_h, y \in R^n$, are derived by approximating the BVP $Au = F$ under the assumptions $V(\bar{\Omega})$, $V(A,F)$ and $V(u)$ from Section 2.1. Then, the properties of the difference operator A_h, a priori estimates and the convergence $y \to u$ as $h \to 0$ are studied in the R^n, $n = n(h) := |\bar{\omega}|$. This discussion is partially based on some papers of the author, cf. HEINRICH [5,6,7,8,9,10].

Let u be the solution of the BVP $Au = F$, and, let Assumptions 2.1., 2.2. and 2.3. be satisfied, i.e. $V(\bar{\Omega})$, $V(A,F)$ and $V(u)$. Firstly, we assume that $u \in W_2^2(\Omega)$, cf. (2.1.9). The approximation of $Au = F$ on $\gamma_1 \subset \Gamma_1$ is fairly simple, because $lu := u = g$ holds on Γ_1 in the sense of $C(\Gamma_1)$, cf. (2.1.11b). Therefore,

$$u(x) = g(x) \quad \text{for } x \in \gamma_1 \tag{3.1.1}$$

makes sense and acts as a restriction to u on γ_1. Now, finite difference approximations of $Au = F$ on $\omega + \gamma_{23} \subset \Omega \cup \Gamma_{23}$ are to be constructed, i.e. for $Lu = f$ on ω and $lu = g$ on γ_{23}. Integrating $Lu = f$ from (2.1.1a) over the box $\varkappa(x)$ of the secondary triangulation $\mathcal{T}_{\varkappa}$ from (2.3.3a), we get, for each grid point $x \in \omega + \gamma_{23}$, the integral identity ("o" over $\varkappa$ is omitted)

$$\int_{\varkappa(x)} (L_o u)(\bar{x})\, d\bar{x} + \int_{\varkappa(x)} (cu)(\bar{x})\, d\bar{x} = \int_{\varkappa(x)} f(\bar{x}) d\bar{x}, \quad x \in \omega + \gamma_{23}, \tag{3.1.2}$$

with L_o from (2.1.3) and k_{ij}, c, f from $V(A,F)$, cf. Assumption 2.2. By Gauss' formula from Appendix DI, the relation

$$\int_{\varkappa(x)} (L_o u)\, d\bar{x} = - \int_{\partial\varkappa(x)} (K\nabla u, n)\, ds, \quad x \in \omega + \gamma_{23}, \tag{3.1.3}$$

holds at least for functions $u \in W_2^2(\varkappa)$ and boxes $\varkappa(x)$, with $\partial\varkappa(x) \in C^{0,1} \cap PC^2$ and $(K\nabla u, n) \in L_2(\Gamma_{23})$ by the trace theorem, cf. Appendix IM. The main part $L_o u$ of Lu and the conormal derivative $(K\nabla u, n)$ are given in (2.1.3). Inserting (3.1.3) into (3.1.2), the identity (3.1.2) is now of the form

$$- \int_{\partial\varkappa(x)} (K\nabla u, n)\, ds + \int_{\varkappa(x)} cu\, d\bar{x} = \int_{\varkappa(x)} f\, d\bar{x}, \quad x \in \omega + \gamma_{23}. \tag{3.1.4}$$

In comparison with (3.1.2), the area integral of the main part $L_o u$ is now replaced by the line integral of the conormal derivative of u along the boundary $\partial\varkappa$. Now, this boundary $\partial\varkappa(x)$ satisfies the relation

$$\partial\mathcal{X}(x)\cap\Gamma = \begin{cases} \emptyset & x\in\omega, \\ & \text{for} \\ \mathcal{S}(x) & x\in\gamma_{23}, \text{ where } \mathcal{S}(x)\subset\Gamma_{23}; \end{cases} \tag{3.1.5}$$

for $\mathcal{S}(x)$ see (2.3.4) and (2.3.16). Therefore, using $lu = g$ on Γ_{23} in the sense of $L_2(\Gamma_{23})$, cf. also (2.1.11c), the line integral of the conormal derivative along $\mathcal{S}(x)\subset\Gamma_{23}$ can by substituted as follows ("o" over $\mathcal{S}(x)$ is omitted):

$$\int_{\mathcal{S}(x)} (K\nabla u,n)\, ds = \int_{\mathcal{S}(x)} (g-\alpha u)ds, \quad x\in\gamma_{23}. \tag{3.1.6}$$

Together with (3.1.4), this leads to the integral identity (3.1.7) called "balance over $\mathcal{X}(x)$"

$$-\int_{\partial\mathcal{X}(x)\setminus\mathcal{S}(x)} (K\nabla u,n)\, ds + \int_{\mathcal{S}(x)} \alpha u\, ds + \int_{\mathcal{X}(x)} cu\, d\bar{x} = \int_{\mathcal{X}(x)} f\, d\bar{x} + \int_{\mathcal{S}(x)} g\, ds, \tag{3.1.7}$$

for $x\in\omega+\gamma_{23}$, with $\mathcal{S}(x) = \emptyset$ for $x\in\omega$ and $\mathcal{S}(x) = \partial\mathcal{X}(x)\cap\Gamma$ for $x\in\gamma_{23}$.

The relation (3.1.7) noted for all grid points $x\in\omega+\gamma_{23}$ is called the "system of balance equations" on the grid $\omega+\gamma_{23}$ and is to be taken under the restriction $u = g$ on γ_1.

The splitting of the contour $\partial\mathcal{X}(x)\setminus\mathcal{S}(x)$ into several polygonal subsets $\mathcal{S}(x,\xi)$ and $\mathcal{S}(x,x'')$, cf. (2.3.4) and (2.3.16), yields a sum representation of the leading term in (3.1.7) (the flux crossing $\partial\mathcal{X}(x)\setminus\mathcal{S}(x)$):

$$-\int_{\partial\mathcal{X}(x)\setminus\mathcal{S}(x)} (K\nabla u,n)ds = -\sum_{\xi\in S'(x)} \int_{\mathcal{S}(x,\xi)} (K\nabla u,n)ds$$
$$= \sum_{x''\in S''(x)} \int_{\mathcal{S}(x,x'')} (K\nabla u,n)ds \tag{3.1.8}$$

for $x\in\omega+\gamma_{23}$, with $\mathcal{S}(x) = \emptyset$ for $x\in\omega$.

The system of balance equations (3.1.7) is, under the restriction $u(x) = g(x)$ for $x\in\gamma_1$ and together with the splitting (3.1.8), the starting point for the construction of finite difference approximations, which are determined by the choice of $\mathcal{X}(x)$ and of quadrature, cubature and finite difference formulas for approximating (3.1.7).

Unfortunately, the global regularity hypothesis $u\in W_2^2(\Omega)$ is not fulfilled for all BVPs $Au = F$. Now we shall check whether the system of balance equations (3.1.7) may be derived under the weaker assumptions (2.1.10), (2.1.11), cf. V(u). Again we have $u = g$ on Γ_1 in the sense of $C(\Gamma_1)$ and, thus, (3.1.1) is justified. Identity (3.1.2) can be noted as previously. The integral theorem (3.1.3) holds obviously for all $\mathcal{X}(x)$, where $x\in\omega+(\gamma_{23}\setminus\gamma_{23}^s)$, and γ_{23}^s denotes the set of singular points belonging to γ_{23}. Relation (3.1.3) is then a consequence of $u\in W_2^2(\Omega\setminus\mathcal{X})$, cf. (2.1.10). For the singular grid points $x\in\gamma_{23}^s$, there is no contact of $\mathcal{S}(x) = \partial\mathcal{X}(x)\cap\Gamma$ with Γ_1 ($\mathcal{S}(x)\cap\Gamma_1 = \emptyset$), since collision points $P_c\in\Gamma_{23}\cap\Gamma_1$ belong to γ_1, by

Definition 2.7. Therefore, according to (2.1.10), $(K\nabla u,n)\in L_2(\mathfrak{S}(x))$ holds for $x\in\gamma^s_{23}$. In such cases, (3.1.3) is a consequence of Assumption 2.3. - V(u) or can be proved directly for important special cases. Thus, (3.1.3) can be verified for the operator $L_0: = -\Delta$ (Δ: Laplacian) and polygonal boundaries Γ_{23}, because the singular part u_s is explicitly known (possibly without the constants in the finite singular expansion) and (3.1.3) can be confirmed for u_s. For general operators L_0 and $\Gamma_{23}\in C^{0,1}\cap PC^2$, i.e. $\partial\mathcal{R}\in C^{0,1}\cap PC^2$, the integral relation (3.1.3) follows from Green's formula (2.1.16), which is proved in COSTABEL [1] at least for analytical coefficients k_{ij}. Then, taking into account $\bar{\Omega}=\mathcal{R}$ (momentarily), $v\equiv 1$ and $(K\nabla u,n)\in L_2(\mathfrak{S}(x))$ as in (2.1.10), identity (3.1.3) holds also in the case $x\in\gamma^s_{23}$. This proves relation (3.1.4) for all $x\in\omega+\gamma_{23}$, and, (3.1.6) is a direct consequence of $(K\nabla u,n)$, αu, $g\in L_2(\mathfrak{S}(x))$ and (2.1.11c). This implies that the weaker smoothness condition (2.1.10), together with (2.1.11) and a generalization of Green's formula (2.1.16), also admits the existence of the system of balance equations (3.1.7).

We shall now make some remarks on the finite difference approaches which are based on the secondary networks (PB), (MD) and are called method (PB), method (MD), respectively. In the literature, these methods are almost exclusively encountered in the work of engineers for special classes of elliptic problems, but rarely in that of mathematicians. In our work, the approaches made in the engineering literature from a physical point of view are revised and viewed mathematically for more general situations, e.g. for mixed boundary conditions, piecewise smooth boundary, weaker assumptions on A, F and generalized regular solutions u.

The method (PB) will be considered for $k := k_{11} \equiv k_{22}$ and $k_{12}\equiv 0$ only, since the special choice of quadrature points $x'_{\xi}\in h(x,\xi)$ ($h(x,\xi)$: mesh segment) for approximating the line integral $\int(K\nabla u,n)ds$ gives symmetric finite difference operators A_h even in this case. An extension of the method (PB) to anisotropic potential fields (in general: $k_{11}\neq k_{22}$, $k_{12}\neq 0$) with favourable properties of A_h is possible, if the circumcentres in each element $e\in\{\Delta,\Box\}$ are taken as quadrature points. But in this case, the method (PB) is not so advantageous as the method (MD), where the centres of gravity are used as quadrature points to $\int(K\nabla u,n)ds$, and the algorithm of constructing A_h would be more convenient. In the author's view, the origin of the concept (PB) was given by McNEAL [1] (1953).

The method (MD) seems to have appeared first in a paper by WINSLOW [1] (1967) as a method of constructing finite difference approximations of the elliptic differential expression $Lu \equiv -\operatorname{div}(k \operatorname{grad} u)$, with applications to the numerical solution of field problems in magnetostatics, cf. also REISSMANN [2,3]. On the other hand, there are obviously no papers in the literature describing and justifying this method for a sufficiently large class of BVPs $Au = F$ in the sense of numerical analysis. In this monograph, a class of finite difference approximations via the method (MD) is proposed for relatively general BVPs $Au = F$. At the same time, the properties of the resulting operators A_h are investigated systematically, and, finally, a priori estimates and error estimates taking into account generalized solutions $u \in W_2^2(\Omega)$ are derived on the basis of new techniques in finite difference methods. This is done for difference operators via method (MD) as well as (PB).

Some variations in the methods (PB) and (MD) are essentially due to the distinct definition of the boxes $\mathfrak{X}, \tilde{\mathfrak{X}}$ and to the choice of the quadrature points for approximating $\int (K\nabla u,n)ds$. Since the latter fact has a great effect on the finite difference approximation, one can say that the methods (PB) and (MD) are significant approaches concerning the choice of quadrature points. Therefore, we have to take into account different types of the splitting of error functionals and modified versions of some proofs. Nevertheless, there are also many aspects, basic techniques and results involved in the method (PB) as well as in the method (MD), which are quite similar.

3.2. Finite difference schemes via method (PB)

3.2.1. The approximation of the balance equations. For the approximation of the BVP $Au = F$ from (2.1.1), (2.1.2) in the case $k_{11} \equiv k_{22}$, $k_{12} \equiv 0$, i.e.

$$Lu \equiv - \sum_{i=1}^{2} \frac{\partial}{\partial x_i} \left(k \frac{\partial u}{\partial x_i}\right) + cu = f \text{ in } \Omega, \quad k := k_{11} \equiv k_{22},$$

$$lu \equiv \begin{cases} u = g & \Gamma_1, \\ k \dfrac{\partial u}{\partial n} = g & \text{on } \Gamma_2, \quad \Gamma = \bigcup_{i=1}^{3} \Gamma_i, \\ k \dfrac{\partial u}{\partial n} + \alpha u = g & \Gamma_3, \end{cases} \tag{3.2.1}$$

some topics of the FDM via method (PB) were already considered by the author in HEINRICH [5,6], especially for locally irregular grids. In this monograph, we allow the networks to be, in general, irregular in the whole domain. Furthermore, the smoothness assumptions on A, F and u are essentially weakened.

The system of balance equations assigned to (3.2.1) is a special case of (3.1.7) and, by splitting $\partial\mathfrak{X}(x) \setminus \mathfrak{S}(x)$ into the perpendicular bisectors $\mathfrak{S}(x,\xi)$ according to (3.1.8) and (2.3.4), it can be rewritten in the form

$$-\sum_{\xi \in S'(x)} \int_{\mathfrak{S}(x,\xi)} k\frac{\partial u}{\partial n} ds + \int_{\mathfrak{S}(x)} \alpha u ds + \int_{\mathfrak{X}(x)} cu \, d\bar{x} = \int_{\mathfrak{X}(x)} f \, d\bar{x} + \int_{\mathfrak{S}(x)} g \, ds \tag{3.2.2}$$

for $x \in \omega + \gamma_{23}$, with $\mathfrak{S}(x) = \emptyset$ for $x \in \omega$,

where $\mathfrak{X}(x) := \mathfrak{X}_p(x)$ is taken from Definition 2.15. We now approximate the terms of (3.2.2) under the assumptions $V(\bar{\Omega})$, $V(A,F)$, $V(u)$ and

VO(PB,ϖ) for the BVP Au = F and the network, respectively, cf. Assumption 2.1., 2.2., 2.3. and Condition 2.16. First, the special smoothness assumptions

$$k,\ c,\ f \in C^{s}(\bar{\Omega}) \text{ and } \alpha,\ g \in PC^{s}(\Gamma_{23}) \text{ for } s \in \{1,2\},\ g \in C(\Gamma_{1}), \tag{3.2.3a}$$

$$u \in C^{1+s}(\bar{\Omega}) \text{ for } s \in \{1,2\}, \tag{3.2.3b}$$

are considered, where the local approximation errors can be given by powers of h as it is done in the following. The approximation of the flux through the edge $\mathfrak{S}(x,\xi)$ is defined by the one-point quadrature rule

$$\int_{\mathfrak{S}(x,\xi)} r(s)ds = r(x'_{\xi})s(x'_{\xi}) + O(h^{1+l}) \text{ for } r \in C^{l}(\mathfrak{S}(x,\xi)), \tag{3.2.4}$$

with $r := k\frac{\partial u}{\partial n}$ (the specific flux). Defining the approximate values $k_h(x'_{\xi})$ by

$$k_h(x'_{\xi}) := \tfrac{1}{2}\{k(x) + k(\xi)\} \approx k(x'_{\xi}) \text{ or } k_h(x'_{\xi}) := k(x'_{\xi}) \text{ (exact value)} \tag{3.2.5}$$

and the central difference quotient $u_{\hbar}(x'_{\xi})$ by

$$u_{\hbar}(x'_{\xi}) := \frac{u(\xi)-u(x)}{h(x,\xi)} \approx \frac{\partial u}{\partial n}(x'_{\xi}), \tag{3.2.6}$$

we finally obtain

$$\int_{\mathfrak{S}(x,\xi)} k\frac{\partial u}{\partial n}ds = k_h(x'_{\xi})\, u_{\hbar}(x'_{\xi})\, s(x'_{\xi}) + O(h^{1+l}),$$

$$l = \begin{cases} 2 & \text{for } x \in \omega,\ x \text{ regular}, \\ 1 & \text{for } x \in \omega + \gamma_{23}, \end{cases} \quad k \in C^{l}(\bar{\Omega}),\ u \in C^{1+l}(\bar{\Omega}), \tag{3.2.7}$$

cf. HEINRICH [5, Abb. 3a, b], [6, pointer (16)]. Here and throughout this paper, the subscript $\hbar$ (script letter) denotes difference quotients, but h (typescript letter) indicates that a grid function is taken into account, e.g. see (3.2.7). In order to approximate the source terms $\int_{\mathfrak{S}(x)} r\, ds$ of (3.2.2), for $x \in \gamma_{23}$, $r := \alpha u$ and $r := g$, the quadrature formulas

$$\int_{\mathfrak{S}(x)} r(s)ds = (s^{-}r^{-}+s^{+}r^{+})(x) + j\left\{-\frac{(s^{-})^{2}}{2}\frac{dr(x_{\Gamma}^{-})}{ds} + \frac{(s^{+})^{2}}{2}\frac{dr(x_{\Gamma}^{+})}{ds}\right\}$$

$$+ O(h^{2+j}) \text{ for } x \in \gamma_{23},\ r \in C^{1+j}(\mathfrak{S}^{\pm}),\ \mathfrak{S}^{\pm}(x) \text{ from (2.3.4)}, \tag{3.2.8}$$

$$r^{\pm}(x) := r(x^{\pm}0),$$

are used alternatively for j = 0 or j = 1, cf. Fig. 5b and HEINRICH [5, Section 5.]. Therefore, in the case j = 0 in (3.2.8), a one-point quadrature formula on the arcs $\mathfrak{S}^{\pm}(x)$ ($x^{\pm}0$: quadrature point) is involved. On the other hand, for j = 1, we have a two-point quadrature formula with Hermite-character and with the same asymptotic accuracy as the trapezoid rule applied to a function on the arcs $\mathfrak{S}^{\pm}(x)$. The derivative $\frac{dr}{ds}$ at x_{Γ}^{+} will be approximated by

$$\frac{\delta r}{\delta \mathfrak{b}}(x'_{\xi_{\Gamma}^{+}}) := \frac{r(\xi_{\Gamma}^{+}-0)-r(x+0)}{\text{mes}\ \mathfrak{b}(x,\xi_{\Gamma}^{+})} \approx \frac{dr(x_{\Gamma}^{+})}{ds}, \quad \mathfrak{b}(x,\xi_{\Gamma}^{+}) := \widehat{x\xi_{\Gamma}^{+}}, \tag{3.2.9}$$

and at x_Γ^- analogously, i.e., the values of r at the grid points $x, \xi_\Gamma^\pm \in \gamma$ are employed. Especially in the case $r' = (\alpha u)' = \alpha' u + \alpha u'$ at the points $x_\Gamma^\pm$, the values of α, u are approximately replaced by

$$\alpha(x_\Gamma^\pm) \approx \frac{1}{2}\left\{\alpha(x\pm 0) + \alpha(\xi_\Gamma^\pm \mp 0)\right\} =: \alpha_h(x'_{\xi_\Gamma^\pm}), \quad u(x_\Gamma^\pm) \approx u(x) \qquad (3.2.10)$$

Due to $\Gamma \in C^{0,1} \cap PC^2$ and α, $g \in PC^1(\Gamma_{23})$, Assumption 2.4. ensures that $\widehat{x\xi_\Gamma^\pm} \in C^2$ and $\alpha, g \in C^1(\widehat{x\xi_\Gamma^\pm})$ hold for $\widehat{x\xi_\Gamma^\pm} \subset \Gamma_{23}$, i.e., we have smooth functions α, g on continuously curved arcs $\widehat{x\xi_\Gamma^\pm}$. Jumps of α, g appear at grid points only, where the left- and right-hand limit will be taken.
The source terms $\int_{\mathcal{X}(x)} r\, d\bar{x}$ in (3.2.2), with $r := cu$ or $r := f$, will be approximated by the one-point cubature formula

$$\int_{\mathcal{X}(x)} r(\bar{x})d\bar{x} = r(x)\, H(x) + O(h^{2+l}), \quad l = \begin{cases} 1 & \text{for } x \in \omega + \gamma_{23},\ r \in C^1(\mathcal{X}), \\ 2 & \text{for } x \in \omega,\ x \text{ regular},\ r \in C^2(\mathcal{X}). \end{cases} \qquad (3.2.11)$$

Up to now, the secondary triangulation $\mathcal{T}_{\mathcal{X}}$ has been the underlying subdivision of $\bar{\Omega}$, cf. (2.3.3a). In general, the boxes $\mathcal{X}(x)$, $x \in \gamma$, have a curved boundary part $\partial\mathcal{X}(x) \cap \Gamma$. They can be approximately replaced by the polygonal boxes $\tilde{\mathcal{X}}(x)$, whose parameters are readily available, cf. $\mathcal{T}_{\tilde{\mathcal{X}}}$ from (2.3.3b, c). Clearly because of (3.2.2) and (2.3.3c), only the boxes for $x \in \gamma_{23}$ are of interest here. For grid points $x \in \gamma_{23}$, the measures $s(x, \xi_\Gamma^\pm)$, $s^\pm(x)$, H(x) of the box $\mathcal{X}(x)$, cf. (2.3.13), are now substituted by the corresponding approximations $\tilde{s}(x, \xi_\Gamma^\pm)$, $\tilde{s}^\pm(x) := h(x, \xi_\Gamma^\pm)/2$, $\tilde{H}(x)$ of the properly polygonally bounded box $\tilde{\mathcal{X}}(x)$, cf. (2.3.5), (2.3.13) and Fig. 5c. Instead of $b(x, \xi_\Gamma^+) := \text{mes}\, \mathfrak{b}(x, \xi_\Gamma^+)$ in (3.2.9), the approximate value $h(x, \xi_\Gamma^+)$ will be taken. As a consequence of $\Gamma \in C^{0,1} \cap PC^2$ and of the definition of γ, the errors are now

$$|H(x) - \tilde{H}(x)| \leq K_0 h^3, \quad |s(x, \xi_\Gamma^\pm) - \tilde{s}(x, \xi_\Gamma^\pm)| \leq K_1 h^2, \qquad (3.2.12)$$
$$|s^\pm(x) - \tilde{s}^\pm(x)| \leq K_2 h^3 \text{ for } x \in \gamma, \text{ expecially } x \in \gamma_{23}.$$

Instead of $\tilde{s}^\pm(x)$ and $\tilde{s}(x)$, we shall from now on write $\frac{1}{2}\, h(x, \xi_\Gamma^\pm)$ and h(x), respectively; cf. (2.3.13).

The substitution of the approximations suggested in (3.2.4) - (3.2.12), together with the total error p_E made, into the balance equations (3.2.2) yields an exact approximation of these equations in (3.2.2) in terms of some "point functionals" p_A^u and p_F defined on the grid $\omega + \gamma_{23}$, that is

$$p_A^u(x) = p_F(x) + p_E(x) \text{ for } x \in \omega + \gamma_{23}, \quad u(x) = g(x) \text{ for } x \in \gamma_1. \quad (3.2.13)$$

The functional p_A^u requires the values of u, k, c, α at the grid points $x \in \bar{\omega}$ (perhaps k at $x'_\xi \in \bar{\omega}'$) and approximates the left-hand side of (3.2.2). Analogously, the functional $p_F(x)$, with the values of f, g at the grid points, does the same for the right-hand side of (3.2.2). The error $p_E(x)$ depends on u, k, c, α, f, g and on $\bar{\Omega}, \bar{\omega}$, h. Neglecting the unknown error $p_E(x)$ at $x \in \omega + \gamma_{23}$, the equality in (3.2.13) is lost and an inequality $p_A^u(x) \approx p_F(x)$ remains. We now introduce an unknown grid function $y \in D$, D from (2.2.8), such that after replacing the grid function $u|_{\bar\omega} \in D$ by $y \in D$, the equality is maintained, that is

$$p_A^y(x) = p_F(x) \text{ for } x \in \omega + \gamma_{23},\ y(x) = g(x) \text{ for } x \in \gamma_1. \tag{3.2.14}$$

The equations $p_A^y(x) = p_F(x)$, $x \in \omega + \gamma_{23}$, constitute discrete analogues of the balance equations (3.2.2) under the restriction $y(x) = g(x)$ for $x \in \gamma_1$. Beside (3.2.14), we often consider the modified system

$$\tilde{p}_A^y(x) = \tilde{p}_F(x) \quad \text{for } x \in \omega + \gamma_{23}, \tag{3.2.15}$$

which results from (3.2.14) by replacing the values $y(x)$ for $x \in \gamma_1$ by $g(x)$ and writing the known terms with $g(x)$, $x \in \gamma_1$, on the right-hand side of the equation.

Furthermore, we scale the balance equations in (3.2.14) by the factor $1/H(x)$ for $x \in \omega$ and $1/h(x)$ for $x \in \gamma_{23}$ and immediately obtain the FDS $A_h y = F_h$ as a discrete analogue of the BVP $Au = F$, i.e.

$A_h y = F_h$ in the sense of $(A_h y)(x) = F_h(x)$ for $x \in \bar\omega$,

$y \in D$ from (2.2.8), with p_A^y, p_F from (3.2.14) and

$$(A_h y)(x) := \begin{cases} y(x) \\ \frac{1}{h(x)} p_A^y(x) \\ \frac{1}{H(x)} p_A^y(x) \end{cases} \text{ and } F_h(x) := \begin{cases} g(x) & x \in \gamma_1, \\ \frac{1}{h(x)} p_F(x) & \text{for } x \in \gamma_{23}, \\ \frac{1}{H(x)} p_F(x) & x \in \omega. \end{cases} \tag{3.2.16}$$

The unknown grid function $y \in D$, which is to be determined as the solution of the linear system of equations (3.2.14) or (3.2.16), is an approximation of the exact solution u of the BVP $Au = F$, as will be shown below. Beside (3.2.16), we shall later on take into account a modified FDS $\tilde{A}_h y = \tilde{F}_h$ on $\omega + \gamma_{23}$ which arises from (3.2.15) by the same way as (3.2.16) from (3.2.14):

$\tilde{A}_h y = \tilde{F}_h$ in the sense of $(\tilde{A}_h y)(x) = \tilde{F}_h(x)$ for $x \in \omega + \gamma_{23}$,

$y \in \tilde{D}$ from (2.2.8), with $\tilde{p}_A^y$, $\tilde{p}_F$ from (3.2.15) and

$$(\tilde{A}_h y)(x) := \begin{cases} \frac{1}{h(x)} \tilde{p}_A^y(x) \\ \frac{1}{H(x)} \tilde{p}_A^y(x) \end{cases} \text{ and } \tilde{F}_h(x) := \begin{cases} \frac{1}{h(x)} \tilde{p}_F(x) & \text{for } x \in \gamma_{23}, \\ \frac{1}{H(x)} \tilde{p}_F(x) & x \in \omega. \end{cases} \tag{3.2.17}$$

The FDS $A_h y = F_h$ or $\tilde{A}_h y = \tilde{F}_h$ derived under assumption (3.2.3) can be also obtained for more general parameters k, c, α, f, g and solutions u, for as long as the functionals p_A^u, p_F from (3.2.13) are correctly defined. Taking u from (2.1.9) or (2.1.10) in V(u), we see that $u|_{\bar\omega}$ makes sense because of $u \in C(\bar\Omega)$. Examining k, c, α, f and g from (2.1.6) in V(A,F), it will become clear that the values of k, c, α, f and g can be taken pointwise on the grid, with the exception of $f \in W_2^s(\Omega)$, $s \in \{0,1\}$; cf. the imbedding theorems from Appendix IM. The approximations of $\int f\, d\bar{x}$ for $f \notin C(\bar\Omega)$ will be discussed separately in Section 5.4.3. If V(A,F) is taken into account instead of (3.2.3), then the error of the quadrature, cubature and difference formulas must be revised, which will be done in Chapter 5.

3.2.2. Finite difference schemes of the type (PB). In the author's earlier papers HEINRICH [5,6], two classes of FDSs $A_h^j y = F_h^j$ $(j=0,1)$ were constructed as discrete analogues of the BVP $Au = F$ (3.2.1) on locally irregular grids and under some grid regularity condition $V(\mathfrak{X},\bar{\omega})$ from HEINRICH [5]. The following Theorem 3.1. contains a more general assertion for irregular networks $\mathfrak{T}_\Delta$ from (2.2.3b) and the relatively weak grid regularity condition $VO(PB,\bar{\omega})$ from (2.3.8) as well as for parameters from $V(A,F)$.

Theorem 3.1. Let (i), (ii), (iii) be given, that is
(i) the BVP $Au = F$ from (2.1.2) for the case (3.2.1) with $V(\bar{\Omega})$, $V(A,F)$ and $V(u)$, cf. Assumption 2.1., 2.2., 2.3.,
(ii) grids $\bar{\omega}$ from (2.2.6) based on the general triangulation $\mathfrak{T}_\Delta$ from (2.3.3), for $h \leq h_0$ and sufficiently small h_0, with arbitrary triangles Δ and (or) rectangles $\square$, but under the Condition 2.16. - $VO(PB,\bar{\omega})$,
(iii) secondary triangulations $\mathfrak{T}_{\tilde{\mathfrak{X}}}$ of the type (PB) from Definition 2.15.
Then, there are FDSs $A_h^j y = F_h^j$ $(j=0,1)$ on $\bar{\omega}$ as discrete analogues of the BVP $Au = F$ via method (PB), for $h \leq h_0$ and sufficiently small h_0. The FDSs

$$A_h^j y = F_h \text{ on } \bar{\omega} = \gamma_1 + \gamma_{23} + \omega, \tag{3.2.18a}$$
$$\text{with } A_h^j = (l_h, l_h^j, L_h)^T,\quad F_h^j = (g_h, g_h^j, f_h)^T,$$

will be described for $j \in \{0,1\}$ as follows:

$$(l_h y)(x) := y(x) = g_h(x) \text{ for } x \in \gamma_1,\quad g_h(x) := g(x),$$

$$(l_h^j y)(x) := -\frac{1}{h(x)} \sum_{\xi \in S'(x)} s^j(x'_\xi)\, k_h(x'_\xi) y_{\hat{x}}(x'_\xi) + \alpha^j(x) y(x) = \tag{3.2.18b}$$
$$= g_h^j(x),\quad x \in \gamma_{23},$$

$$(L_h y)(x) := -\frac{1}{H(x)} \sum_{\xi \in S'(x)} s(x'_\xi) k_h(x'_\xi) y_{\hat{x}}(x'_\xi) + c(x)y(x) = f_h(x),\quad x \in \omega,$$

with the following notations for $j \in \{0,1\}$:

$$y_{\hat{x}}(x'_\xi) := \frac{y(\xi)-y(x)}{h(x'_\xi)},\quad k_h(x'_\xi) := \frac{1}{2}\{k(x)+k(\xi)\},\quad f_h(x) := f(x)$$
for $f \in C(\bar{\Omega})$, and $f_h(x)$ from Section 5.4.3. for $f \notin C(\bar{\Omega})$,

$$s^j(x'_\xi) := \begin{cases} s(x'_\xi) & \text{for } x \in \omega + \gamma_{23},\ \xi \in \omega, \\ \tilde{s}(x'_{\xi_\Gamma^\pm}) - \delta_{1j} \dfrac{h^2(x'_{\xi_\Gamma^\pm})}{8} \dfrac{\alpha_h(x'_{\xi_\Gamma^\pm})}{k_h(x'_{\xi_\Gamma^\pm})} & \text{for } x \in \gamma_{23}, \\ & \xi = \xi_\Gamma^\pm \in S'(x) \cap \gamma, \end{cases} \tag{3.2.18c}$$

$$\alpha_h(x'_{\xi_\Gamma^\pm}) := \frac{1}{2}\{\alpha(x \pm 0) + \alpha(\xi_\Gamma^\pm \mp 0)\},$$

$$\alpha^j(x) := \left(\frac{3}{4} + \frac{\delta_{oj}}{4}\right) \tilde{\alpha}(x) + \frac{\delta_{1j}}{4} \bar{\alpha}(x) + \frac{\tilde{H}(x)}{h(x)} c(x)$$
$$g_h^j(x) := \left(\frac{3}{4} + \frac{\delta_{oj}}{4}\right) \tilde{g}(x) + \frac{\delta_{1j}}{4} \bar{g}(x) + \frac{\tilde{H}(x)}{h(x)} f_h(x) \qquad \text{for } x \in \gamma_{23}, \qquad (3.2.18c)$$

$\tilde{\alpha}(x) := \frac{1}{2h(x)}\{h(x'_{s_\Gamma}-)\, \alpha(x-0) + h(x'_{s_\Gamma}+)\, \alpha(x+0)\}$, $\tilde{g}(x)$ analogously,

$\bar{\alpha}(x) := \frac{1}{2h(x)}\{h(x'_{s_\Gamma}-)\, \alpha(s_\Gamma^-+0) + h(x'_{s_\Gamma}+)\, \alpha(s_\Gamma^+-0)\}$, $\bar{g}(x)$ analogously,

$\delta_{1j} := j$, $\delta_{oj} := 1-j$; $h(x)$, $H(x)$, $\tilde{H}(x)$, $h(x'_s)$, $s(x'_s)$ and $\tilde{s}(x'_s)$ are taken from (2.3.13). ≡

Proof: The proof follows the line of the constructive approximation of the balance equations described in Section 3.2.1. Due to VO(PB,$\bar{\omega}$), the secondary triangulation $\mathcal{T}_{\tilde{\mathfrak{X}}}$ is correctly defined. Therefore, the system of balance equations (3.2.2) and $u = g$ on γ_1 can be derived. Under the assumptions V($\bar{\Omega}$), V(A,F), V(u) and $h \leq h_o$, for sufficiently small h_o, the point functionals p_A^u, p_F and the unknown error p_E are well-defined. Therefore, the discrete analogues (3.2.14), (3.2.16) and (3.2.18) can be formulated. Some details of the approximations used will be given explicitly in Chapter 5.

Remark 3.2. The assertion that the error p_E in (3.2.13) is arbitrarily small for sufficiently small h is not included in Theorem 3.1., but will be investigated in Chapter 5., together with the convergence $y \to u$ for h tending to zero and under certain assumptions on $\bar{\omega}, \bar{\Omega}$ and A, u, F. The networks employed in Theorem 3.1. consist, in general, of triangles and rectangles. Clearly, the special cases "triangles only", "rectangles only" and "locally irregular networks" from Definition 2.10., 2.11., 2.12. and 2.13. are also included in Theorem 3.1. Moreover, the FDSs proposed for Au = F can also be obtained on other networks, if $\mathcal{T}_{\tilde{\mathfrak{X}}}$ is correctly obtainable via method (PB). Naturally, general convex quadrangles cannot be utilized. ≡

3.3. Finite difference schemes via method (MD)

3.3.1. Quadrature formulas for the balance equations. We now consider the BVP Au = F from (2.1.1), (2.1.2) and secondary triangulations with boxes $\mathfrak{X}(x) := \mathfrak{X}_m(x)$, $\tilde{\mathfrak{X}}(x) := \tilde{\mathfrak{X}}_m(x)$ of the type (MD) from Definition 2.25. The starting point of the approximations to be carried out is the system of balance equations (3.1.7). According to the index set S"(x), the leading term $\int(K\nabla u,n)ds$ in (3.1.7) will now be split in the same manner as in (3.1.8). Consequently, we have

$$-\sum_{x'' \in S''(x)} \int_{\mathcal{S}(x,x'')} (K\nabla u,n)ds + \int_{\mathcal{S}(x)} \alpha u ds + \int_{\mathfrak{X}(x)} cu d\bar{x} = \int_{\mathfrak{X}(x)} f d\bar{x} + \int_{\mathcal{S}(x)} g ds \qquad (3.3.1)$$

for $x \in \omega + \gamma_{23}$, with $\mathcal{S}(x) = \emptyset$ for $x \in \omega$, and with $\mathfrak{S}(x,x'')$, $\mathfrak{S}(x)$ from (2.3.15), (2.3.16). Next we approximate the terms in (3.3.1) by quadrature, cubature and finite difference formulas, where the last ones are treated in more detail in Section 3.3.2. Initially and for indicating local approximation errors, the special smoothness assumptions

$$k_{ij}, c, f \in C^s(\bar{\Omega})\ (i,j=1,2),\ \alpha, g \in PC^s(\Gamma_{23})\ \text{for}\ s \in \{1,2\},\ g \in C(\Gamma_1), \tag{3.3.2a}$$

$$u \in C^{1+s}(\bar{\Omega})\ \text{for}\ s \in \{1,2\}, \tag{3.3.2b}$$

are taken into account. For the approximation of the flux through $\mathfrak{S}(x,x'')$, i.e. $\int_{\mathfrak{S}(x,x'')}(K\nabla u,n)ds$, we consider the subsets $\mathfrak{S}^{\pm}(x,x''), \tilde{\mathfrak{S}}^{\pm}(x,x'')$ of $\mathfrak{S}(x,x'')$, $\tilde{\mathfrak{S}}(x,x'')$, and the outer normals $n_{\mathfrak{S}\pm}, n_{\tilde{\mathfrak{S}}\pm}$ to $\mathfrak{S}^{\pm}$, $\tilde{\mathfrak{S}}^{\pm}$, respectively, cf. Definition 2.23., (2.3.15), (2.3.16) and Figs. 6d, e. The partial medians $\tilde{\mathfrak{S}}^{\pm}(x,x'')$ of Δ originate from $\mathfrak{S}^{\pm}(x,x'')$ in $\triangle$ by shortening or lengthening of $\mathfrak{S}^{\pm}(x,x'')$ as described in Definition 2.23. Obviously, the error of the corresponding measures can be estimated by

$$|s^{\pm}(x,x'') - \tilde{s}^{\pm}(x,x'')| \leq K_1 h^2 \quad \text{for}\ x \in \gamma_{23}. \tag{3.3.3}$$

For the approximation of $\int_{\mathfrak{S}(x,x'')} r ds$, where r may have a jump at x", we choose quadrature formulas on the parts $\mathfrak{S}^{\pm}(x,x'')$, with one quadrature point only, viz. $x'' \in \mathfrak{S}^{\pm}(x,x'')$, that is

$$\int_{\mathfrak{S}(x,x'')} r ds \approx r^-(x'')s^-(x,x'') + r^+(x'')s^+(x,x''),\quad r^{\pm}(x''):=r(x''\pm 0). \tag{3.3.4}$$

In triangles Δ, the partial medians $\tilde{\mathfrak{S}}^{\pm}(x,x'')$ of Δ are used for approximating $\mathfrak{S}^{\pm}(x,x'')$ in $\triangle$. For $u \in C^2(\bar{\Omega}), k_{ij} \in C^1(\bar{\Omega})$ and $r:=(K\nabla u,n)$, this leads to the relations

$$\int_{\mathfrak{S}(x,x'')} (K\nabla u,n)ds = \begin{cases} (K(x'')\nabla u(x''),\ n_{\mathfrak{S}-}s^-(x,x'')+n_{\mathfrak{S}+}s^+(x,x''))+O(h^2) \\ \text{for}\ e:=\Delta(x,\mathfrak{s},\mathfrak{s}^+), & (3.3.5a) \\ (K(x'')\nabla u(x''),\ n_{\tilde{\mathfrak{S}}-}\tilde{s}^-(x,x'')+n_{\tilde{\mathfrak{S}}+}\tilde{s}^+(x,x''))+O(h^2) \\ \text{for}\ e:=\triangle(x,\mathfrak{s},\mathfrak{s}^+), & (3.3.5b) \end{cases}$$

where $x \in \omega + \gamma_{23}$. Bearing in mind (2.3.22), we see that

$$\frac{h^{\oplus} - h^{\circ}}{2} = \begin{cases} \mathfrak{S}^{\ominus} + \mathfrak{S}^{\oplus} = n_{\mathfrak{S}-}s^-(x,x'')+n_{\mathfrak{S}+}s^+(x,x'') & e = \Delta, \quad (3.3.6a) \\ & \text{for} \\ \tilde{\mathfrak{S}}^{\ominus} + \tilde{\mathfrak{S}}^{\oplus} = n_{\tilde{\mathfrak{S}}-}\tilde{s}^-(x,x'')+n_{\tilde{\mathfrak{S}}+}\tilde{s}^+(x,x'') & e = \triangle, \quad (3.3.6b) \end{cases}$$

which simplifies (3.3.5) considerably. Thus,

$$\int_{\mathfrak{S}(x,x'')} (K\nabla u,n)ds = (K\nabla u,\ \frac{h^{\oplus}-h^{\circ}}{2})(x'')+O(h^2)\ \text{for}\ x \in \omega+\gamma_{23}. \tag{3.3.7}$$

If we take into account rectangles $\square\,(\square \subset \bar{\Omega})$, then relations (3.3.5a), (3.3.6a) and (3.3.7) are also valid, if the vectors h, h^+ are arranged as shown in Fig. 6d. Therefore, relation (3.3.7) defines an approximation of the line integral $\int_{\mathfrak{S}(x,x'')} (K\nabla u,n)ds$ for the general case $e \in \{\Delta, \square, \triangle\}$.

For the approximation of the integrals $\int_{\mathfrak{S}(x)} r\,ds$ from (3.3.1), with $r := \alpha u$ or $r := g$, we use here the same quadrature formulas as described in (3.2.8) for $j \in \{0,1\}$, with the only difference that now $\mathfrak{S}(x)$, $\mathfrak{S}^{\pm}(x)$, $x_{\Gamma}^{\pm}$, $s^{\pm}(x)$ are to be understood in the sense of Definition 2.23. (MD-boxes). The approximation of the integrals $\int_{\mathcal{H}(x)} r\,d\bar{x}$ in (3.3.1), with $r := cu$ or $r := f$, is made in analogy to (3.2.11), but here with $\mathcal{H}(x)$, $H(x)$ from method (MD). The parameters of the boxes $\mathcal{H}(x)$, $x \in \gamma$, are replaced by those of the polygonal modifications $\tilde{\mathcal{H}}(x)$, $x \in \gamma$, since $\tilde{s}^{\pm}(x,x'')$ $\approx s^{\pm}(x,x'')$, $\tilde{s}^{\pm}(x) \approx s^{\pm}(x)$, $\tilde{H}(x) \approx H(x)$ hold, and the values with "$\sim$" are readily available; cf. (2.3.15), (2.3.16), (2.3.18), and Fig. 6c in comparison with Fig. 6b. Estimating the error, we get (3.3.3) and

$$|H(x) - \tilde{H}(x)| < K_0 h^3, \quad |s^{\pm}(x) - \tilde{s}^{\pm}(x)| \leq K_2 \begin{cases} h^2 & \text{case a)} \\ & \text{for} \qquad \text{of (2.2.1)}, \\ h^3 & \text{case b)} \end{cases} \tag{3.3.8}$$

with $\tilde{\tilde{s}}^{\pm}(x) = h(x, \xi_{\Gamma}^{\pm})/2$, $x \in \gamma$. Comparing the estimates for $|s^{\pm}(x) - \tilde{s}^{\pm}(x)|$ with respect to the methods (PB) and (MD), we see that PB-boxes produce an error which is asymptotically smaller than for MD-boxes, cf. (3.3.8) with (3.2.12).

Finally, $K(x'')$ needed in (3.3.7) will be calculated exactly at x'' or approximated by

$$K_h(x'') := \tfrac{1}{3}\{K(x)+K(\xi)+K(\xi^+)\} \text{ for } x'' \in \Delta(x,\xi,\xi^+) \text{ and } x'' \in \triangle(x,\xi,\xi^+), \tag{3.3.9a}$$

$$K_h(x'') := \tfrac{1}{4}\{K(x)+K(\xi)+K(\eta)+K(\xi^+)\} \text{ for } x'' \in \square(x,\xi,\eta,\xi^+), \tag{3.3.9b}$$

cf. Figs. 6e, d. The values of r and $\frac{dr}{ds}$ at the quadrature points x and $x_{\Gamma}^{\pm}$ from formula (3.2.8), with $r := \alpha u$ or $r := g$, will be taken as in (3.2.9), (3.2.10). What remains to be done now is to choose appropriate finite difference quotients for the approximation of the first order derivative ∇u in (3.3.7).

<u>3.3.2. Finite difference quotients.</u> In this section, a simple approach to the construction of finite difference quotients on finite elements e will be given. Additionally, the notations of some well-known finite difference quotients are introduced.

Let us consider a finite region e, with $e = \bar{e}$ and $e \subset \bar{\Omega}$, and e satisfies the same assumption V(e) as $\bar{\Omega}$ in $V(\bar{\Omega})$, cf. Assumption 2.1. Furthermore, let n be the outer normal to ∂e and $u \in C^1(\bar{\Omega})$ or $u \in W_2^1(\Omega)$. By Gauss' formula (cf. Appendix DI), we get

$$\int_e \frac{\partial u}{\partial x_i}\,dx = \int_{\partial e} u n_i\,ds = \begin{cases} \int_{\partial e} u\,dx_2 & i = 1, \\ & \text{for} \\ -\int_{\partial e} u\,dx_1 & i = 2. \end{cases} \tag{3.3.10}$$

For sufficiently smooth functions u, for instance $u \in C^1(\bar{\Omega})$, the one-point cubature formula

$$\frac{\partial u}{\partial x_i}(\tilde{x}) = \frac{1}{\text{mes } e} \int_e \frac{\partial u}{\partial x_i}\, dx + q_i(\partial^1 u, \tilde{x}), \quad \tilde{x} \in e, \qquad (3.3.11)$$

holds for an arbitrary point $\tilde{x} \in e$. Now, by means of (3.3.11), (3.3.10) and quadrature formulas for approximating the line integrals $\int_{\partial e} u\, dx_i$, we are able to construct simple difference analogues to ∇u at $\tilde{x} \in e$, where e may be curvilinear. We confine ourselves to the cases $e \in \{\Delta, \square, \mathcal{\Delta}\}$ For $e := \Delta(x, \xi, \xi^+)$ and sufficiently smooth functions u, using the trapezoid rule on each of the sides Δ and collecting the single expressions, we obtain the relations

$$\frac{\partial u}{\partial x_i}(\tilde{x}) = \frac{(-1)^{1+i}}{\text{mes } \Delta} \int_{\partial\Delta} u\, dx_{3-i} + q_i(\partial^1 u, \tilde{x}) = (\nabla_h u(\tilde{x}), e_i) + d_i(\partial^1 u, \tilde{x}), \qquad (3.3.12)$$

with $e_1 = (1,0)^T$, $e_2 = (0,1)^T$, q_i from (3.3.11) and ∇_h defined in (3.3.14). Clearly, the difference analogue ∇_h depends on the geometry of Δ only, i.e. ∇_h does not depend on a special $\tilde{x} \in \Delta$. For $e := \mathcal{\Delta}(x, \xi, \xi^+)$, the same difference quotient from (3.3.14) arises, if $\mathcal{\Delta}$ is approximated by Δ. The following relation holds for triangles $\Delta, \Delta \subset \Omega_h$:

$$|\nabla u(\tilde{x}) - \nabla_h u(\tilde{x})| =: |d(\partial^1 u, \tilde{x})| = O(h) \text{ for } u \in C^2(\Delta \cap \bar{\Omega}), \qquad (3.3.13)$$

$\tilde{x} \in \Delta(x, \xi, \xi^+) \cap \bar{\Omega}$, $x \in \omega + \gamma_{23}$. Relation (3.3.13) can be verified by Taylor's expansion and by taking into account the $O(h^3)$-asymptotic of the arc b on the segment h, cf. Appendix GE. The cubature error q_i, cf. (3.3.11) and (3.3.12), can be estimated as in (3.2.11). In particular, we have $q_i = O(h^2)$ for $u \in C^3$, $e = \Delta$ and $\tilde{x} = x''$.
Furthermore, the following difference quotients are defined, which will be used here and in several places of this monograph.

<u>Definition 3.3.</u> The following difference quotients will be introduced.

a) the difference quotient $\nabla_h u(\tilde{x})$ for $\nabla u(\tilde{x})$, $\tilde{x} \in \Delta$ or $\tilde{x} \in \mathcal{\Delta}$:

$$\nabla_h u(\tilde{x}) := \left(\frac{\delta u}{\delta x_1}, \frac{\delta u}{\delta x_2}\right)^T(\tilde{x}) := \frac{[u(\xi)-u(x)]\, h^{\oplus} - [u(\xi^+)-u(x)]\, h^{\circ}}{(h^{\oplus}, h)}, \qquad (3.3.14)$$

where $h^{\oplus}$ and h° are taken from (2.3.20);

b) the difference quotient $\frac{\delta u}{\delta \mu}(\tilde{x})$ for $\frac{\partial u}{\partial \mu}(\tilde{x})$ (directional derivative), $\tilde{x} \in e$:

$$\frac{\delta u}{\delta \mu}(\tilde{x}) := (\nabla_h u(x), \mu), \; |\mu| = 1, \text{ especially: } \mu = e_i \; (i = 1,2), \qquad (3.3.15)$$
with $e_1 = (1,0)^T$, $e_2 = (0,1)^T$,

c) the difference quotient $\frac{\delta u}{\delta h}(\tilde{x})$ for the directional derivative $\frac{\partial u}{\partial h}(\tilde{x})$, where $\tilde{x} \in h(x, \xi)$:

$$\frac{\delta u}{\delta h}(\tilde{x}) := u_h(\tilde{x}) := \frac{u(\xi)-u(x)}{h(x,\xi)}, \text{ with } h(x,\xi) := \text{mes } h(x,\xi), \qquad (3.3.16)$$

where $\tilde{x}$ is often the midpoint x'_ξ of the mesh segment $h(x,\xi)$; then $\frac{\delta u}{\delta h}(x'_\xi)$ is the central difference quotient;

d) the difference quotient $\frac{\delta u}{\delta b}(\tilde{x})$ for the derivative $\frac{\partial u}{\partial s}(\tilde{x})$, $\tilde{x} \in b(x,\xi)$, along the arc $b(x,\xi)$ over the mesh segment $h(x,\xi)$, cf. Fig. 3a:

$$\frac{\delta u}{\delta b}(\tilde{x}) := u_b(\tilde{x}) := \frac{u(\xi)-u(x)}{b(x,\xi)}, \text{ with } b(x,\xi) := \text{mes } b(x,\xi), \tag{3.3.17}$$

because of $b(x,\xi) - h(x,\xi) = O(h^3)$, $\frac{\delta u}{\delta b}$ is frequently replaced by $\frac{\delta u}{\delta h}$ from (3.3.16);

e) the difference quotient $\nabla_h u(\tilde{x}) = (\frac{\delta u}{\delta x_1}, \frac{\delta u}{\delta x_2})^T(\tilde{x})$ for $\nabla u(\tilde{x})$, $\tilde{x} \in \Box$, where $\Box$ is provided with sides parallel to the coordinate axes, cf. Fig. 3b:

$$\frac{\delta u}{\delta x_1}(\tilde{x}) := \frac{u(\xi)-u(x)+u(\eta)-u(\xi^+)}{2h(x,\xi)}, \tag{3.3.18}$$

$$\frac{\delta u}{\delta x_2}(\tilde{x}) := \frac{u(\eta)-u(\xi)+u(\xi^+)-u(x)}{2h(x,\xi^+)}. \equiv$$

The difference quotients in (3.3.18) are well known and can be derived via (3.3.10), (3.3.11) and (3.3.12), where e or Δ must be replaced by $\Box$. The difference quotients encountered previously do not depend on the position of $\tilde{x}$ in e, h or b, but the special choice of $\tilde{x}$ (for ∇u, $\frac{\partial u}{\partial \mu}$, etc.) is essential for the error estimation. The error of the difference quotients is of the order $O(h^l)$, $l \in \{1,2\}$, where l depends on the formula used, especially on $\tilde{x}$, and u is assumed to be sufficiently smooth (maximally $u \in C^3$). A more detailed treatment of this can be found in Chapter 5.

Substituting the approximations described by (3.3.7) - (3.3.9), (3.3.14), (3.3.16), (3.3.18) and by (3.2.8), (3.2.11) into the system of balance equations (3.3.1), we now obtain, with the reasoning of method (PB) transferred to method (MD), the equations (3.2.13), (3.2.14), (3.2.15), (3.2.16) and (3.2.17), but now for the method (MD). Thus some FDSs $A_h y = F_h$ for the BVP $Au = F$ are constructed. For functions k_{ij}, c, f, g, α from V(A,F) and solutions u from V(u), the composed approximations are also defined, since the point functionals of these functions on the grid $\bar{\omega}$ are well-defined.

<u>3.3.3. Finite difference schemes of the type (MD).</u> The following Theorem 3.4. summarizes the approximations proposed in Sections 3.3.1. and 3.3.2. Two FDSs $A_h^j y = F_h^j$ (j=0,1) will be noted, which differ from each other in the approximation of the line integral $\int_{S(x)} r\,ds$ according to (3.2.8) for j=0,1 and method (MD).

<u>Theorem 3.4.</u> Let (i), (ii), (iii) be given, that is
(i) the BVP $Au = F$ (2.1.1), (2.1.2), with $V(\bar{\Omega})$, V(A,F) and V(u) from Assumption 2.1., 2.2., 2.3.,
(ii) grids $\bar{\omega}$ from (2.2.6) based on the general triangulation $\mathcal{T}_\Delta$ from (2.3.3), for $h \leq h_0$ and sufficiently small h_0, with arbitrary triangles,

but under the Condition 2.24. - VO(MD, $\bar{\omega}$),

(iii) secondary triangulations $\mathcal{T}_{\tilde{\mathcal{H}}}$ of the type (MD) from Definition 2.23.

Then, there are FDSs $A_h^j y = F_h^j$ $(j=0,1)$ on $\bar{\omega}$ as discrete analogues of the BVP $Au = F$ via method (MD), for $h \leq h_0$ and sufficiently small h_0. The FDSs

$$A_h^j y = F_h^j \text{ on } \bar{\omega} = \gamma_1 + \gamma_{23} + \omega, \text{ with } A_h^j = (l_h, l_h^j, L_h)^T,\ F_h^j = (g_h, g_h^j, f_h)^T, \tag{3.3.19a}$$

will be described for $j \in \{0,1\}$ as follows:

$$(l_h y)(x) := y(x) = g_h(x) \text{ for } x \in \gamma_1,\ g_h(x) := g(x),$$

$$(l_h^j y)(x) := -\frac{1}{h(x)} \sum_{x'' \in S''(x)} (K_h \nabla_{\hbar} y, \frac{\hbar^{\bullet} - \hbar^{o}}{2})(x'') \tag{3.3.19b}$$

$$+ \frac{j}{h(x)} \sum_{\xi \in S'(x) \cap \gamma} \frac{h^2(x_\xi')}{8} \alpha_h(x_\xi') y_{\hbar}(x_\xi') + \alpha^j(x) y(x) = g_h^j(x) \quad \text{for } x \in \gamma_{23},$$

$$(L_h y)(x) := -\frac{1}{H(x)} \sum_{x'' \in S''(x)} (K_h \nabla_{\hbar} y, \frac{\hbar^{\bullet} - \hbar^{o}}{2})(x'') + c(x) y(x) = f_h(x) \quad \text{for } x \in \omega,$$

with the following notations:

$(.,.)$ scalar product of R^2, $K_h(x'')$ from (3.3.9a), $\nabla_{\hbar} y(x'')$ from (3.3.14), $y_{\hbar}(x_\xi')$ from (3.3.16) or (3.2.18c), $f_h(x) := f(x)$ for $f \in C(\bar{\Omega})$, and $f_h(x)$ from Section 5.4.3. for $f \notin C(\bar{\Omega})$,

$$\alpha_h(x'_{\xi_\Gamma^\pm}) := \frac{1}{2}\{\alpha(x \pm 0) + \alpha(\xi_\Gamma^\pm \mp 0)\}, \tag{3.3.19c}$$

$$\alpha^j(x) := (\tfrac{3}{4} + \frac{\delta_{0j}}{4}) \tilde{\alpha}(x) + \frac{\delta_{1j}}{4} \bar{\alpha}(x) + \frac{\tilde{H}(x)}{h(x)} c(x),\quad \tilde{\alpha}, \bar{\alpha} \text{ from (3.2.18c)},$$

$$g_h^j(x) := (\tfrac{3}{4} + \frac{\delta_{0j}}{4}) \tilde{g}(x) + \frac{\delta_{1j}}{4} \bar{g}(x) + \frac{\tilde{H}(x)}{h(x)} f_h(x),\quad \tilde{g}, \bar{g} \text{ from (3.2.18c)},$$

$\delta_{0j} := 1-j$, $\delta_{1j} := j$; $h(x)$, $h(x_\xi')$, $H(x)$ and $\tilde{H}(x)$ are taken from (2.3.18). ≡

<u>Proof:</u> The argumentation is quite similar to that of the proof of Theorem 3.1.

Theorem 3.4. is formulated for networks of triangles, but it holds also for networks of triangles and (or) rectangles. The formulas of the FDSs $A_h^j y = F_h^j$ obtained for such networks are identical to (3.3.19). But for rectangles □, the segments $\hbar$, $\hbar^+$ and $\nabla_{\hbar} y(x'')$ must be taken from Fig. 3b and (3.3.18), respectively, $K_h(x'')$ from (3.3.9b) (case □). If rectangles □ are used, then extended difference stars $S_e(x)$ mentioned in Section 2.2.1. must be considered, and mesh segments $\hbar(x,\xi)$ exist only for $\xi \in S_e'(x) \cap S'(x)$. For the sake of brevity, we restrict ourselves to networks of triangles, if method (MD) is discussed. But it should be pointed out that all final results concerning the symmetry and positive definiteness of the difference operators (Chapter 4) and the convergence (Chapter 5) are maintained for networks involving rectangles. For FDSs $A_h y = F_h$ with extended difference stars $S_e(x)$ and networks of rectangles only, we

refer to the monographs of SAMARSKIJ [1], SAMARSKIĬ/ANDREEV [1].
In analogy to the method (PB), Remark 3.2. is to add here for Theorem 3.4., with the only exception that convex quadrangles could be used for method (MD), too, cf. SAMARSKIĬ/TISHKIN et al. [1]. The assumptions on the grid regularity are here considerably weaker than in Theorem 3.1., since VO(MD, $\bar{\omega}$) is only a condition concerning the asymptotic behaviour of the grid $\bar{\omega}$.

Remark 3.5. It is often suitable that the summation in (3.3.19b) is written with respect to the grid points $\xi \in S'(x)$ instead of $x'' \in S''(x)$. Using triangles only, we can easily rewrite the FDSs $A_h^j y = F_h^j$ ($j=0,1$) from (3.3.19) as follows:

$$(l_h y)(x) := y(x) = g_h(x) \text{ for } x \in \gamma_1,$$

$$(l_h^j y)(x) := -\frac{1}{h(x)} \sum_{\xi \in S'(x)} h(x'_\xi)\, K_h^j(x'_\xi) y_{\hbar}(x'_\xi) + \alpha^j(x) y(x) = g_h^j(x) \quad \text{for } x \in \gamma_{23}, \tag{3.3.20a}$$

$$(L_h y)(x) := -\frac{1}{H(x)} \sum_{\xi \in S'(x)} h(x'_\xi)\, K_h(x'_\xi) y_{\hbar}(x'_\xi) + c(x) y(x) = f_h(x) \quad \text{for } x \in \omega .$$

Here, $H(x)$, $h(x)$, $h(x'_\xi)$, $\alpha^j(x)$, $g_h^j(x)$, $f_h(x)$, $y_{\hbar}(x'_\xi)$ are to be taken from (3.3.19), but $K_h(x'_\xi)$, $K_h^j(x'_\xi)$ are defined by ($j=0,1$)

$$K_h(x'_\xi) := K_h^j(x'_\xi) := \frac{1}{2}\left\{ \frac{(K_h^+ \hbar^\oplus, \hbar^\oplus - \hbar^\circ)}{(\hbar^\oplus, \hbar)} + \frac{(K_h^- \hbar^\ominus, \hbar^\ominus - \hbar^\circ)}{(\hbar^\circ, \hbar^-)} \right\}$$

for $x \in \omega$ and $\xi \in S'(x)$, $x \in \gamma_{23}$ and $\xi \in S'(x) \cap \omega$, $K_h^\pm := K_h(x''_\pm)$,

$$K_h^j(x'_\xi) := \frac{1}{2}\left\{ \frac{(K_h^+ \hbar^\oplus, \hbar^\oplus - \hbar^\circ)}{(\hbar^\oplus, \hbar)} - j\,\frac{h(x'_\xi)}{4}\,\alpha_h(x'_\xi) \right\} \quad \text{for } x \in \gamma_{23},\ \xi = \xi_\Gamma^+ \in S'(x), \tag{3.3.20b}$$

$$K_h^j(x'_\xi) := \frac{1}{2}\left\{ \frac{(K_h^- \hbar^\ominus, \hbar^\ominus - \hbar^\circ)}{(\hbar^\circ, \hbar^-)} - j\,\frac{h(x'_\xi)}{4}\,\alpha_h(x'_\xi) \right\} \quad \text{for } x \in \gamma_{23},\ \xi = \xi_\Gamma^- \in S'(x).$$

The symbols $\hbar$, $\hbar^\pm$, $\hbar^\circ$ and $\hbar^\oplus, \hbar^\ominus$ are defined in (2.3.19), (2.3.20); and $K_h^\pm := K_h(x''_\pm)$ is taken from (3.3.9a), where $x''_\pm$ are the centres of gravity in the triangles $\Delta(x, \xi, \xi^\pm)$. ■

4. ANALYTICAL AND MATRIX PROPERTIES OF THE DIFFERENCE OPERATORS A_h

<u>4.1. General remarks and notations.</u> It is well known that the operator A of the BVP (2.1.1), (2.1.2) is an operator "of monotonic type", cf. COLLATZ [1], GILBARG/TRUDINGER [1], HEINRICH [3], MEYER [1]. Furthermore, some linear operator $\tilde{A}:V \rightarrow V^*$ (V^*: dual space), can be associated with A (case $g = 0$ on Γ_1) by means of the variational formulation from (2.1.12), where $\tilde{A}$ is symmetric, positive definite, V-elliptic (V-coercive) and V-continuous on $V = \mathring{W}^1_{2,\Gamma_1}$, cf. AZIZ [1], CIARLET [2], GAJEWSKI et al. [1], GILBARG/TRUDINGER [1], LADYSHENSKAYA [1], MICHLIN/SMOLIZKI [1], NEČAS [1], ODEN/REDDY [1], WLOKA [1]. For the positive definiteness and the V-ellipticity, condition (2.1.8) is essential, and either of these properties can be deduced from the other one. The properties mentioned above are often taken advantage of, when a priori estimates are derived, which are the basis of assertions on the existence and uniqueness of a solution to the BVP, sometimes for inclusion theorems for the solution u. Indeed, if the properties "of monotonic type", on the one hand, and the symmetry, positive definiteness on the other hand can be preserved for difference operators A_h and $\tilde{A}_h$, respectively, then we are often able to derive a priori estimates, error estimates, existence, uniqueness and inclusion assertions for the solution y of the FDS $A_h y = F_h$ (or $\tilde{A}_h y = \tilde{F}_h$) and efficient methods for solving the system of linear algebraic equations $\mathcal{B}_h y = G_h$ assigned to $A_h y = F_h$.
Now the FDSs $A_h y = F_h$ constructed in Chapter 3. will be investigated, and it will be shown how the properties "of monotonic type", "symmetry" and "positive definiteness" can be proved under certain assumptions. Additionally, a discrete analogue of the "V-ellipticity" will be obtained.

We consider the difference operator A_h in its "canonical representation" (4.1.1), with real coefficients $a(x,\xi)$ and the index set S(x) ("full difference star") from (2.2.2), i.e.

$$(A_h y)(x) := \sum_{\xi \in S(x)} a(x, \xi)\, y(\xi) \quad \text{for } x \in \bar{\omega} \quad , \; S(x) = S'(x) + \{x\}. \tag{4.1.1}$$

Moreover, further operators $\tilde{A}_h$, B_h and matrices $\mathcal{A}_h$, $\tilde{\mathcal{A}}_h$, $\mathcal{B}_h$ will be associated with A_h and defined subsequently. After numbering the n grid points of $\bar{\omega}$ in a one-to-one fashion with the integers 1,2,...,n, we have $\bar{\omega} := \{P_1, P_2, \dots, P_n\}$. On the basis of this ordering, a square matrix $\mathcal{A}_h = (a_{kl})_{n\times n}$ and a system of linear algebraic equations $\mathcal{A}_h y = F_h$ are associated with the operator A_h from (4.1.1) and with the FDS $A_h y = F_h$, respectively, where $\mathcal{A}_h$ and $\mathcal{A}_h y = F_h$ are defined as follows:

$$\mathcal{A}_h := (a_{kl})_{n\times n},\ a_{kl} := \begin{cases} a(P_k,P_l) & P_l \in S(P_k) \\ & \text{for} \\ 0 & P_l \notin S(P_k) \end{cases},\ a(P_k,P_l) \text{ from (4.1.1)}, \tag{4.1.2}$$

with $k,l = 1,2,\ldots,n$, $n := |\bar{\omega}|$ from (2.2.8),

$\mathcal{A}_h y = F_h$, with vectors y, $F_h \in R^n$,

$y = (y_1,\ldots,y_n)^T$, $F_h = (F_{h1},\ldots,F_{hn})^T$, $y_i = y(P_i)$, $F_{hi} = F_h(P_i)$.

Henceforth, we shall always be able to alternatively use the operator A_h or its matrix representation $\mathcal{A}_h$, the grid functions y, F_h on $\bar{\omega}$ or the corresponding vectors y, $F_h \in R^n$ (which are not distinguished in their symbols). In doing so, several results of general and special matrix theory, e.g. from ORTEGA/RHEINBOLDT [1] and VARGA [1], can be utilized here. For the vector $y \in R^n$ of the unknown grid function of the FDS $A_h y = F_h$, we actually have $n_{\gamma_1} := |\gamma_1|$ known components. Therefore, the first $\tilde{n}$ components (without loss of generality), $\tilde{n} := n - n_{\gamma_1} = |\omega + \gamma_{23}|$, are really unknown and have to be calculated. After replacing the components $y_{\tilde{n}+1},\ldots,y_n$ by the given values $g_i = g(P_i)$ ($i=\tilde{n}+1,\ldots,n$, and $P_i \in \gamma_1$), the following system of equations defining the unknown reduced vector $y \in R^{\tilde{n}}$ (the symbol y will not be changed) is obtained:

$$\tilde{\mathcal{A}}_h y = \tilde{F}_h,\ \tilde{\mathcal{A}}_h = (\tilde{a}_{ij})_{\tilde{n}\times\tilde{n}} \text{ and } \tilde{F}_h \in R^{\tilde{n}} \text{ from (4.1.4)}. \tag{4.1.3}$$

System (4.1.3) is associated with a FDS $\tilde{A}_h y = \tilde{F}_h$ on $\omega + \gamma_{23}$, which is defined by

$$(\tilde{A}_h y)(x) := \sum_{\xi \in S'(x)\setminus\gamma_1} a(x,\xi)y(\xi) = F_h(x) - \sum_{\xi \in S(x)\cap\gamma_1} a(x,\xi)g(\xi) =: \tilde{F}_h(x),$$
$$x \in \omega + \gamma_{23}. \tag{4.1.4}$$

Finally, we take into account the linear system of algebraic equations $\mathcal{B}_h y = G_h$ which is associated with the discrete analogue $B_h y = G_h$ of the system of balance equations (3.1.7) for $x \in \omega + \gamma_{23}$, with the inserted restriction $y(x) = g(x)$ for $x \in \gamma_1$, cf. (3.2.15). That is, we consider the FDS

$$B_h y = G_h \text{ on } \omega + \gamma_{23}, \text{ where } (B_h y)(x) := \tilde{p}_A^y(x),\ G_h(x) := \tilde{p}_F(x)$$
$$\text{for } x \in \omega + \gamma_{23}, \tag{4.1.5}$$

and $\tilde{p}_A^y$, $\tilde{p}_F$ are taken from (3.2.15) considered for method (PB) and method (MD). The matrix $\mathcal{B}_h$ is derived from the difference operator B_h in the same way as $\tilde{\mathcal{A}}_h$ from $\tilde{A}_h$. Obviously, the only difference between the matrices $\mathcal{B}_h$ and $\tilde{\mathcal{A}}_h$ as well as B_h and $\tilde{A}_h$ rests on the scaling by the factors $1/h(x)$ for $x \in \gamma_{23}$ and $1/H(x)$ for $x \in \omega$. Therefore, the following representation of $\mathcal{B}_h y = G_h$ via $\tilde{\mathcal{A}}_h y = \tilde{F}_h$ from (4.1.3) can be given, cf. (3.2.15) with (3.2.17):

$\mathcal{B}_h y = G_h$, with $\mathcal{B}_h = (b_{kl})_{\tilde{n}x\tilde{n}}$, $G_h = (G_{h1},\dots,G_{h\tilde{n}})^T \in R^{\tilde{n}}$,

$$\mathcal{B}_h := \tilde{H}_h \tilde{\mathcal{A}}_h,\ G_h := \tilde{H}_h \tilde{F}_h,\ \tilde{H}_h := \begin{pmatrix} H_h(P_1) & & \mathbf{0} \\ & \ddots & \\ \mathbf{0} & & H_h(P_{\tilde{n}}) \end{pmatrix}, \tag{4.1.6}$$

$$H_h(P_k) := \begin{cases} h(P_k) & P_k \in \gamma_{23} \\ & \text{for} \\ H(P_k) & P_k \in \omega \end{cases},\ H(x),\ h(x) \text{ from (2.3.13) or (2.3.18).}$$

The system $\mathcal{B}_h y = G_h$ (4.1.6) of $\tilde{n}$ simultaneous equations is that system which must be actually solved on a computer. It arises immediately by discretization of the system of balance equations (3.1.7), without the detour of using $\tilde{\mathcal{A}}_h y = \tilde{F}_h$. Clearly, $\mathcal{B}_h y = G_h$ also defines an operator equation $B_h y = G_h$ on $\omega + \gamma_{23}$, which was already noted in (4.1.5). It should be mentioned that a slight modification of the symbol B_h is used in HEINRICH [5,7] as compared with HEINRICH [8] and the work presented here.

4.2. Monotonicity and other matrix properties

4.2.1. Operators A_h and matrices via method (PB). The property of a difference operator A_h to be "of monotonic type" and its utilization for deriving a priori estimates with respect to the C-norm (maximum-norm) have been discussed by BRAMBLE/HUBBARD [1], CIARLET/RAVIART [1], COLLATZ [1,2], LORENZ [1], MICHLIN/SMOLIZKI [1], TÖRNIG [1], SAMARSKIĬ/ANDREEV [1], but also by HEINRICH [2,3,4,6] and in a great number of other papers. It is well known that the following Condition 4.1. - $M(A_h,\bar{\omega})$ is sufficient for a difference operator A_h to be "of monotonic type" (if A_h^{-1} exists: "inverse isotone"), i.e., that $A_h y \geq 0$ on $\bar{\omega}$ implies $y \geq 0$ on $\bar{\omega}$. Furthermore, $M(A_h,\bar{\omega})$ guarantees that $\mathcal{A}_h$ is an M-matrix, cf. VARGA [1].

Condition 4.1. - $M(A_h,\bar{\omega})$. Let the difference operator A_h from (4.1.1) be equipped with the following properties:

(i) The real coefficients $a(x,\xi)$ satisfy the sign restrictions

$$a(x,\xi) \begin{cases} > 0 & \xi = x \\ & \text{for} \\ < 0 & \xi \in S'(x) \end{cases},\ x \in \bar{\omega}. \quad \text{(for case "} \leq 0\text{", see Remark 4.2.)} \tag{4.2.1a}$$

(ii) For each $x \in \bar{\omega}$, the inequality $(A_h 1)(x) := \sum_{\xi \in S(x)} a(x,\xi) \geq 0$ (4.2.1b) holds.

(iii) At least at one point $x \in \bar{\omega}$, $(A_h 1)(x) > 0$ is satisfied. (4.2.1c)

(iv) If $(A_h 1)(x) = 0$ holds at a point $x \in \bar{\omega}$, then there is a finite sequence of grid points $x^{(o)}, x^{(1)}, \dots, x^{(m)}$ taken from $\bar{\omega}$ with $x^{(o)} := x$, $x^{(j)} \in S'(x^{(j-1)})$ for $j = 1,2,\dots,m$ $(m \geq 1)$ and $(A_h 1)(x^{(m)}) > 0$, $(A_h 1)(x^{(j)}) = 0$ for $j = 0,1,\dots,m-1$. ≡ (4.2.1d)

Remark 4.2. If $a(x,\xi) \leq 0$ holds for $\xi \in S'(x)$, but $a(x,\xi) < 0$ for $\xi \in S_r'(x) \subset S'(x)$, then in $M(A_h,\bar{\omega})$, $S'(x)$ and $S(x)$ are to be replaced by $S_r'(x)$ and $S_r(x) := S_r'(x) + \{x\}$, respectively. Here, the set $S_r'(x)$ is defined by the neighbouring points of x, which are genuinely connected with x, i.e. coupled with x by coefficients $a(x,\xi) \neq 0$. ≡

It should be mentioned that there are also other conditions on A_h, which guarantee the property of A_h to be "of monotonic type" and which are weaker than $M(A_h,\bar{\omega})$, cf. LORENZ [1].

In the following Theorem 4.3., the difference operators A_h, $\tilde{A}_h$, B_h and the associated matrices $\mathcal{A}_h$, $\tilde{\mathcal{A}}_h$, $\mathcal{B}_h$ are characterized with respect to several properties which result partly from $M(A_h,\bar{\omega})$ and are important for further considerations.

Theorem 4.3. Let the difference operators $A_h := A_h^j$ $(j=0,1)$ from Theorem 3.1. be given, cf. (3.2.18), which are constructed via method (PB) for the BVP $Au = F$ from (3.2.1), under the condition $VO(PB,\bar{\omega})$ and for $h \leq h_o$, h_o sufficiently small. Then, the following assertions can be made:

1. The operator A_h fulfils the monotonicity condition $M(A_h,\bar{\omega})$, cf. (4.2.1).
2. a) The operator A_h is "of monotonic type" on D from (2.2.8), that is, $A_h y \geq 0$ on $\bar{\omega}$ implies $y \geq 0$ on $\bar{\omega}$.

2. b) The assertion a) holds in the strict sense, too, that is, from $A_h y > 0$ on $\dot{\omega}$ we can conclude $y > 0$ on $\bar{\omega}$.

3. The operators $\tilde{A}_h$, B_h defined on $\tilde{D}$ from (2.2.8) are provided with the same properties 1., 2., which are stated for A_h with respect to D. For $\tilde{A}_h$, $\tilde{B}_h$, see (4.1.4), (4.1.5).

4. The matrices $\mathcal{A}_h$, $\tilde{\mathcal{A}}_h$ and $\mathcal{B}_h$ are matrices "of monotonic type" in the sense of COLLATZ [1]. They are also irreducibly diagonally dominant matrices (ORTEGA/RHEINBOLDT [1], VARGA [1]) with positive diagonal entries and nonpositive off-diagonal entries.

5. The matrices $\mathcal{A}_h$, $\tilde{\mathcal{A}}_h$ and $\mathcal{B}_h$ are M-matrices in the sense of VARGA [1]. There exists the inverse matrix to each of them, and we have $\mathcal{A}_h^{-1} > 0$, $\tilde{\mathcal{A}}_h^{-1} > 0$, $\mathcal{B}_h^{-1} > 0$, that is, all entries of the inverse matrices are positive.

6. The $\tilde{n} \times \tilde{n}$ matrix $\mathcal{B}_h$ is symmetric in the sense of $\mathcal{B}_h = \mathcal{B}_h^T$ and positive definite in $R^{\tilde{n}}$. In general, the matrices $\mathcal{A}_h$, $\tilde{\mathcal{A}}_h$ are not symmetric. The matrix $\tilde{\mathcal{A}}_h$ is symmetric in the special case of $\tilde{H}_h = \varkappa I_h$, with $\varkappa > 0$, I_h: identity matrix, $\tilde{H}_h$ from (4.1.6). The matrix $\tilde{\mathcal{A}}_h$ is positive definite with respect to some scalar product having appropriate weights. ≡

Proof: For the secondary networks $\mathfrak{J}_{\varkappa}, \mathfrak{J}_{\tilde{\varkappa}}$ (cf. (2.3.3)) constructed by the method (PB) under the Condition 2.16. - $VO(PB,\bar{\omega})$, the measures of the box boundaries $\partial\mathcal{X}, \partial\tilde{\mathcal{X}}$ satisfy the following relations:

$$\text{mes } \partial\mathcal{K}(x) = \sum_{\xi\in S'(x)} s(x,\xi) \geq 2\pi \min_{\xi\in S'(x)} \frac{h(x_\xi^!)}{2} \geq \pi\varepsilon_0 h > 0 \text{ for } x\in\omega,$$
$$\text{mes } \partial\tilde{\mathcal{K}}(x) > \text{mes } (\partial\tilde{\mathcal{K}}(x) \cap \bar{\Omega}_h) = \sum_{\xi\in S'(x)} \tilde{s}(x_\xi^!) \geq \pi_0 h > 0 \text{ for } x\in\gamma, \tag{4.2.2}$$

with $s(x_\xi^!)$, $\tilde{s}(x_\xi^!)$ from (2.3.13), with the natural constant π and another positive constant π_0 which does not depend on h, $h \leq h_0$. That such a constant π_0 exists is due to the construction of γ and the Assumption 2.1. - $V(\bar{\Omega})$, especially to (2.1.4) and the positive angle at the corner points $P_k \in \gamma \subset \Gamma$, cf. Fig. 1a. Additionally, there is a universal upper bound $\pi_1 h$ ($\pi_1 > 0$) for mes $\partial\mathcal{K}(x)$, mes $\partial\tilde{\mathcal{K}}(x)$ in (4.2.2), and we can also note inclusions for mes $\mathcal{K}(x)$ and mes $\tilde{\mathcal{K}}(x)$ such as $\pi_2 h^2 \geq \text{mes } \mathcal{K}(x) \geq \pi_3 h^2 > 0$ for $x\in\bar{\omega}$, analogously for mes $\tilde{\mathcal{K}}(x)$, $x\in\gamma$. Taking into account these relations for $\mathcal{K}(x)$, $\tilde{\mathcal{K}}(x)$ and recalling TO from VO(PB, $\bar{\omega}$), we easily see that the primary and secondary networks contract uniformly for $h \to 0$. Because some mesh segments $\hbar(x,\xi)$ are affixed to each grid point $x\in\bar{\omega}$, there is a nonempty subset $\{\xi\}$ of neighbouring grid points with $s(x_\xi^!) > 0$ or $\tilde{s}(x_\xi^!) > 0$. Therefore, every two points of $\bar{\omega}$ can be connected by a polygonal line $\bigcup_{x,\xi}\hbar(x,\xi)$ with some $x,\xi\in\bar{\omega}$, where $s(x_\xi^!) > 0$ or $\tilde{s}(x_\xi^!) > 0$ holds for $h \leq h_0$, if h_0 is chosen sufficiently small. Because of VO(PB, $\bar{\omega}$), the inequalities $s(x_\xi^!) \geq 0$ and $s^j(x_\xi^!) \geq 0$ (j=0 and j=1) are fulfilled for $x\in\omega+\gamma_{23}$ and all $\xi\in S'(x)$. Additionally, $s(x_\xi^!) > 0$ and $s^j(x_\xi^!) > 0$ (j=0,1; $s^0(x_\xi^!) = \tilde{s}(x_\xi^!)$, $s^1(x_\xi^!)$ from (3.2.18c)) hold for $\xi\in S_r'(x) \subset S'(x)$ ($S_r'(x) \neq 0$), which results from (4.2.2). The representation of the operator $A_h = A_h^j$ (j=0,1; cf. (3.2.18)) in the canonical form (4.1.1) leads to the following relations for the coefficients $a(x,\xi)$ and their sums:

$$x\in\gamma_1:\ a^j(x,x) = 1,\quad S'(x) = \emptyset;$$

$$x\in\gamma_{23}:\ a^j(x,\xi) = -\frac{1}{h(x)}\,\frac{s^j(x_\xi^!)k_h(x_\xi^!)}{h(x_\xi^!)} < 0 \text{ for } \xi\in S_r'(x)\subset S'(x),\quad S_r'(x) \neq \emptyset,$$

$$a^j(x,x) = -\sum_{\xi\in S'(x)} a^j(x,\xi) + \alpha^j(x) > 0 \text{ for } \alpha^j \geq 0, \tag{4.2.3}$$

$$(A_h 1)(x) = \alpha^j(x)\begin{cases} \geq 0 & \text{for } x\in\gamma_2, \\ \geq \frac{\alpha_0}{2} > 0 & \text{for } x\in\gamma_3,\ \alpha_0 \text{ from (2.1.7b)}; \end{cases}$$

$$x\in\omega:\ a^j(x,\xi) = -\frac{1}{H(x)}\,\frac{s(x_\xi^!)k_h(x_\xi^!)}{h(x_\xi^!)} \text{ for } \xi\in S_r'(x)\subset S'(x),\ S_r'(x)\neq\emptyset,$$

$$a^j(x,x) = -\sum_{\xi\in S'(x)} a^j(x,\xi) + c(x) > 0 \text{ for } c \geq 0,$$

$$(A_h 1)(x) = c(x)\begin{cases} > 0 & \text{for } x\in\omega_0,\ \omega_0 \text{ from (4.2.4)}, \\ = 0 & \text{for } x\in\omega\setminus\omega_0. \end{cases}$$

To prove $a^j(x,x) > 0$, the relations (4.2.2) are again taken into consideration. For $h \leq h_o$, h_o sufficiently small, there is at least one grid point $\hat{x} \in \bar{\omega}$ with $(A_h 1)(\hat{x}) > 0$, since $V(\bar{\Omega})$, assumption (2.1.8) and TO in VO(PB, $\bar{\omega}$) imply the relations

$$\omega_o + \gamma_1 + \gamma_3 \neq \emptyset \text{ and } \sum_{x \in \omega_o} H(x) + \sum_{x \in \gamma_1 + \gamma_3} h(x) \geq C_o > 0, \qquad (4.2.4)$$

with $\omega_o := \{x \in \omega : c(x) > 0\}$ and γ_1, γ_3 from (2.2.6).

Here, C_o does not depend on h, $h \leq h_o$. If grid points $\check{x} \in \bar{\omega}$ with $(A_h 1)(\check{x}) = 0$ occur, then there is a sequence of grid points described in (4.2.1d), where $x^{(1)} \in S_r'(\check{x}), \ldots, \hat{x} \in S_r'(x^{(m-1)}, (A_h 1)(\hat{x}) > 0$. This assertion is an obvious consequence of the considerations made at the beginning of this proof (cf. the text between (4.2.2) and (4.2.3)). Therefore, Condition 4.1. - $M(A_h, \bar{\omega})$ is verified, and the first assertion of Theorem 4.3. is proved. Assertion 2.a) follows from $M(A_h, \bar{\omega})$ and, e.g., from Theorem 3 in HEINRICH [3], and 2.b) results from Theorem 3.11. with Corollary 1 from p. 85 in VARGA [1]. The properties of $\tilde{A}_h$, B_h stated in 3. can be proved in an entirely similar way as for A_h. In assertion 4., the properties presented for A_h, $\tilde{A}_h$ and B_h in 1. and 3. will now be formulated for the associated matrices $\mathcal{A}_h$, $\tilde{\mathcal{A}}_h$ and $\mathcal{B}_h$ in the language adapted to the matrix theory. Therefore, the arguments used above can be easily transferred to this situation (for the terminology, cf. VARGA [1]). The existence of $\mathcal{A}_h^{-1}$, $\tilde{\mathcal{A}}_h^{-1}$ and $\mathcal{B}_h^{-1}$ is a consequence of 4., where the entries of these inverse matrices are nonnegative due to 2.a) and, therefore, $\mathcal{A}_h$, $\tilde{\mathcal{A}}_h$ and $\mathcal{B}_h$ are M-matrices. From 2.b) we see that the entries of the inverse matrices are even positive. The symmetry of the matrix $\mathcal{B}_h$ stated in 6. follows from the symmetry of $s(x,\xi)$, $s^j(x,\xi)$, $h(x,\xi)$ and $k_h(x,\xi)$ with respect to x and ξ. According to Theorem 1.5 in SCHWARZ et al. [1] or Theorem 3.11. (Corollary 2) in VARGA [1], the matrix $\mathcal{B}_h$ is also positive definite, at least for fixed h ($h \leq h_o$), since 4. holds and $\mathcal{B}_h$ is symmetric. The remaining assertions can be concluded from the representation $\mathcal{B}_h = \tilde{H}_h \tilde{\mathcal{A}}_h$ in (4.1.6). The constant in the inequality which describes the positive definiteness of $\mathcal{B}_h$ will be investigated later, in Section 4.4.3., with respect to its asymptotic behaviour.

<u>4.2.2. MD-operators A_h in comparison with PB-operators.</u> We now consider FDSs $A_h^j y = F_h^j$ (j=0,1) from Theorem 3.4., cf. (3.3.19). The MD-operators A_h have the "element representation" (3.3.19) and the "flux representation" (3.3.20), where the relation $K_h^j(x_\xi') > 0$ is basic to the property $M(A_h, \bar{\omega})$.

<u>Lemma 4.4.</u> For the coefficients k_{il} from (2.1.1), let the special case $k_{11} \equiv k_{22} =: k$, $k_{12} \equiv 0$ be given. Then the following assertions hold.

(i) The coefficients $K_h^j(x_\xi')$ from (3.3.20) fulfil the inequality

$$K_h^j(x_\xi') \geq \tilde{k}_o > 0 \text{ for } \begin{cases} \theta_- + \theta_+ \leq \pi - \theta_o & \text{and } x_\xi' \in \omega', \\ \theta_\pm \leq \pi/2 - \theta_o & \text{and } x'_{\xi_\Gamma^\pm} \in \gamma'_{23}, \end{cases} \quad j=0,1, \qquad (4.2.5)$$

where $\tilde{k}_o$ does not depend on h, $h \leq h_o$, if h_o is chosen sufficiently small.

(ii) For $k \equiv const \geq k_o > 0$, we have (4.2.5) for $j=0$ without further restrictions on h_o. Additionally, in place of (4.2.5) the following relations hold for $K_h^o(x_\xi')$ from (3.3.20):

$$K_h^o(x_\xi') \geq 0 \quad \text{for} \quad \begin{cases} \theta_- + \theta \leq \pi & \text{and } x_\xi' \in \omega', \\ \theta_\pm \leq \pi/2 & \text{and } x'_{\xi_\Gamma^\pm} \in \gamma'_{23}, \end{cases} \qquad (4.2.6)$$

where $K_h^o(x_\xi') = 0$ corresponds only to the cases $\theta_- + \theta_+ = \pi$ and $\theta_\pm = \pi/2$, respectively. ∎

<u>Proof:</u> The proof is based on the detailed analysis of the coefficients $K_h^j(x_\xi')$ from (3.3.20) using goniometric transformations and can be found in HEINRICH [7].

If the coefficients k_{il} $(i,l=1,2)$ from (2.1.1) are such that $|k_{12}(x)|$ and $|k_{11}(x) - k_{22}(x)|$ are sufficiently small for all $x \in \bar{\Omega}$, and, if all angles θ satisfy the more restrictive condition $0 < \theta_o \leq \theta_\pm \leq \pi/2 - \theta_o$, then the inequality for K_h^j in (4.2.5) remains valid. Further assertions similar to Lemma 4.4. can be stated for coefficients k_{il} and networks of practical interest, where the angles $\theta_\pm(x,\xi)$ and $\measuredangle(x_1, h^\pm)$, $h^\pm := h(x, \xi^\pm)$, are balanced with respect to k_{il}.

We now consider the relationship between the discrete analogues derived by the methods (PB) and (MD) for an essential special case of the operator L in the BVP Au = F, namely $L := -\Delta$ (Laplacian) and corresponding mixed boundary conditions. In Lemma 4.5., the indices p and m introduced in Assumption 2.14. indicate the method underlying the construction of the difference operators A_h, i.e., (<u>P</u>B) or (<u>M</u>D).

<u>Lemma 4.5.</u> Let the grid $\bar{\omega}$ of the primary triangulation T_Δ be given, where T_Δ consists of triangles only, cf. (2.2.4). Furthermore, let Condition 2.16. - VO(PB, $\bar{\omega}$) be satisfied. For the BVP Au = F from (2.1.1), (2.1.2), the following special case is considered:

$$k := k_{11} \equiv k_{22} \equiv 1,\ k_{12} \equiv 0,\ c \equiv 0,\ \alpha \equiv 0, \text{ i.e.}$$

$$Au = (-\Delta u, \frac{\partial u}{\partial n}, u)^T \text{ with respect to the subsets } \Omega, \qquad (4.2.7)$$

$\Gamma \setminus \Gamma_1$, Γ_1 of $\bar{\Omega}$.

Then, the discrete analogues $\Lambda_h^p := A_h^o(PB)$ by method (PB), A_h^o from (3.2.18), and $\Lambda_h^m := A_h^o(MD)$ by method (MD), A_h^o from (3.3.19), can be represented in the following form:

$$\Lambda_h^p y := \begin{cases} y(x), & x \in \gamma_1, \\ -\frac{1}{h(x)} \sum\limits_{\xi \in S'(x)} s^o(x'_\xi) y_{\hbar}(x'_\xi), & x \in \gamma_{23}, \\ -\frac{1}{H_p(x)} \sum\limits_{\xi \in S'(x)} s(x'_\xi) y_{\hbar}(x'_\xi), & x \in \omega, \end{cases} \tag{4.2.8}$$

with $s^o(x'_\xi)$, $s(x'_\xi)$, $y_{\hbar}(x'_\xi)$, $h(x)$, $H_p(x) := H(x)$ from (3.2.18),

$$\Lambda_h^m := \begin{cases} y(x), & x \in \gamma_1, \\ -\frac{1}{h(x)} \sum\limits_{\xi \in S'(x)} h(x'_\xi) K_h^o(x'_\xi) y_{\hbar}(x'_\xi), & x \in \gamma_{23}, \\ -\frac{1}{H_m(x)} \sum\limits_{\xi \in S'(x)} h(x'_\xi) K_h(x'_\xi) y_{\hbar}(x'_\xi), & x \in \omega, \end{cases} \tag{4.2.9}$$

with $h(x'_\xi)$, $K_h^o(x'_\xi)$, $K_h(x'_\xi)$, $H_m(x) := H(x)$ from (3.3.20).

Moreover, the coefficients of the difference quotient $y_{\hbar}(x'_\xi)$ in (4.2.8) and (4.2.9) are related by

$$\begin{aligned} s^o(x'_\xi) &= h(x'_\xi) K_h^o(x'_\xi) & x \in \gamma_{23} \\ & & \text{for} \qquad\qquad \text{and} \quad \xi \in S'(x). \\ s(x'_\xi) &= h(x'_\xi) K_h(x'_\xi) & x \in \omega \end{aligned} \tag{4.2.10}$$

Hence, the difference operators Λ_h^p, Λ_h^m from (4.2.8), (4.2.9) differ at most in the factors $H_p^{-1}(x)$, $H_m^{-1}(x)$ for $x \in \omega$, where $H_p(x) \neq H_m(x)$ holds, in general. ■

Proof: The representations (4.2.8), (4.2.9) are derived from A_h^o by setting $k = 1$, $c = 0$, $\alpha = 0$. For triangles $\Delta(x,\xi,\xi^{\pm})$ with the joint side $\hbar(x,\xi)$ and angles $\theta_{\pm}(x,\xi)$ opposite to $\hbar(x,\xi)$, the following relations between $\cot\theta_{\pm}$ and the usual scalar product $(.,.)$ in R^2, with vectors $\hbar$, $\hbar^{\pm}$, $\hbar^o$, $\hbar^{\ominus}$ and $\hbar^{\oplus}$ from (2.3.19), (2.3.20), can be easily verified:

$$\cot\theta_- = \frac{(\hbar^-, \hbar^- - \hbar)}{(\hbar^o, \hbar^-)}, \quad \cot\theta_+ = \frac{(\hbar^+, \hbar^+ - \hbar)}{(\hbar^{\oplus}, \hbar)} \text{ for } 0 < \theta_{\pm} < \pi. \tag{4.2.11}$$

Therefore, using (3.3.20b) and (4.2.11), we get

$$h(x'_\xi) K_h(x'_\xi) = \tfrac{1}{2} h(x'_\xi) \left\{ \cot\theta_- + \cot\theta_+ \right\} \quad \text{for } x \in \omega + \gamma_{23} \tag{4.2.12a}$$

and mesh segments $\hbar(x'_\xi)$ with two adjacent triangles $\Delta(x,\xi,\xi^{\pm})$, otherwise

$$h(x'_{\xi^{\pm}_{\Gamma}}) K_h^o(x'_{\xi^{\pm}_{\Gamma}}) = \tfrac{1}{2} h(x'_{\xi^{\pm}_{\Gamma}}) \cot\left\{\theta_{\pm}(x,\xi^{\pm}_{\Gamma})\right\} \text{ for } x \in \gamma_{23}, \ \xi^{\pm}_{\Gamma} \in S'(x) \cap \gamma. \tag{4.2.12b}$$

Using relation (6) from Appendix GE, we see that the measure $s(x'_{\xi})$ of the perpendicular bisector segment $\mathfrak{S}(x'_{\xi}) := \mathfrak{S}(x,\xi)$ is equal to the right-hand side of (4.2.12a) and, for a single triangle touching the boundary Γ_h, $s(x'_{\xi\pm\Gamma})$ coincides with the right-hand side of (4.2.12b), cf. pointer (5) in Appendix GE. Thus, (4.2.10) is verified. It is also easy to see that, in general, $H_p(x) \neq H_m(x)$ holds.

Remark 4.6. For the case $k \equiv const \geq k_o > 0$, $c \neq 0$, $\alpha \neq 0$ of the operator $A = (L,l)^T$ from (3.2.1), the discrete analogues $A_h^p := A_h^o(PB)$ from (3.2.18) and $A_h^m := A_h^o(MD)$ from (3.3.19) differ only in the factors $H_p^{-1}(x)$; $H_m^{-1}(x)$ for $x \in \omega$ and in the value $\alpha^j(x)$ for $x \in \gamma_{23}$. ≡

Now, an analogue of Theorem 4.3. with respect to method (MD) will be formulated.

Theorem 4.7. Let the difference operators $A_h := A_h^j$ $(j=0,1)$ defined in Theorem 3.4. and constructed on secondary networks (MD) be given, cf. (3.3.19) or (3.3.20), under the following special assumptions:

(i) $k_{11}(x) = k_{22}(x) =: k(x)$, $k_{12}(x) = 0$ for $x \in \bar{\Omega}$ and Condition 2.17. - VO"(PB,$\bar{\omega}$), or instead of (i),

(ii) $k_{11} \equiv k_{22} \equiv k \equiv const \geq \varepsilon_o > 0$, $k_{12} \equiv 0$ on $\bar{\Omega}$ and Condition 2.16. - VO(PB,$\bar{\omega}$).

Then, for (i) or (ii), the monotonicity condition $M(A_h,\bar{\omega})$ from (4.2.1) is fulfilled and the assertions 1., 2. ..., 6. from Theorem 4.3. hold verbatim for the operators A_h, $\tilde{A}_h$, B_h and the corresponding matrices $\mathcal{A}_h$, $\tilde{\mathcal{A}}_h$, $\mathcal{B}_h$, which are derived from (3.3.19) ≡

Proof: Due to Lemma 4.5. and Remark 4.6., the discrete analogues $A_h^p := A_h^o(PB)$ and $A_h^m := A_h^o(MD)$ taken for $k \equiv const$ differ at the grid points $x \in \omega$ in the factors $H_p^{-1}(x)$, $H_m^{-1}(x)$ and at $x \in \gamma_{23}$ in the values $\alpha_p^j(x)$, $\alpha_m^j(x)$, where both values $\alpha_p^j(x)$, $\alpha_m^j(x)$ are either positive or equal to zero. Therefore, the assertion of Theorem 4.7. (ii), $j = 0$, follows immediately from Theorem 4.3. In the case (ii), $j = 1$, we take into account that VO(PB,$\bar{\omega}$) implies both $K_h^o(x'_{\xi}) > 0$ and $K_h^1(x'_{\xi}) > 0$ for $x \in \gamma_{23}$, $\xi \in S'(x) \cap \gamma$, if h_o is sufficiently small, cf. Lemma 4.4. If $k(x)$ is variable, cf. case (i), the coefficients in $A_h^j(PB)$, $A_h^j(MD)$ $(j=0,1)$ do not only differ in the factors $H^{-1}(x)$ and $\alpha^j(x)$ mentioned previously. But due to VO"(PB,$\bar{\omega}$) and (4.2.5), the coefficients $K_h^j(x'_{\xi})$ from (3.3.20) are positive for all mesh segments $h(x,\xi)$, $x \in \omega + \gamma_{23}$ and $j = 0,1$. Finally, the remaining conclusions for completing the proof can be made as in the proof of Theorem 4.3.

Remark 4.8. Comparing Theorems 4.3. and 4.7., we see that, in the case of variable $k(x)$, the difference operators A_h^p from method (PB) satisfy $M(A_h,\bar{\omega})$ under the angle restrictions W1, W2 from VO(PB,$\bar{\omega}$), which are weaker than W1", W2" needed for MD-operators A_h^m. ≡

4.3. Scalar products, norms and a trace theorem

4.3.1. Notations for scalar products and norms. We now consider "discrete forms" of scalar products and norms for grid functions, which are used to prove analytical results regarding difference operators A_h on grids $\bar{\omega}$. These discrete forms can be understood in most cases - but not in all - as discrete analogues of the "continuous forms" of scalar products and norms of function spaces.
We again take into account the grids $\bar{\omega}$, $\bar{\omega}'$ and $\bar{\omega}''$, the subsets ω, γ_{23}, ω' and γ'_{23}, the set of grid functions D_o, D' etc., the primary and secondary triangulations $\mathfrak{T}_\sigma$, $\mathfrak{T}_\Delta$ and $\mathfrak{T}_{\varkappa}$, $\mathfrak{T}_{\tilde{\varkappa}}$, always under the grid regularity conditions VO(PB,$\bar{\omega}$) for method (PB) and VO(MD,$\bar{\omega}$) for method (MD). The subscript r indicates the origin of the boxes $\varkappa_r(x)$, $\tilde{\varkappa}_r(x)$, with $r \in \{p,m\}$, and p, m correspond to (PB), (MD), respectively. For these and some other notations employed here, see Sections 2.2., 2.3.

Definition 4.9. - Scalar products. Let $(.,.)_r$ and $(.,.)_{\hat{r}}$ denote the following scalar products for some grid functions defined on the primary grids $\bar{\omega}$ or $\omega+\gamma_{23}$:

$$(y,v)_r := \sum_{x\in\omega} y(x)v(x)H_r(x) + \sum_{x\in\gamma_{23}} y(x)v(x)h(x) \quad \text{for } y,v \in D_o \text{ or } y,v\in\tilde{D}, \tag{4.3.1}$$

$$(y,v)_{\hat{r}} := \sum_{x\in\omega+\gamma_{23}} y(x)v(x)\tilde{H}_r(x) \quad \text{for } y,v\in D_o \text{ or } y,v\in\tilde{D}, \tag{4.3.2}$$

with r = p or r = m, i.e. $r\in\{p,m\}$, and $H_r(x)$, $\tilde{H}_r(x)$ from (2.3.13) (r=p) or (2.3.18) (r=m), $h(x) := \frac{1}{2}\{h(x,\xi_r^-) + h(x,\xi_r^+)\}$, and D_o, $\tilde{D}$ are taken from (2.2.8). Furthermore, $(.,.)'$ and $(.,.)''$ denote scalar products for functions defined on the secondary grids $\omega'+\gamma'_{23}$ and $\bar{\omega}''$:

$$(y,v)' := \sum_{x'\in\omega'+\gamma'_{23}} y(x')v(x')H'(x') \quad \text{for } y,v\in\tilde{D}' \text{ from (2.2.8)}, \tag{4.3.3}$$

with $H'(x') := h(x')s^o(x')$, $s^o(x')$ from (2.3.13) or (3.2.18c),

$$(y,v)'' := \sum_{x''\in\bar{\omega}''} y(x'')v(x'')H''(x'') \quad \text{for } y,v\in D'' \text{ from (2.2.8)}, \tag{4.3.4}$$

with $H''(x'') := \text{mes } e(x'')$, $e\in\{\Delta,\Box\}$. ≡

In Definition 4.9., the quantity $H'(x')$ means the double measure of that portion of $\varkappa_p(x)$ $(x\in\omega)$ or $\tilde{\varkappa}_p(x)$ $(x\in\gamma_{23})$, which is assigned to the mesh segment $h(x')$, cf. Fig. 9f. The scalar products $(.,.)'$ and $(.,.)''$ are used preferentially for the methods (PB) and (MD), respectively. Obviously, all "scalar products" from Definition 4.9. satisfy the usual properties of a scalar product. Therefore, discrete forms of L_2-norms can be introduced. The maximum norm (C-norm) will also be defined below.

<u>Definition 4.10. - Norms.</u> Let $\|\cdot\|_{o,r}$, $\|\cdot\|_{o,\hat{r}}$, $\|\cdot\|_{o,\hat{r},\omega_T}$ and $\|\cdot\|_{o,\gamma_t}$ denote the following norms for some grid functions defined on the primary grids $\bar{\omega}$ or $\omega+\gamma_{23}$, on ω_T and γ_t:

$$\|y\|^2_{o,r} := (y,y)_r = \sum_{x\in\omega} y^2(x)H_r(x) + \sum_{x\in\gamma_{23}} y^2(x)h(x) \text{ for } y\in D_o \text{ or } y\in\tilde{D}, \quad (4.3.5)$$

$$\|y\|^2_{o,\hat{r}} := (y,y)_{\hat{r}} = \sum_{x\in\omega+\gamma_{23}} y^2(x)\tilde{H}_r(x) \text{ for } y\in D_o \text{ or } y\in\tilde{D}, \quad (4.3.6)$$

with $(.,.)_r$, $(.,.)_{\hat{r}}$ from Definition 4.9. and $r\in\{p,m\}$,

$$\|y\|^2_{o,\hat{r},\omega_T} := \|y\|^2_{o,r,\omega_T} := \sum_{x\in\omega_T} y^2(x)H_r(x) \text{ for } y\in D_o \text{ and } \omega_T\subset\omega, \; r\in\{p,m\}, \quad (4.3.7)$$

$$\|y\|^2_{o,\gamma_t} := \sum_{x\in\gamma_t} y^2(x)h(x) \text{ for } y\in D_o \text{ and } \gamma_t\subset\gamma_{23}, \quad (4.3.8)$$

$$h(x) := \frac{1}{2}\left\{h(x,\xi_\Gamma^-) + h(x,\xi_\Gamma^+)\right\}.$$

Moreover, the discrete C-norm will be designated by

$$\|y\|_{C(\omega_T)} := \max |y(x)| \text{ for } \omega_T\subset\bar{\omega} \text{ and } y\in D \text{ from (2.2.8).} \quad (4.3.9)$$

Obviously, the norm $\|\cdot\|_{o,\hat{r}}$ is really a discrete analogue of the $L_2(\Omega_h)$-norm. Furthermore, discrete L_2-norms $\|\cdot\|'_o$, $\|\cdot\|''_o$ using the secondary network can be introduced:

$$\|v\|'^2_o := (v,v)' \text{ for } v\in\tilde{D}', \quad \|v\|''^2_o := (v,v)'' \text{ for } v\in D'', \quad (4.3.10)$$

cf. (4.3.3) and (4.3.4). These norms are used preferably as discrete L_2-norms for the difference quotients y_h and $\nabla_h y$, and, therefore, they are a seminorm $|y|_1$ for the grid function y itself.

<u>Definition 4.11. - Seminorms.</u> Let $\|\cdot\|'_o$ and $\|\cdot\|''_o$ denote the following norms of the first order difference quotients y_h and $\nabla_h y$, respectively:

$$|y|^2_{1,p} := \|y_h\|'^2_o := (y_h,y_h)' := \sum_{x'\in\omega'+\gamma'_{23}} y_h^2(x')H'(x') \text{ for } y\in D_o \quad (4.3.11)$$

and secondary networks (PB), $(.,.)'$ from (4.3.3),

$$|y|^2_{1,m} := \||\nabla_h y|\|''^2_o := (|\nabla_h y|,|\nabla_h y|)'' := \sum_{x''\in\bar{\omega}''} |\nabla_h y(x'')|^2 \operatorname{mes} e(x'') \text{ for } y\in D_o \quad (4.3.12)$$

and secondary networks (MD), $(.,.)''$ from (4.3.4),

with

$$y_{\hbar}(x'_{\xi}) := \frac{y(\xi)-y(x)}{h(x,\xi)} \quad \text{and} \quad \nabla_{\hbar} y(x'') := \frac{[y(\xi)-y(x)]\hbar^{\bullet} - [y(\xi^+)-y(x)]\hbar^{o}}{2 \text{ mes } \Delta(x'')}$$

for $e(x''):=\Delta(x'')$, or $\nabla_{\hbar} y$ from (3.3.18) for $e:=\square$,

$$|\nabla_{\hbar} y|^2 := (\nabla_{\hbar} y, \nabla_{\hbar} y)_{R^2}.$$ ■

Due to the Definitions 4.10., 4.11., we are in position to introduce discrete forms of W_2^1- and W_2^{-1}-norms.

<u>Definition 4.12. - W_2^1-and W_2^{-1}-norms.</u> For $r \in \{p,m\}$, let $\|.\|_{1,r}$ and $\|.\|_{1,\hat{r}}$ denote discrete W_2^1-norms of the following type:

$$\left.\begin{aligned} \|y\|^2_{1,p} &:= \|y\|^2_{o,p} \\ \|y\|^2_{1,\hat{p}} &:= \|y\|^2_{o,\hat{p}} \end{aligned}\right\} + |y|^2_{1,p} \quad \text{for } y \in D_o \text{ and secondary networks (PB),} \tag{4.3.13}$$

$$\left.\begin{aligned} \|y\|^2_{1,m} &:= \|y\|^2_{o,m} \\ \|y\|^2_{1,\hat{m}} &:= \|y\|^2_{o,\hat{m}} \end{aligned}\right\} + |y|^2_{1,m} \quad \text{for } y \in D_o \text{ and secondary networks (MD),} \tag{4.3.14}$$

where $\|.\|_{o,r}$, $\|.\|_{o,\hat{r}}$, $r \in \{p,m\}$, are taken from (4.3.5), (4.3.6), and $|\cdot|_{1,r}$, $r \in \{p,m\}$, from (4.3.11), (4.3.12). Furthermore, let $\|.\|_{-1,r}$, $\|\cdot\|_{-1,\hat{r}}$ be given by

$$\|y\|_{-1,r} := \sup_{\substack{v \in D_o \\ v \neq 0}} \frac{|(y,v)_r|}{\|v\|_{1,r}} \quad \text{for } y \in D_o,\ r \in \{p,m\}, \text{ or } \hat{r} \text{ instead of } r, \tag{4.3.15}$$

where $(.,.)_r$ and $(.,.)_{\hat{r}}$ are taken from (4.3.1) and (4.3.2), respectively. ■

Obviously, the norms $\|.\|_{i,r}$ and $\|.\|_{i,\hat{r}}$ differ for fixed $i \in \{0,1\}$ and $r \in \{p,m\}$ in the L_2-term only, and here exclusively in the weights $h(x)$ and $\widetilde{H}_r(x)$ for boundary points $x \in \gamma_{23}$. The discrete norms $\|\cdot\|_{i,\hat{r}}$, for $i \in \{0,1\}$ and with the symbol "$\wedge$", are adapted to the well-known L_2- and W_2^1-norm as used in Appendix DI. The norms $\|.\|_{i,r}$, for $i \in \{0,1\}$ and without "$\wedge$", are more convenient for the discrete analogues $A_h y = F_h$ of the BVP $Au = F$. That all "norms" ("seminorms") introduced in the Definitions 4.10., 4.11. and 4.12. are equipped with the usual properties of a norm (seminorm) can be proved easily by verifying the norm axioms. Although the "discrete spaces" W_2^1, W_2^{-1}, etc., depend on D_o, the index D_o will not be written. For simplicity, we shall also drop the subscripts r or $\hat{r}$ in some places, where the context clearly yields the correct subscript $r = p$ or $r = m$. Thus, we shall use the short notations

$$\|y\|_o^2 := (y,y),\ \|y\|_1^2 := \|y\|_o^2 + |y|_1^2,\ \|y\|_{-1} := \sup_{\substack{v \in D_o \\ v \neq 0}} \frac{|(y,v)|}{\|v\|_1},\ y \in D_o. \tag{4.3.16}$$

4.3.2. $\Omega - \Omega_h$-equivalence for norms of special functions. The domain $\bar{\Omega}_h$ with the representation (2.2.3b) can be described by triangles Δ only, cf. (2.2.4), if each rectangle $\square$ is divided correctly into two triangles Δ. Analogously, $\bar{\Omega}$ can be represented by triangles only, namely by straight triangles Δ in the interior of Ω and by curved triangles at most at the boundary Γ. The index set $\bar{\omega}''$ in (2.2.3) is now substituted by another set $\bar{\omega}''$ (without change of notation) which comprises the centres of gravity of the original triangles and the triangles obtained after subdivision of each $\square$ into two triangles Δ. Obviously, we are now in position to note relations (4.3.17). Each $x' \in \gamma'$, γ' from (2.3.14), is associated with a mesh segment $\mathfrak{h}(x') := \mathfrak{h}(x,\xi)$ and an arc $\mathfrak{b}(x') := \mathfrak{b}(x,\xi)$, where the former is a subset of Γ_h and the latter is a subset of Γ. The arc $\mathfrak{b}(x')$ coincides with a curved triangle edge, cf. Figs. 7a, b. According to assumption (2.2.1), the two cases of Figs. 7a, b are typical situations for the correspondence of $\mathfrak{h}(x,\xi)$ and $\mathfrak{b}(x,\xi)$, or we have $\mathfrak{b}(x,\xi) = \mathfrak{h}(x,\xi)$ for straight parts of Γ. Although we principally consider triangles $\varDelta$ with one curved edge $\mathfrak{b}(x,\xi)$ on Γ, the results obtained are also true if two edges of some triangles $\varDelta$ lie on Γ. Then, the proofs require only slight modification. We now consider the triangulations mentioned previously and some subsets of Γ, Γ_h, γ and γ'.

Assumption 4.13. Let $\bar{\Omega}_h$, $\bar{\Omega}$ be triangulated such that the relations

$$\mathcal{T}_\Delta : \bar{\Omega}_h = \bigcup_{x'' \in \bar{\omega}''} \Delta(x''), \quad \mathcal{T}_\varDelta : \quad \bar{\Omega} = \bigcup_{x'' \in \bar{\omega}''} \varDelta(x'') \quad \text{for } h \leq h_0 \tag{4.3.17}$$

hold. Furthermore, let the subsets $\gamma'_T \subset \gamma'$, $\bar{\gamma}_T \subset \gamma$, $\Gamma_T \subset \Gamma$ and $\Gamma_{Th} \subset \Gamma_h$ be related as follows:

$$\Gamma_T = \bigcup_{x' \in \gamma'_T} \mathfrak{b}(x'), \quad \Gamma_{Th} = \bigcup_{x' \in \gamma'_T} \mathfrak{h}(x'), \quad \bar{\gamma}_T := \{x \in \gamma : x \in \Gamma_T\}, \tag{4.3.18}$$

where $\mathfrak{b}(x')$ is, in general, a curved triangle edge associated with the mesh segment $\mathfrak{h}(x')$, cf. Figs. 7a, b. ■

Assumption 4.13. implies that Γ_{Th} is the polygonal approximation of Γ_T and we have $\partial\Gamma_T = \partial\Gamma_{Th} \subset \bar{\gamma}_T \subset \gamma$, i.e. the boundary points of Γ_T, Γ_{Th} coincide (if $\partial\Gamma_T \neq \emptyset$) and are contained in $\bar{\gamma}_T$. For instance, Γ_1 and Γ_{1h} can be represented via $\gamma'_1 := \{x' = x'_\xi \in \gamma' : x \in \gamma_1, \xi \in \gamma_1\}$ and γ_1 as follows:

$$\Gamma_1 = \bigcup_{x' \in \gamma'_1} \mathfrak{b}(x') = \bigcup_{x,\xi \in \gamma_1} \mathfrak{b}(x,\xi), \quad \Gamma_{1h} = \bigcup_{x' \in \gamma'_1} \mathfrak{h}(x') = \bigcup_{x,\xi \in \gamma_1} \mathfrak{h}(x,\xi). \tag{4.3.19}$$

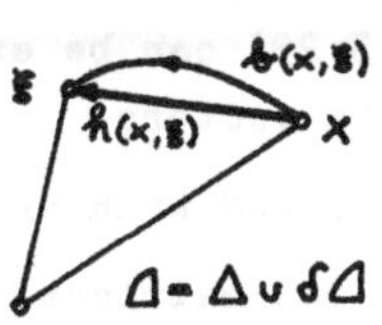

Fig. 7a

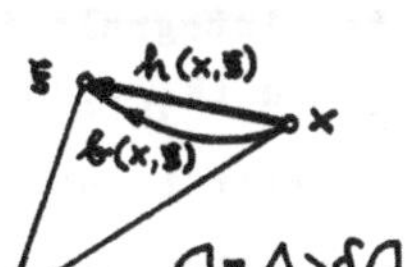

Fig. 7b

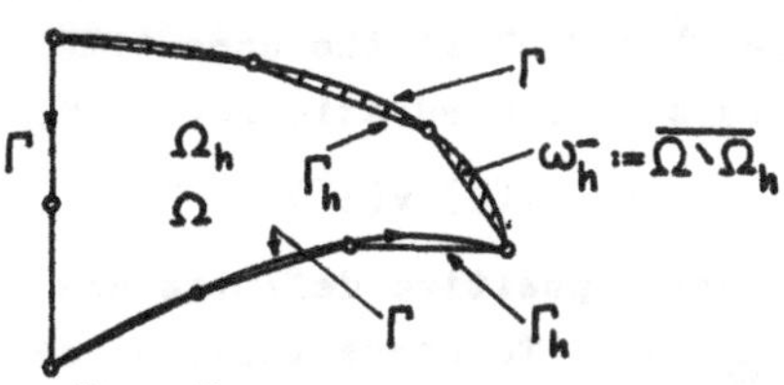

Fig. 7c

We now introduce special functions $\hat{y}$ and $\tilde{y}$ defined on $\bar{\Omega}_h$ and $\bar{\Omega} \cup \bar{\Omega}_h$, respectively.

Definition 4.14. Let the grid function $y \in D_o$, D_o from (2.2.8), be associated with the piecewise linear and continuous interpolant $\hat{y}$ given by

$$\hat{y} \in PP_1(\bar{\Omega}_h) := \left\{ \hat{v} \in C(\bar{\Omega}_h) : \hat{v}|_\Delta \in P_1(\Delta), \Delta \subset \bar{\Omega}_h, \ \hat{v}|_{\bar{\omega}} = v \in D_o \right\}, \quad (4.3.20)$$

where $P_1(\Delta)$ is the set of all polynomials of degree ≤ 1, $\hat{v}|_\Delta$ and $\hat{v}|_{\bar{\omega}}$ the restrictions of $\hat{v}$ on Δ and $\bar{\omega}$, respectively. Moreover, a function $\tilde{y}$ associated with $\hat{y}$ from (4.3.20) will be introduced. The function $\tilde{y}$ is continuous and piecewise linear on $\bar{\Omega} \cup \bar{\Omega}_h$, with $\tilde{y} \equiv \hat{y}$ on $\bar{\Omega}_h$, and, $\tilde{y}$ will be defined as follows:

$$\tilde{y}(\bar{x}) := \begin{cases} \hat{y}(\bar{x}) & \bar{x} \in \bar{\Omega}_h, \\ \hat{y}^*(\bar{x}) & \text{for } \bar{x} \in \delta\Delta \subset \omega_h^-, \ b(x,\xi) \subset \Gamma_{23}, \\ 0 & \bar{x} \in \delta\Delta \subset \omega_h^-, \ b(x,\xi) \subset \Gamma_1. \end{cases} \quad (4.3.21)$$

Here, $\omega_h^- := \overline{\Omega \setminus \Omega_h}$ is used, and $\hat{y}^*$ is the extension of the function $\hat{y}$ defined on Δ to the set $\Delta = \Delta \cup \delta\Delta$ via linear extrapolation; $b(x,\xi)$ denotes the Γ-arc of $\delta\Delta$, cf. Figs. 7a, c. ■

Evidently, for the functions $\hat{y}$ and $\tilde{y}$ from Definition 4.14., where $\tilde{y}$ is taken on $\bar{\Omega}$, the relations

$$\hat{y} \in \overset{\circ}{W}{}^1_{2,\Gamma_{1h}}(\Omega_h) := \left\{ v \in W_2^1(\Omega_h) : v|_{\Gamma_{1h}} = 0 \right\} \text{ and } \tilde{y} \in W_2^1(\Omega) \quad (4.3.22)$$

hold. Hence, in general, $\tilde{y}$ does not vanish on Γ_1, but $\tilde{y}$ satisfies nearly the zero boundary condition on Γ_1.

Now we turn to the investigation of several equivalence relations for the continuous functions $\hat{y}$, $\tilde{y}$ and the grid function $y \in D_o$, which are connected by Definition 4.14. For $\tilde{y}$ from (4.3.21) and triangles Δ, Δ from (4.3.17), with $\Delta = \Delta \cup \delta\Delta$ or $\Delta = \Delta \setminus \delta\Delta$, cf. Figs. 7a, b, the following "local equivalence" of L_2-norms holds:

$$C_o^- \int_{\Delta(x")} \tilde{y}^2(\bar{x}) d\bar{x} \leq \int_{\Delta(x")} \tilde{y}^2(\bar{x}) d\bar{x} \leq C_o^+ \int_{\Delta(x")} \tilde{y}^2(\bar{x}) d\bar{x} \text{ for } y \in D_o, \ x" \in \bar{\omega}". \quad (4.3.23)$$

The positive constants $C_o^\pm$ do not depend on $\tilde{y}$, x" and are of the type $C_o^\pm = 1 \pm C_o' h$, $C_o' = O(1)$, if $h \leq h_o$ is satisfied and h_o is sufficiently small. If $\tilde{y}$ is not linear on Δ and Δ at the same time, then $\tilde{y} = 0$ holds on $\delta\Delta$ and (4.3.23) is straightforward. Otherwise, if $\tilde{y}$ is linear on Δ and Δ at the same time, then the integrals in (4.3.23) can be expressed by quadratic forms $(M_\Delta y, y)_{R^3}$ and $(M_\Delta y, y)_{R^3}$, with vectors $y = (y(x), y(\xi), y(\xi^+))^T \in R^3$, $(.,.)_{R^3}$: scalar product in R^3, and with symmetric, positive definite and "energetically" equivalent 3x3-matrices M_Δ, M_Δ. To prove this, we denote the vertices x, ξ, ξ^+ of the triangle

$\Delta(x''):=\Delta(x,\xi,\xi^+)$ now by $x^{(1)}$, $x^{(2)}$, $x^{(3)}$ and introduce functions φ_i, $i\in\{1,2,3\}$, as follows:

$$\varphi_i(\bar{x}):=\begin{cases}1 & \text{at } \bar{x}=x^{(i)},\\ \text{linear on } \Delta \text{ and } \square, & \\ 0 & \text{at } \bar{x}=x^{(j)},\ j\neq i,\ j\in\{1,2,3\},\end{cases} \qquad i=1,2,3. \quad (4.3.24)$$

The functions $\hat{y}$ or $\tilde{y}$ can be represented by

$$\hat{y}(\bar{x})=\sum_{i=1}^{3} y_i\,\varphi_i(\bar{x}) \text{ for } \bar{x}\in\Delta,\ \tilde{y}(\bar{x})=\sum_{i=1}^{3} y_i\,\varphi_i(\bar{x}) \text{ for } \bar{x}\in\square, \quad (4.3.25)$$

with $y_i:=y(x^{(i)}),i\in\{1,2,3\}$, $y\in D_o$. Using (4.3.25) we get the relations

$$\int_\Delta \hat{y}^2(\bar{x})d\bar{x}=\sum_{i,j=1}^{3} y_i y_j\int_\Delta \varphi_i\,\varphi_j d\bar{x}=:(M_\Delta y,y)_{R^3}, \quad (4.3.26)$$

where the matrix entries m_{ij} of $M_\Delta:=(m_{ij})_{3x3}$ are defined by $m_{ij}:=\int_\Delta \varphi_i\,\varphi_j d\bar{x}$. Analogously, for $\square$ we obtain

$$\int_\square \tilde{y}^2(\bar{x})d\bar{x}=(M_\square y,y)_{R^3},\ M_\square:=(\tilde{m}_{ij})_{3x3},\ \tilde{m}_{ij}:=\int_\square \varphi_i\,\varphi_j d\bar{x}. \quad (4.3.27)$$

The matrices M_Δ, $M_\square$ are obviously symmetric and positive definite. The latter property is becoming clear from the fact that the integrals of $\hat{y}^2$ or $\tilde{y}^2$ from (4.3.26) and (4.3.27) are nonnegative and equal to zero for $\hat{y}\equiv 0$ or $\tilde{y}\equiv 0$ only, i.e. for $y=0\in R^3$. Together with the symmetric matrix $\delta M=(\delta m_{ij})_{3x3}$, $\delta m_{ij}:=\int_{\delta\square}\varphi_i\,\varphi_j d\bar{x}$, we have $(M_\square y,y)_{R^3}=(M_\Delta y,y)_{R^3}\pm(\delta M y,y)_{R^3}$ $(\square=\Delta\uplus\delta\square)$.
Hence,

$$\frac{(M_\square y,y)_{R^3}}{(M_\Delta y,y)_{R^3}}=1\pm\frac{(\delta M y,y)_{R^3}}{(M_\Delta y,y)_{R^3}}\begin{cases}\leq 1+\dfrac{\|\delta M\|}{\lambda_{min}}\\ \geq 1-\dfrac{\|\delta M\|}{\lambda_{min}}\end{cases} \quad \text{for } y\neq 0\in R^3,$$

with $0\leq\|\delta M\|\leq Ch^3$ due to mes $\delta\square\leq C_1h^3$, and $\lambda_{min}=$ mes $\Delta/12$ for the smallest eigenvalue $\lambda_{min}(M_\Delta)$. Thus (4.3.23) is proved. Calculating the greatest eigenvalue of M_Δ, we obtain $\lambda_{max}=$ mes $\Delta/3$. Hence, the inequalities

$$\frac{1}{4}\,\frac{\text{mes}\,\Delta}{3}(y,y)_{R^3}\leq\int_\Delta \hat{y}^2 d\bar{x}\leq\frac{\text{mes}\,\Delta}{3}(y,y)_{R^3} \text{ for } y\in D_o \quad (4.3.28)$$

and $\Delta=\Delta(x'')$, $x''\in\bar{\omega}''$, are satisfied.

For a function $\tilde{y}$ linear both on Δ and $\square$, the identities

$$\int_{\square(x'')}|\nabla\tilde{y}|^2d\bar{x}=\{|\nabla\tilde{y}|^2 \text{ mes }\square\}(x''),\ \int_{\Delta(x'')}|\nabla\tilde{y}|^2d\bar{x}=\{|\nabla\tilde{y}|^2 \text{ mes }\Delta\}(x'') \quad (4.3.29)$$

for $y \in D_o$, $x'' \in \bar{\omega}''$, are evident. Using mes $\delta\square \leq C_1 h^3$ and (4.3.29), we are able to establish the inequalities

$$C_1^- \int_\Delta |\nabla \tilde{y}|^2 d\bar{x} \leq \int_\square |\nabla \tilde{y}|^2 d\bar{x} \leq C_1^+ \int_\Delta |\nabla \tilde{y}|^2 d\bar{x} \text{ for } y \in D_o, \tag{4.3.30}$$

$\Delta = \Delta(x'')$, $\square = \square(x'')$, $x'' \in \bar{\omega}''$, with constants $C_1^\pm = 1 \pm C_1' h$.

For $\tilde{y} = 0$ on $\delta\square$ or $\delta\square = \emptyset$, (4.3.23) and (4.3.30) are fulfilled trivially.

As introduced in Appendix DI, we consider here the usual (semi-)norms $\|.\|_{o,G}$, $|.|_{1,G}$ and $\|.\|_{1,G}$ for functions $u \in W_2^1(G)$ and domains $G \in \{\Omega, \Omega_h\}$, i.e.

$$\|u\|_{o,G}^2 = \int_G u^2 dx,\ |u|_{1,G}^2 := \int_G |\nabla u|^2 dx,\ \|u\|_{1,G}^2 := \|u\|_{o,G}^2 + |u|_{1,G}^2. \tag{4.3.31}$$

By means of (4.3.17) we can note the following relations between the "local" and "global" norms:

$$\|\hat{y}\|_{i,\Omega_h}^2 = \sum_{x'' \in \bar{\omega}''} \|\hat{y}\|_{i,\Delta(x'')}^2,\ \|\tilde{y}\|_{i,\Omega}^2 = \sum_{x'' \in \bar{\omega}''} \|\tilde{y}\|_{i,\square(x'')}^2 \text{ for } i = 0,1 \tag{4.3.32}$$

and analogous identities for the seminorms $|\hat{y}|_{1,\Omega_h}$, $|\tilde{y}|_{1,\Omega}$.

<u>Lemma 4.15.</u> For the functions $\hat{y}$, $\tilde{y}$ assigned to $y \in D_o$ by Defintion 4.14., the following "global equivalence" of the seminorms $|.|_1$ and the L_2- and W_2^1-norms hold:

$$\begin{aligned} &C_1^- |\hat{y}|_{1,\Omega_h}^2 \leq |\tilde{y}|_{1,\Omega}^2 \leq C_1^+ |\hat{y}|_{1,\Omega_h}^2,\ C_1^\pm \text{ from } (4.3.30), \\ &\bar{C}_i^- \|\hat{y}\|_{i,\Omega_h}^2 \leq \|\tilde{y}\|_{i,\Omega}^2 \leq \bar{C}_i^+ \|\hat{y}\|_{i,\Omega_h}^2 \text{ for } i \in \{0,1\}, \end{aligned} \tag{4.3.33}$$

with $\bar{C}_o^\pm := C_o^\pm$, $\bar{C}_1^- := \min\{C_o^-, C_1^-\}$, $\bar{C}_1^+ := \max\{C_o^+, C_1^+\}$.

The constants noted in (4.3.33) are positive and independent of $\hat{y}$, $\tilde{y}$ (also y) and h, for $h \leq h_o$ and sufficiently small h_o. ■

<u>Proof:</u> The assertion is an obvious consequence of (4.3.23), (4.3.30) and (4.3.32).

<u>4.3.3. Γ-Γ_h-relations for norms of special functions.</u> We now revert to the sets Γ_T, Γ_{Th}, $\mathfrak{b}(x')$, $\mathfrak{h}(x')$ from (4.3.18) and define the L_2-norm $\|.\|_{o,\mathfrak{s}}$ on curves $\mathfrak{s}$ by

$$\|u\|_{o,\mathfrak{s}}^2 := \int_\mathfrak{s} u^2 ds \text{ for } u \in L_2(\mathfrak{s}) \text{ and } \mathfrak{s} \in \{\Gamma_T, \Gamma_{Th}, \mathfrak{b}, \mathfrak{h}\},\ \mathfrak{s} \in PC^2. \tag{4.3.34}$$

Let the arc $\mathfrak{b}(x') \subset \partial\square(x'')$ and the segment $\mathfrak{h}(x') \subset \partial\Delta(x'')$ be in correspondence by $x' \in \gamma'$, cf. Figs. 7a, b and (4.3.18). Then for $\hat{y}$, $\tilde{y}$

from Definition 4.14., and with positive constants $M_i' = O(1)$, $M_i'' = O(1)$ and $M_i = 1 + M_i''h$, $i \in \{1,2\}$, the following local norm relations can be verified:

$$\left.\begin{aligned} \|\tilde{y}\|^2_{o,\mathfrak{b}(x')} &\le M_1\|\hat{y}\|^2_{o,\mathfrak{h}(x')} + M_1'h^2|\hat{y}|^2_{1,\Delta(x'')} \\ \|\hat{y}\|^2_{o,\mathfrak{h}(x')} &\le M_2\|\tilde{y}\|^2_{o,\mathfrak{b}(x')} + M_2'h^2|\tilde{y}|^2_{1,\varDelta(x'')} \end{aligned}\right\} \begin{array}{l}\text{for all } x'\in\gamma' \\ \text{and } h\le h_o, \\ \|\cdot\|_{o,s} \text{ from (4.3.34).}\end{array} \quad (4.3.35)$$

Relation (4.3.35) compares the L_2-norms $\|\tilde{y}\|_{o,\mathfrak{b}(x')}$ and $\|\hat{y}\|_{o,\mathfrak{h}(x')}$, but there is no equivalence of these norms. If $\tilde{y} = 0$ on $\delta\varDelta$ or $\delta\varDelta = \emptyset$ holds, then (4.3.35) is clear. For the essential case where $\tilde{y}$ is linear on Δ and $\varDelta$ at the same time, we must prove (4.3.35) by bearing in mind int $\mathfrak{h}(x') \cap \bar{\Omega} = \emptyset$ or $\mathfrak{h}(x') \subset \bar{\Omega}$ from (2.2.1), which can be handled analogously. Let us consider the case $\mathfrak{h}(x') \subset \bar{\Omega}$, with $\mathfrak{b}(x') \subset \Gamma_{23}$, $\mathfrak{h}(x') \subset \partial\Delta(x'')$, $\mathfrak{b}(x') \subset \partial\varDelta(x'')$, cf. Fig. 7a. We take a point $\bar{x}_o \in \mathfrak{h}(x')$, the straight line l spanned by the normal $n(\bar{x}_o)$ to $\mathfrak{h}(x')$ and the point of intersection $\bar{x} := l \cap \mathfrak{b}(x')$. Applying Taylor's expansion of $\tilde{y}(\bar{x})$ at the point $\bar{x}_o$ with respect to the direction $n(\bar{x}_o)$, taking into account the constant vectors $\nabla\hat{y}$, $\nabla\tilde{y}$, with $\nabla\hat{y} = \nabla\tilde{y}$, squaring $\tilde{y}(\bar{x})$ and estimating $\tilde{y}^2(\bar{x})$ from above, we get the inequality

$$\tilde{y}^2(\bar{x}) \le M_1\hat{y}^2(\bar{x}_o) + \tilde{M}_1h^3|\nabla y(x'')|^2, \quad M_1 = 1 + O(h), \quad \tilde{M}_1 = O(1). \quad (4.3.36)$$

Integrating (4.3.36) and inserting the norm $|\hat{y}|_{1,\Delta}$, the first inequality in (4.3.35) is verified. Analogously, the second inequality can be proved. The "local" and "global" norms on the boundaries are connected by

$$\|\tilde{y}\|^2_{o,\Gamma_T} = \sum_{x'\in\gamma_T'} \|\tilde{y}\|^2_{o,\mathfrak{b}(x')}, \quad \|\hat{y}\|^2_{o,\Gamma_{Th}} = \sum_{x'\in\gamma_T'} \|\hat{y}\|^2_{o,\mathfrak{h}(x')}, \quad (4.3.37)$$

cf. (4.3.18). Thus, summing up the local norms via (4.3.37) and using (4.3.35), we obtain the global norm relations

$$\|\tilde{y}\|^2_{o,\Gamma_T} \le M_1\|\hat{y}\|^2_{o,\Gamma_{Th}} + M_1'h^2|\hat{y}|^2_{1,\Omega_\Delta}, \quad \text{with } \Omega_\Delta := \bigcup_{x''\in\omega_t''} \Delta(x''), \quad (4.3.38)$$

$$\|\hat{y}\|^2_{o,\Gamma_{Th}} \le M_2\|\tilde{y}\|^2_{o,\Gamma_T} + M_2'h^2|\tilde{y}|^2_{1,\Omega_\varDelta}, \quad \text{with } \Omega_\varDelta := \bigcup_{x''\in\omega_t''} \varDelta(x''),$$

where ω_t'' is - somewhat more roughly than necessary - defined by $\omega_t'' := \{x'' \in \bar{\omega}'' : \mathfrak{h}(x') \subset \partial\Delta(x''), \ x' \in \gamma_T'\}$, and $\Omega_\Delta, \Omega_\varDelta$ denote subsets of $\bar{\Omega}_h, \bar{\Omega}$ near the boundaries Γ_h, Γ, respectively. For the continuous function $\hat{y} \in PP_1(\bar{\Omega}_h)$ associated with the grid function $y \in D_o$, cf. (4.3.20), the following local identity (4.3.39) and estimate (4.3.40) hold:

$$\int_{\hbar(x')} \hat{y}^2 ds = \frac{\text{mes } \hbar(x')}{3} \{y^2(x) + y(x)y(\xi) + y^2(\xi)\}, \text{ with } x,\xi \in \partial\hbar(x'), \tag{4.3.39}$$

$$\frac{1}{3} \frac{\text{mes } \hbar(x')}{2} (y,y)_{R^2} \le \int_{\hbar(x')} \hat{y}^2 ds \le \frac{\text{mes } \hbar(x')}{2} (y,y)_{R^2}, \text{ with} \tag{4.3.40}$$

$$y = (y(x),y(\xi))^T \in R^2.$$

4.3.4. Relations between "continuous" and "discrete" norms

We now describe the connection between norms for the continuous function $\hat{y} \in \overset{\circ}{W}{}^1_{2,\Gamma_{1h}}(\Omega_h)$ assigned to $y \in D_o$ and grid norms for $y \in D_o$.

Lemma 4.16. For grid functions $y \in D_o$ and the function $\hat{y} \in PP_1(\bar{\Omega}_h)$ from (4.3.20), the following relations hold:

$$|\hat{y}|_{1,\Omega_h} = |y|_{1,m}, \; \frac{1}{4}\|y\|_{i,\hat{m}} \le \|\hat{y}\|_{i,\Omega_h} \le \|y\|_{i,\hat{m}} \text{ for } i = 0,1, \tag{4.3.41}$$

with $|\cdot|_{1,\Omega_h}, \|\cdot\|_{i,\Omega_h}$ from (4.3.31) and $|\cdot|_{1,m}, \|\cdot\|_{i,\hat{m}}$ from (4.3.12), (4.3.14).≡

Proof: The identity in (4.3.41) is an obvious consequence of summing up the local norm relations

$$\int_{\Delta(x'')} |\nabla\hat{y}|^2 d\bar{x} = \{|\nabla_{\hbar} y|^2 \text{ mes}\Delta\}(x'') \text{ for all } x'' \in \bar{\omega}'' \tag{4.3.42}$$

due to the identity $\nabla\hat{y}|_{\Delta(x'')} = \nabla_{\hbar} y(x'')$, $\nabla_{\hbar}$ from (4.3.12).
The norm inequalities in (4.3.41) are based essentially on the identity

$$\sum_{x''\in\bar{\omega}''} \{y^2(x) + y^2(\xi) + y^2(\xi^+)\} \frac{\text{mes } \Delta(x'')}{3} = \sum_{x\in\omega+\gamma_{23}} y^2(x)\tilde{H}_m(x) =: \|y\|^2_{o,\hat{m}}, \tag{4.3.43}$$

which can be proved via summation by parts, cf. also (4.3.6).

Identity (4.3.43) combined with the local norm relation (4.3.28) yields the proof of the inequalities in (4.3.41), case $i = 0$. The proof for $i = 1$ is completed by combining the case $i = 0$ and the left-hand identity in (4.3.41). In the following lemma, L_2-norms of functions defined on the boundaries Γ_{Th}, Γ_{23h} and their discrete analogues are involved.

Lemma 4.17. For the trace $y|_{\gamma}$ of the grid function $y \in D_o$ and the trace $\hat{y}|_{\Gamma_h}$ of the continuous function $\hat{y} \in PP_1(\bar{\Omega}_h)$ from (4.3.20), the norm relations

$$\|\hat{y}\|_{o,\Gamma_{Th}} \le \|y\|_{o,\bar{\gamma}_T}, \; \frac{1}{3}\|y\|^2_{o,\gamma_{23}} \le \|\hat{y}\|^2_{o,\Gamma_{23h}} \le \|y\|^2_{o,\gamma_{23}} \tag{4.3.44}$$

hold, with norms $\|\cdot\|_{o,\cdot}$ defined in (4.3.8) and (4.3.34) for the sets Γ_{Th}, Γ_{23h}, $\bar{\gamma}_T$, γ_{23} according to (4.3.18), (2.2.6) (γ'_{23} from (2.3.14)).≡

Proof. Due to the inequalities (4.3.40) and $y|_{\gamma_1} = 0$, the following relations are evident

$$\|\hat{y}\|^2_{0,\Gamma_{Th}} = \sum_{x' \in \gamma'_T} \int_{h(x')} \hat{y}^2 ds \le \sum_{x' \in \gamma'_T} \{y^2(x) + y^2(\xi)\}\frac{h(x')}{2} \le \|y\|^2_{0,\bar{\gamma}_T}, \tag{4.3.45}$$

$$\|\hat{y}\|^2_{0,\Gamma_{23h}} = \sum_{x' \in \gamma'_{23}} \int_{h(x')} \hat{y}^2 ds \ge \sum_{x' \in \gamma'_{23}} \{y^2(x)+y^2(\xi)\}\frac{h(x')}{6} = \frac{1}{3}\|y\|^2_{0,\gamma_{23}}.$$

For $y \in D_0$, we always have $\|y\|_{0,\gamma_{23}} = \|y\|_{0,\bar{\gamma}_{23}}$. Hence by setting $\Gamma_{Th} := \Gamma_{23h}$, $\bar{\gamma}_T := \bar{\gamma}_{23}$, the third inequality in (4.3.44) is a consequence of the first. Therefore, all inequalities in (4.3.44) are verified by (4.3.45).

4.3.5. A trace theorem and the equivalence of certain grid norms

For several purposes, e.g. for the equivalence of certain grid norms, it is desirable to have a discrete analogue of the well-known trace theorem. For functions $u \in W_2^1(\Omega)$, the trace of u on Γ is defined and the inequality

$$\|u\|_{0,\Gamma} \le C_E \|u\|_{1,\Omega}, \quad \text{for } u \in W_2^1(\Omega), \tag{4.3.46}$$

holds, where the constant C_E is independent of u, cf. Appendix IM. Such a theorem will now be formulated for grid functions $y \in D_0$ with traces on γ or on subsets $\gamma_t \subset \gamma$.

Theorem 4.18. (Trace Theorem). Let $\bar{\omega}$ be the grid assigned to the triangulation $\mathcal{T}_\Delta$ from (4.3.17), under the weak grid regularity condition $VO(MD,\bar{\omega})$ from (2.3.23). Then the inequality

$$\|y\|_{0,\gamma_t} \le C_0 \|y\|_{1,\hat{m}}, \quad \text{for } y \in D_0 \text{ and any } \gamma_t \subset \gamma, \tag{4.3.47}$$

holds, where C_0 is a positive constant independent of y and $h (h \le h_0)$. The norms $\|\cdot\|_{0,\gamma_t}$, $\|\cdot\|_{1,\hat{m}}$ are taken from (4.3.8), (4.3.14). ■

Proof: Since $y \in D_0$, we have $\|y\|_{0,\gamma_t} \le \|y\|_{0,\gamma_{23}}$ for any subset $\gamma_t \subset \gamma$. For subsets $\Gamma_T \subset \Gamma$ and the function $\tilde{y} \in W_2^1(\Omega)$ from (4.3.21), cf. (4.3.22), the inequality $\|\tilde{y}\|_{0,\Gamma_T} \le C_E\|\tilde{y}\|_{1,\Omega}$ holds due to (4.3.46) and $\|y\|_{0,\Gamma_T} \le \|y\|_{0,\Gamma}$, especially for $\Gamma_T = \Gamma_{23} \subset \Gamma$. Taking into account (4.3.44), (4.3.38), (4.3.33) and (4.3.41), in this order, we obtain the estimates

$$\begin{aligned} \|y\|^2_{0,\gamma_t} \le \|y\|^2_{0,\gamma_{23}} &\le 3\|\hat{y}\|^2_{0,\Gamma_{23h}} \le 3\, M_2\|\tilde{y}\|^2_{0,\Gamma_{23}} + 3\, M_2' h^2 |\tilde{y}|^2_{1,\Omega_\Delta} \\ &\le C\|\tilde{y}\|^2_{1,\Omega} \le C_0^2\|\hat{y}\|^2_{1,\Omega_h} \le C_0^2\|y\|^2_{1,\hat{m}}, \text{ with } C_0^2 = 3\, C_E^2 + C'h, \end{aligned} \tag{4.3.48}$$

$C' = O(1)$, that is (4.3.47) with $C_0 = \sqrt{3}\, C_E + O(h)$ for $h \leq h_0$ and sufficiently small h_0. Thus, the proof of (4.3.47) has been reduced to (4.3.46) and several norm relations shown in the preceding sections.

By means of Theorem 4.18. and other norm relations, we are able to prove the equivalence of some grid norms for the secondary triangulations (PB) and (MD), which are summarized in the next two lemmas.

<u>Lemma 4.19.</u> Let VO(PB, $\bar{\omega}$) from (2.3.8) be satisfied for grids $\bar{\omega}$ which are based on networks of triangles. Then, for the seminorms $|\cdot|_{1,p}$, $|\cdot|_{1,m}$ from (4.3.11), (4.3.12) and for the norms $\|\cdot\|_{i,r}$, $\|\cdot\|_{i,\hat{r}}$ from (4.3.5), (4.3.6), (4.3.13) and (4.3.14), with $r \in \{p,m\}$ and $i \in \{0,1\}$, the following relations hold:

$$|y|_{1,p} = |y|_{1,m} \quad \text{for } y \in D_0, \tag{4.3.49}$$

$$\left.\begin{array}{l} C^-_{mp}\|y\|^2_{i,p} \leq \|y\|^2_{i,m} \leq C^+_{mp}\,\|y\|^2_{i,p} \\ \hat{C}^-_{mp}\|y\|^2_{i,\hat{p}} \leq \|y\|^2_{i,\hat{m}} \leq \hat{C}^+_{mp}\,\|y\|^2_{i,\hat{p}} \end{array}\right\} \text{for } y \in D_0 \text{ and } i = 0,1. \tag{4.3.50}$$

The positive constants $C^{\pm}_{mp}$, $\hat{C}^{\pm}_{mp}$ do not depend on y and h ($h \leq h_0$). Thus, the discrete L_2- and W^1_2-norms defined analogously (type r or $\hat{r}$) for secondary networks (PB) and (MD) are equivalent in the sense of (4.3.50). ≡

<u>Proof:</u> VO(PB, $\bar{\omega}$) implies that VO(MD, $\bar{\omega}$) is fulfilled and the boxes $\mathcal{K}_r$ and $\tilde{\mathcal{K}}_r$, $r \in \{p,m\}$, are correctly defined for $h \leq h_0$ and sufficiently small h_0. The identity of the seminorms (4.3.49) is shown in Section 4.4.1., cf. (4.4.12), by means of Green's formula, so that we can refer to this proof. Due to VO(PB, $\bar{\omega}$), the case $i = 0$ (L_2-norm) in (4.3.50) is proved by the fact that there are positive constants $C^{\pm}_{m} = O(1)$ such that the inequalities $C^-_{mp}H_p(x) \leq H_m(x) \leq C^+_{mp}H_p(x)$, $x \in \omega$, hold, and similar relations for $\tilde{H}_p(x)$, $\tilde{H}_m(x)$, $x \in \omega + \gamma_{23}$. The case $i = 1$ in (4.3.50) is an obvious consequence of (4.3.49) and (4.3.50), case $i = 0$.

<u>Lemma 4.20.</u> Let VO(MD, $\bar{\omega}$) from (2.3.23) be satisfied. Then, for $\|\cdot\|_{1,m}$ and $\|\cdot\|_{1,\hat{m}}$ from (4.3.14), the following inequalities hold for secondary networks (MD):

$$\|y\|_{1,\hat{m}} \leq \|y\|_{1,m} \leq C_{m\hat{m}}\|y\|_{1,\hat{m}} \quad \text{for } y \in D_0. \tag{4.3.51}$$

Moreover, if VO(PB, $\bar{\omega}$) from (2.3.8) is fulfilled and secondary networks (PB) are taken into account, then we have

$$\|y\|_{1,\hat{p}} \leq \|y\|_{1,p} \leq C_{p\hat{p}}\,\|y\|_{1,\hat{p}} \quad \text{for } y \in D_0, \tag{4.3.52}$$

with $\|\cdot\|_{1,\hat{p}}$, $\|\cdot\|_{1,p}$ from (4.3.13). The positive constants $C_{m\hat{m}}$, $C_{p\hat{p}}$ do not depend on y and h ($h \leq h_0$). Thus, the W^1_2-norm analogues $\|\cdot\|_{1,r}$ and $\|\cdot\|_{1,\hat{r}}$ are equivalent on D_0. ≡

Proof: For sufficiently small h_o, we get $\tilde{H}_r(x) \leq h(x)$ for $r \in \{p,m\}$, $x \in \gamma_{23}$ and $h \leq h_o$. Then, the inequalities on the left-hand side in (4.3.51), (4.3.52) are evident. The Trace Theorem 4.18., with inequality (4.3.47) for $\gamma_t := \gamma_{23}$, yields the inequality on the right-hand side of (4.3.51), since

$$\|y\|^2_{1,m} := \sum_{x \in \omega} y^2(x) H_m(x) + \|y\|^2_{o,\gamma_{23}} + |y|^2_{1,m} \leq (1+C_o^2)\, \|y\|^2_{1,\hat{m}}$$

holds for $y \in D_o$ and C_o from (4.3.47). Thus, we have $C^2_{m\hat{m}} := 1 + C_o^2$. The inequality on the right-hand side of (4.3.52) is an immediate consequence of (4.3.51) combined with (4.3.50).

Remark 4.21. Contrary to the equivalence of the norms $\|.\|_{1,r}$ and $\|.\|_{1,\hat{r}}$ in Lemma 4.20., we generally have no equivalence of the norms $\|.\|_{o,r}$ and $\|.\|_{o,\hat{r}}$. This is caused by the weights $h(x)$ and $\tilde{H}_r(x)$ for $x \in \gamma_{23}$, which are of the exact order $h(x) = O(h)$ and $\tilde{H}_r(x) = O(h^2)$. Nevertheless, this fact does not disturb the equivalence of $\|.\|_{1,r}$ and $\|.\|_{1,\hat{r}}$, which is essentially due to the fact that the Trace Theorem 4.18. holds. Moreover, on the basis of Lemma 4.19. and 4.20., we draw the conclusion that the four discrete W_2^1-norms $\|.\|_{1,r}$, $\|.\|_{1,\hat{r}}$ defined in (4.3.14), for $r \in \{p,m\}$, are mutually equivalent on D_o and on grids $\bar{\omega}$, if the secondary networks are correctly defined. ≡

4.4. Green's formula, inequalities of Friedrichs-Poincaré type and the positive definiteness of A_h

4.4.1. Green's formula, the symmetry and the "energy" of A_h. We now study the question whether the difference operators A_h via method (PB) and (MD) are symmetric operators in the sense of scalar products introduced in Definition 4.9. To prove this property and related assertions, it is helpful to derive discrete analogues of Green's first formula, which will be done in what follows.

Lemma 4.22. (PB-operators A_h). Let A_h^j ($j=0,1$) denote the difference operators from Theorem 3.1., cf. (3.2.18), which are constructed via method (PB) on irregular networks of triangles and (or) rectangles, with a secondary network of the type (PB) and under the Condition 2.16. - VO(PB, $\bar{\omega}$). Then, a discrete analogue (4.4.2) of Green's formula will be valid, and the operators A_h^j are symmetric in the sense of

$$(A_h^j y, v)_p = (y, A_h^j v)_p \quad \text{for } y,v \in D_o,\ h \leq h_o \text{ and } j = 0,1, \qquad (4.4.1)$$

with the scalar product $(.,.)_p$ from (4.3.1). ≡

Proof: To prove the analogue of Green's formula, we consider $(A_h^j y, v)_p$ and $A_h^j y = (0, l_h^j y, L_h y)^T$ from (3.2.18a) for $y \in D_o$. Replacing the summation index $x \in \omega + \gamma_{23}$ by $x' \in \omega' + \gamma'_{23}$ by means of "summation by parts" within the main part of L_h, l_h^j, and taking into account the symmetry relations $h(x'_\xi) := h(x,\xi) = h(\xi,x)$, $s^j(x'_\xi) := s^j(x,\xi) = s^j(\xi,x)$, $k_h(x'_\xi) := k_h(x,\xi) = k_h(\xi,x)$ and the skew symmetry of the difference quotient $y_{\hbar}(x'_\xi) := y_{\hbar}(x,\xi) = -y_{\hbar}(\xi,x)$, cf. (3.2.18), we obtain the following identity

$$(A_h^j y, v)_p := \sum_{x\in\omega} \{(L_h y) v H_p\}(x) + \sum_{x\in\gamma_{23}} \{(l_h^j y) v h\}(x) \tag{4.4.2}$$

$$= \sum_{x'\in\omega'+\gamma'_{23}} k_h(x') y_{\hbar}(x') v_{\hbar}(x') H^j(x') + \sum_{x\in\omega} c(x) y(x) v(x) H_p(x)$$

$$+ \sum_{x\in\gamma_{23}} \alpha^j(x) y(x) v(x) h(x) \quad \text{for } y,v \in D_o \text{ and } j = 0,1.$$

The meaning of $H^j(x')$ is given by

$$H^j(x') = h(x') s^j(x') \text{ for } j = 0,1, \text{ with } s^j(x') \text{ from (3.2.18c)}, \tag{4.4.3}$$

i.e., $H^0(x') = H'(x')$ holds, cf. (4.3.3), and $\alpha^j(x)$ is defined in (3.2.18c). Finally, the symmetry relation (4.4.1) is a consequence of the apparent symmetry of the right-hand side of (4.4.2).

<u>Lemma 4.23. (MD-operators A_h).</u> Let A_h^j ($j=0,1$) denote the difference operators from Theorem 3.4., cf. (3.3.19), which are constructed via method (MD) and under the Condition 2.24. - VO(MD, $\bar{\omega}$). Then, a discrete analogue (4.4.5) of Green's formula holds, and the operators A_h^j are symmetric on D_o, i.e.

$$(A_h^j y, v)_m = (y, A_h^j v)_m \text{ for } y,v \in D_o,\ h \le h_o \text{ and } j = 0,1, \tag{4.4.4}$$

with the scalar product from (4.3.1).∎

<u>Proof:</u> We consider $(A_h^j y, v)$ for $y,v \in D_o$, with $A_h^j y = (0, l_h^j y, L_h y)^T$ from (3.3.19a), and get by "summation by parts" within the main part of L_h, l_h^j the identity

$$(A_h^j y, v)_m := \sum_{x\in\omega} \{(L_h y) v H_m\}(x) + \sum_{x\in\gamma_{23}} \{(l_h^j y) v h\}(x)$$

$$= \sum_{x''\in\bar{\omega}''} \{(K_h \nabla_{\hbar} y, \nabla_{\hbar} v) H''\}(x'') - \frac{1}{8} \sum_{x'\in\gamma'_{23}} \alpha_h(x') y_{\hbar}(x') v_{\hbar}(x') h^3(x')$$

$$+ \sum_{x\in\omega} c(x) y(x) v(x) H_m(x) + \sum_{x\in\gamma_{23}} \alpha^j(x) y(x) v(x) h(x) \tag{4.4.5}$$

$$\text{for } y,v \in D_o,\ j = 0,1.$$

Here, the formulas of the summation by parts are given by

$$-\sum_{x\in\omega+\gamma_{23}} \{\sum_{x''\in S''(x)} (K_h \nabla_{\hbar} y, \frac{\hbar^{\bullet} - \hbar^{o}}{2})(x'')\} v(x) = \sum_{x''\in\bar{\omega}''} \{(K_h \nabla_{\hbar} y, \nabla_{\hbar} v) H''\}(x), \tag{4.4.6}$$

$$\sum_{x\in\gamma_{23}}\sum_{\xi\in S'(x)\cap\gamma}\alpha_h(x'_\xi)y_{\hbar}(x'_\xi)h^2(x'_\xi)v(x) = -\sum_{x'\in\gamma'_{23}}\alpha_h(x'_\xi)y_{\hbar}(x'_\xi)v_{\hbar}(x'_\xi)h^3(x'_\xi).$$

The meaning of K_h, $\hbar^{\circ}$, $\hbar^{o}$, $\alpha^j(x)$, $\alpha_h(x'_\xi)$, $h(x'_\xi)$ and $y_{\hbar}(x'_\xi)$, $\nabla_{\hbar}y(x'')$ is the same as in (3.3.19). Moreover, $H''(x''):=$ mes $\Delta(x'')$ and $\alpha=0$ on γ_2 are taken into account. The relation (4.4.4) follows from the evident symmetry of the right-hand side of (4.4.5) with respect to y and v.

Remark 4.24. The symmetry relations (4.4.1) and (4.4.4) can also be noted for $\hat{p}$, $\hat{m}$ instead of p, m, i.e.,

$$(A_h^j y,v)_{\hat{r}} = (y,A_h^j v)_{\hat{r}} \text{ for } y,v\in D_o,\ h\le h_o,\ j\in\{0,1\},\ r\in\{p,m\}, \quad (4.4.7)$$

where $(.,.)_{\hat{r}}$ is taken from (4.3.2). The proof is quite similar to that of (4.4.1), (4.4.4). But (4.4.7) can also be verified by the following relation between the scalar products $(.,.)_r$, $(.,.)_{\hat{r}}$ from (4.3.1), (4.3.2):

$$(y,v)_r = (y,v)_{\hat{r}} + \sum_{x\in\gamma_{23}} y(x)v(x)\{h(x) - \tilde{H}_r(x)\} \quad \text{for } y,v\in D_o, \quad (4.4.8)$$

with $h(x) - \tilde{H}_r(x)\ge \varepsilon h > 0$ for $h\le h_o$ and sufficiently small h_o, $x\in\gamma_{23}$. ■

Finally, we shall investigate the relationship between the seminorms $|y|_{1,r}$ and the "energy" $(\Lambda_h^r y,y)_r$ of some simple difference operators Λ_h^r given by

$$\left.\begin{array}{l}\Lambda_h^p := A_h^o(PB) \text{ from } (3.2.18)\\ \Lambda_h^m := A_h^o(MD) \text{ from } (3.3.19)\end{array}\right\} \begin{array}{l}\text{for } k:= k_{11} = k_{22} = 1,\ k_{12} = 0,\\ c = 0 \text{ on } \bar{\Omega},\ \alpha = 0 \text{ on } \Gamma\setminus\Gamma_1,\end{array} \quad (4.4.9)$$

with VO(PB, $\bar{\omega}$) for r = p and VO(MD, $\bar{\omega}$) for r = m.

The operators Λ_h^p, Λ_h^m were discussed earlier in Lemma 4.5.

Lemma 4.25. For Λ_h^r, $r\in\{p,m\}$, from (4.4.9), the following relations hold:

$$(\Lambda_h^p y,v)_p = (y_{\hbar},v_{\hbar})',\quad (\Lambda_h^m y,v)_m = \sum_{i=1}^{2}\left(\frac{\delta y}{\delta x_i},\frac{\delta v}{\delta x_i}\right)'' \text{ for } y,v\in D_o, \quad (4.4.10)$$

$$(\Lambda_h^p y,y)_p = |y|_{1,p}^2,\quad (\Lambda_h^m y,y)_m = |y|_{1,m}^2 \quad \text{for } y\in D_o, \quad (4.4.11)$$

furthermore,

$$(\Lambda_h^p y,y)_p = (\Lambda_h^m y,y)_m, \text{ i.e. } |y|_{1,p} = |y|_{1,m} \quad \text{for } y\in D_o, \quad (4.4.12)$$

where $(.,.)_r$, $(.,.)'$, $(.,.)''$, $|.|_{1,r}$, $r\in\{p,m\}$, are taken from (4.3.1),

(4.3.3), (4.3.4), (4.3.11) and (4.3.12), with $\delta./\delta x_i$ from (3.3.15) for $\mu = e_i$ $(i=1,2)$. ▬

Proof: Green's formulas (4.4.10) are a direct consequence of (4.4.2) and (4.4.5). Therefore, the squares of the seminorms $|.|_{1,r}$ can be represented by the energy of the difference operators Λ_h^r. This follows from (4.4.10) and (4.3.11), (4.3.12). On the basis of Lemma 4.5. about the comparison of Λ_h^p and Λ_h^m on networks of triangles, we see from (4.2.8) - (4.2.10) and the definition of $(.,.)_r$, $r \in \{p,m\}$, that (4.4.12) holds.

Thus, the "discrete energies" of the operators Λ_h^p and Λ_h^m as well as the seminorms $|.|_{1,r}$ coincide on networks of triangles.

4.4.2. Inequalities of Friedrichs-Poincaré-type. We now derive inequalities which are discrete analogues of the corresponding inequalities of Friedrichs and Poincaré within the $W_2^1(\Omega)$-theory. Beside Ω, Γ_1 and ω, γ_1, we take into account subsets Γ_T, $\bar{\gamma}_T$ of Γ, γ according to (4.3.18) and subdomains Ω_T of Ω. Furthermore, $\omega_T := \omega \cap \Omega_T$. Assume that $\Gamma_T \subset \Gamma_{23}$ and

$$\operatorname{mes} \Gamma_1 > 0 \text{ or } \operatorname{mes} \Gamma_T > 0 \text{ or } \operatorname{mes} \Omega_T > 0. \tag{4.4.13}$$

Thus, by virtue of the grid regularity conditions VO(MD,$\bar{\omega}$) or VO(PB,$\bar{\omega}$), we have

$$\sum_{x \in \gamma_{23}} h(x) \geq M_1 > 0 \text{ or } \sum_{x \in \bar{\gamma}_T} h(x) \geq \bar{M}_T > 0 \text{ or } \sum_{x \in \omega_T} H_r(x) \geq M_T > 0, \tag{4.4.14}$$

$r \in \{p,m\}$, where M_1, $\bar{M}_T$ and M_T are independent of h, $h \leq h_o$, and h_o is chosen sufficiently small.

Theorem 4.26. - Friedrichs-Poincaré-type inequality. Under the assumption (4.4.14), which is based on (4.4.13), and the grid regularity conditions VO(PB, $\bar{\omega}$) for r = p and VO(MD, $\bar{\omega}$) for r = m, the following inequalities hold:

$$\| y \|^2_{1,\hat{r}} \leq C^{\hat{r}}_{fp} \{ | y |^2_{1,r} + \| y \|^2_{o,\bar{\gamma}_T} + \| y \|^2_{o,\hat{r},\omega_T} \} \text{ for } y \in D_o \text{ and } r \in \{p,m\}, \tag{4.4.15}$$

and with r instead of $\hat{r}$, too. The positive constants $C^{\cdot}_{fp}$ do not depend on y and h, $h \leq h_o$, for sufficiently small h_o. The (semi-)norms $\|.\|_{1,r}$, $\|.\|_{1,\hat{r}}$, $|.|_{1,r}$, $\|.\|_{o,\bar{\gamma}_T}$, $\|.\|_{o,r,\omega_T}$ and $\|.\|_{o,\hat{r},\omega_T}$ $(\omega_T \subset \omega)$ are taken from (4.3.11) - (4.3.14), (4.3.8), (4.3.7), with $r \in \{p,m\}$. ▬

Proof: The proof will only be carried out for (4.4.15), case $\hat{m}$. Then, the proof of the remaining estimates (4.4.15), viz. the cases m, $\hat{p}$ and p, is a simple consequence of the equivalence of several grid norms. For this, cf. (4.3.49) - (4.3.52), and the norms $\|.\|_{o,l,\omega_T}$, where l stands for $\hat{m}$, m, $\hat{p}$ and p, are equivalent, too. Now, it is well known from the literature that the following inequality of Friedrichs-Poincaré-type holds:

$$\|u\|^2_{1,\Omega} \leq C_{fp}\{|u|^2_{1,\Omega} + \|u\|^2_{o,\Gamma_1} + \|u\|^2_{o,\Gamma_T} + \|u\|^2_{o,\Omega^o_T}\} \quad \text{for } u \in W^1_2(\Omega), \qquad (4.4.16)$$

if mes Γ_1 + mes Γ_T + mes $\Omega^o_T \neq 0$ is satisfied for Γ_1, Γ_T from (4.4.13) and some subdomain $\Omega^o_T \subset \Omega$ characterized below. The positive constant C_{fp} does not depend on u, and the norms used in (4.4.16) are given in (4.3.31) and (4.3.34). For instance, (4.4.16) can be derived from Lemma 1.36 in GAJEWSKI et al. [1, p. 36]. Clearly, (4.4.16) is also true for $u := \tilde{y}$ defined in (4.3.21), cf. (4.3.22). We now estimate $\|\tilde{y}\|^2_{1,\Omega}$ from below and the right-hand side in (4.4.16) for $u := \tilde{y}$ from above. By (4.3.33) and (4.3.41), we have the estimates

$$\|\tilde{y}\|^2_{1,\Omega} \geq \bar{C}^-_1 \|\hat{y}\|^2_{1,\Omega_h} \geq \frac{1}{4} \bar{C}^-_1 \|y\|^2_{1,\hat{m}}, \quad |\tilde{y}|^2_{1,\Omega} \leq C^+_1 |\hat{y}|^2_{1,\Omega_h} = C^+_1 |y|^2_{1,m}. \qquad (4.4.17)$$

Furthermore, due to (4.3.38) (also for Γ_1, Γ_{1h}), $\hat{y}|_{\Gamma_{1h}} = 0$ and (4.3.44), the inequalities

$$\|\tilde{y}\|^2_{o,\Gamma_1} + \|\tilde{y}\|^2_{o,\Gamma_T} \leq M_1\|\hat{y}\|^2_{o,\Gamma_{Th}} + M'_1 h^2 |y|^2_{1,m} \leq M_1\|y\|^2_{o,\bar{\gamma}_T} + M'_1 h^2 |y|^2_{1,m} \qquad (4.4.18)$$

can be inferred. For $y \in D_o$, the following estimates hold (cf. Lemma 4.16. and (4.3.28)):

$$\|y\|^2_{o,\hat{m},\omega_T} \geq \sum_{x'' \in \omega''_T} \{y^2(x) + y^2(\xi) + y^2(\xi^+)\} \frac{\text{mes } \Delta(x'')}{3} \geq \sum_{x'' \in \omega''_T} \int_{\Delta(x'')} \hat{y}^2 d\bar{x} \qquad (4.4.19)$$

$$\geq \int_{\Omega^o_T} \hat{y}^2 d\bar{x} =: \|\hat{y}\|^2_{o,\Omega^o_T}, \quad \text{with } \omega''_T := \{x'' \in \bar{\omega}'' : \Delta(x'') \subset \Omega_T\},$$

and Ω^o_T is a subset of Ω_T, with $\Omega^o_T \subset \bigcap_{h \leq h_o} \{\bigcup_{\omega''_T} \Delta(x'')\}$ and mes $\Omega^o_T \neq 0$, if mes $\Omega_T \neq 0$ is valid and h_o is chosen sufficiently small. The inequalities in (4.4.17) - (4.4.19) yield, together with (4.4.16), the inequality (4.4.15), case $\hat{m}$, with a positive constant $C^{\hat{m}}_{fp}$ independent of $y \in D_o$ and $h \leq h_o$, if h_o is chosen sufficiently small.

Remark 4.27. The Friedrichs-Poincaré-type inequality holds obviously on networks of triangles and (or) rectangles, with secondary networks (PB) and VO(PB, $\bar{\omega}$), i.e.,

$$\|y\|^2_{1,l} \leq C^l_{fp} \{|y|^2_{1,p} + \|y\|^2_{o,\bar{\gamma}_T} + \|y\|^2_{o,l,\omega_T}\} \quad \text{for } y \in D_o,\ l = p, \hat{p}. \qquad (4.4.20)$$

Clearly, (4.4.20) is also valid for regular or irregular networks of rectangles used for domains Ω which are rectangles or are composed of rectangles. ■

The proof of (4.4.20) uses the "complete triangulation" of the given original network such that each rectangle □ is subdivided into two triangles △ by one diagonal. But this diagonal $\hbar(x,\xi)$ is associated with a perpendicular bisector segment $\mathfrak{s}(x,\xi)$ of measure zero, i.e. $s(x'_\xi) = \tilde{s}(x'_\xi) = 0$. Thus, there are no changes in the boxes $\mathfrak{X}_p(x)$, $\tilde{\mathfrak{X}}_p(x)$ and in the norms $\|.\|_{0,l}$, $l \in \{p,\hat{p}\}$, $|.|_{1,p}$, if such diagonals are inserted. It should be noted that in (2.3.8e) from VO(PB,$\bar{\omega}$) the value $j = 0$ can be taken, since no assumptions on FDSs $A_h^1 y = F_h^1$ are used. Therefore, the condition $\theta \leq \pi/2$ for boundary triangles, cf. (2.3.8d), must be replaced by $\theta < \frac{\pi}{2} - \theta_o$, cf. (2.3.8e), only in such cases where int $\hbar(x,\xi) \cap \bar{\Omega} = \emptyset$ holds and where triangles are originally employed in the given network.

Especially, if the domain Ω is a rectangle and networks of rectangles are used, then an inequality related with (4.4.20) is also presented in SAMARSKIĬ/ANDREEV [1]; concerning the FEM cf. ŽENIŠEK [1].

4.4.3. The positive definiteness of A_h. In order to prove the positive definiteness of A_h, we shall use

$$k_h(x') \geq k_o > 0 \text{ and } (K_h(x'')\xi,\xi)_{R^2} \geq k_o(\xi,\xi)_{R^2} \text{ for } \xi \in R^2 \qquad (4.4.21)$$

owing to (2.1.7a) and to the definition of k_h and K_h in (3.2.5) and (3.3.9), respectively. Comparing $H^1(x')$ and $H^o(x')$ from (4.4.3) as well as $\alpha^1(x)$ and $\alpha^o(x)$ from (4.4.2) or (4.4.5), we also see that the following relations hold:

$$H^1(x') = \begin{cases} H^o(x') & x' \in \omega' + \gamma'_2, \\ H^o(x') - \frac{1}{8}\,\frac{\alpha_h(x')h^3(x')}{k_h(x')} & \text{for } \quad x' \in \gamma'_{23} \setminus \gamma'_2, \end{cases} \quad \alpha|_{\gamma_2} = 0, \qquad (4.4.22a)$$

$$\sum_{x \in \gamma_{23}} \alpha^1(x)y^2(x)h(x) - \frac{1}{8}\sum_{x' \in \gamma'_{23}} \alpha_h(x')y_\hbar^2(x')h^3(x') \geq \sum_{x \in \gamma_{23}} \frac{\alpha^o(x)}{2}\, y^2(x)h(x) \qquad (4.4.22b)$$

for $y \in D_o$.

Using (4.4.2), (4.4.5), (4.4.21), (4.4.22) and estimating $(A_h y,y)$ from below, the following inequalities are obtained,

$$\text{case (PB): } (A_h^j y,y)_p \geq k_o(y_\hbar,y_\hbar)' + (dy,y)_p \text{ for } y \in D_o,\ j = 0,1, \qquad (4.4.23a)$$

$$\text{case (MD): } (A_h^j y,y)_m \geq k_o(|\nabla_\hbar y|,|\nabla_\hbar y|)'' + (dy,y)_m \text{ for } y \in D_o,\ j = 0,1, \qquad (4.4.23b)$$

with A_h^j(PB), A_h^j(MD) from (3.2.18), (3.3.19), respectively. The grid function d is specified by

$$d(x) := \begin{cases} c(x) & x \in \omega, \\ \frac{\alpha^o(x)}{2} \quad \text{for} & x \in \gamma_{23}, \quad \alpha^o(x) \text{ from (3.2.18) (PB) or (3.3.19) (MD).} \\ 0 & x \in \gamma_1, \end{cases} \tag{4.4.24}$$

Introducing the difference operator M_h by

$$(M_h y)(x) := \begin{cases} (\Lambda_h y)(x) + k_o^{-1} d(x) y(x) & x \in \omega + \gamma_{23}, \\ \quad\quad\quad\quad \text{for} & \quad\quad \Lambda_h \text{ from (4.4.9),} \\ y(x) & x \in \gamma_1, \end{cases} \tag{4.4.25}$$

it is apparent that M_h is a difference operator which is positive definite and an "energetic minorant" with respect to $k_o^{-1} A_h^j$, i.e.

$$(A_h^j y, y) \geq k_o(M_h y, y) = k_o(\Lambda_h y, y) + (dy, y) = k_o |y|_1^2 + (dy, y) > 0 \quad \text{for } y \in D_o, \tag{4.4.26}$$

with $(.,.)$, $|\cdot|_1$ from (4.3.16). The inequality $k_o\,|y|_1^2 + (dy,y) > 0$ in (4.4.26) is a consequence of the relation $\omega_o + \gamma_1 + \gamma_3 \neq \emptyset$ from (4.2.4) and can be easily discussed by stating three particular cases $\omega_o \neq \emptyset$, $\gamma_1 \neq \emptyset$, and $\gamma_3 \neq \emptyset$, under the restriction $y|_{\gamma_1} = 0$. We pay attention to the fact that $|y|_1 = 0$ is valid only for grid functions $y \equiv$ const on $\bar{\omega}$, $d > 0$ holds on $\omega_o + \gamma_3$ and no function $y = \text{const} \in D_o$ is admissible for $k_o |y|_1^2 + (dy,y) > 0$ in the case $\gamma_1 \neq \emptyset$. Hence, M_h and A_h^j $(j=0,1)$ are positive definite on D_o at least for each fixed h, $h \leq h_o$.
We now formulate Theorem 4.28., where we assert that the positive definiteness is uniform with respect to h, $h \leq h_o$.

Theorem 4.28. Let $A_h := A_h^j$ $(j=0,1)$ be a difference operator constructed in Theorem 3.1. for networks of triangles and (or) rectangles via method (PB) and in Theorem 3.4. for networks of triangles via method (MD), cf. (3.2.18), (3.3.19), where the corresponding grid regularity conditions VO(PB, $\bar{\omega}$) or VO(MD, $\bar{\omega}$) are assumed to be fulfilled.
Then, the difference operators are symmetric, i.e.

$$(A_h y, v) = (y, A_h v) \text{ for } y, v \in D_o \text{ (} r, \hat{r} \text{ are omitted).} \tag{4.4.27}$$

Furthermore, let relation (2.1.8) be satisfied, i.e., mes Γ_1 + mes Γ_3 + + mes $\Omega_c > 0$. Then, the difference operators $A_h^r := A_h$, $r \in \{p,m\}$, are positive definite, uniformly with respect to $h \leq h_o$, i.e.

$$\|y\|_{o,r}^2 \leq C_r (A_h^r y, y)_r \text{ for } y \in D_o,\ r \in \{p,m\}, \tag{4.4.28}$$

where C_r is a positive constant independent of h and y. Additionally, the following a priori estimates, being discrete analogues of the "V-ellipticity", hold:

$$\|y\|^2_{1,r} \le C_r(A_h y,y)_r \text{ and } \|y\|^2_{1,\hat{r}} \le C_{\hat{r}}(A^r_h y,y)_r \text{ for } y\in D_o, \quad r\in\{p,m\}. \tag{4.4.29}$$

Here, the constants C_r and $C_{\hat{r}}$ are of the same type as in (4.4.28). The scalar products $(.,.)_r$ and the norms $\|.\|_{o,r}$, $\|.\|_{1,r}$, $\|.\|_{1,\hat{r}}$ are taken from (4.3.1), (4.3.5), (4.3.13) and (4.3.14) ≡

<u>Proof:</u> The symmetry relation (4.4.27) was already proved, cf. (4.4.1), (4.4.4) and (4.4.7). By virtue of (2.1.8) and the definition of $\bar{\omega} = \omega + \gamma$, we have the relations

$$\sum_{x\in\gamma_1} h(x) \ge M_1 > 0 \text{ or } \sum_{x\in\gamma_3} h(x) \ge M_3 > 0 \text{ or } \sum_{x\in\omega_{c_o}} H(x) \ge M_o > 0, \tag{4.4.30}$$

with $\omega_{c_o} := \omega\cap\Omega_{c_o}$, $\Omega_{c_o} := \{x\in\Omega : c(x) \ge c_o\}$, $c_o(\varepsilon) := \varepsilon \max_{x\in\bar{\Omega}} c(x)$, $0 < \varepsilon < 1$; ε fixed, c from (2.1.7b). The constants M_i, $i\in\{0,1,3\}$, do not depend on h, $h \le h_o$. The inequalities in (4.4.30) can be derived from mes $\Gamma_1 > 0$, mes $\Gamma_3 > 0$ or mes $\Omega_c > 0$, with a continuous function c. The scalar product (dy,y) from (4.4.26) can be bounded from below by

$$(dy,y) \ge c_o \|y\|^2_{o,\omega_{c_o}} + \frac{\alpha_o}{2}\|y\|^2_{o,\gamma_3} \quad \text{for } y\in D_o, \tag{4.4.31}$$

where d is defined in (4.4.24), and the inequalities $c(x) \ge c_o > 0$ on ω_{c_o} as well as $\alpha^o(x) \ge \alpha_o > 0$ on γ_3 are taken into account, α_o from (2.1.7b). In accordance with (4.4.23), (4.4.26), (4.4.31) and $\|y\|_{o,\gamma_3} = \|y\|_{o,\bar{\gamma}_3}$ ($y\in D_o$, cf. (2.2.6) (4.3.18)), we have

$$(A^r_h y,y)_r \ge k_o|y|^2_{1,r} + c_o\|y\|^2_{o,r,\omega_{c_o}} + \frac{\alpha_o}{2}\|y\|^2_{o,\bar{\gamma}_3} \quad \text{for } y\in D_o, \quad r\in\{p,m\}. \tag{4.4.32}$$

Inequality (4.4.32) and the Friedrichs-Poincaré-type inequalities (4.4.15), (4.4.20) lead, together with the trivial relation $\|y\|_{o,r} \le \|y\|_{1,r}$ (or with $\hat{r}$ instead of r), to the estimates

$$\|y\|^2_{o,r} \le \|y\|^2_{1,r} \le C_r(A^r_h y,y)_r, \quad \|y\|^2_{o,\hat{r}} \le \|y\|^2_{1,\hat{r}} \le C_{\hat{r}}(A^r_h y,y)_r \tag{4.4.33}$$

for $y\in D_o$, $r\in\{p,m\}$, $0 < C_r = O(1)$, $C_{\hat{r}}$ analogously, where A^r_h is to be taken for $r\in\{p,m\}$ and includes the cases $j = 0$ and $j = 1$ for A^j_h. The constants $C_r(C_{\hat{r}})$ depend on k_o, c_o, α_o, $C^r_{fp}(C^{\hat{r}}_{fp})$ and h_o, but they are independent of h ($h \le h_o$) and $y\in D_o$. Thus, the a priori estimates (4.4.29) and the positive definiteness of A^r_h in (4.4.28) are both proved by (4.4.33), for $r\in\{p,m\}$.

Remark 4.29. - Friedrichs' inequality. The constants C_r, $C_{\hat{r}}$ are available a priori in many cases. For instance, if $\Gamma = \Gamma_1$ holds and secondary networks (MD) are utilized, then the following discrete analogue of Friedrichs' inequality can be proved:

$$|y|_{1,m}^2 \geq \frac{\pi}{2 \operatorname{mes} \tilde{\Omega}} \|y\|_{0,m}^2 \quad \text{for } y \in D_0, \tag{4.4.34}$$

where $\tilde{\Omega}$ is a domain with the properties $\Omega \subset \tilde{\Omega}$ and $\Omega_h \subset \tilde{\Omega}$ for $h \leq h_0$, cf. HEINRICH [7]. Therefore, we have

$$(A_h^m y, y)_m \geq k_0 |y|_{1,m}^2 \geq \frac{k_0 \pi}{2 \operatorname{mes} \tilde{\Omega}} \|y\|_{0,m}^2 \quad \text{for } y \in D_0,$$

$$\|y\|_{1,m}^2 \leq C_m (A_h^m y, y)_m \text{ for } y \in D_0, \; C_m := \frac{\pi + 2 \operatorname{mes} \tilde{\Omega}}{k_0 \pi} . \tag{4.4.35}$$

Taking into account the smallest circle K_R, with radius R, containing $\Omega \cup \Omega_h$ for each $h \leq h_0$, we can write

$$\|y\|_{0,m}^2 \leq \frac{2R^2}{k_0} (A_h^m y, y)_m \quad \text{and } \|y\|_{1,m}^2 \leq \frac{1+2R^2}{k_0} (A_h^m y, y)_m \quad \text{for } y \in D_0. \tag{4.4.36}$$

Due to the equivalence of the norms for secondary networks (MD) and (PB), estimates of the type (4.4.34) - (4.4.36) can also be noted for p instead of m. ≡

Remark 4.30. The best constant C_1 in the estimate $(A_h y, y) \geq C_1 (y, y)$ for $y \in D_0$, which can be taken uniformly with respect to h, $h \leq h_0$, is defined by $C_1 := \inf\limits_{h \leq h_0} \lambda_0$. Here, λ_0 is the smallest eigenvalue of A_h on D_0, and, due to the symmetry and the positive definiteness of A_h, λ_0 is always real and positive for $h \leq h_0$. For the numerical approximation of such constants C_1, cf. LANGER/JUNG [1]. ≡

The method of proving the positive definiteness of A_h via Friedrichs-Poincaré-type inequalities was presented in the proof of Theorem 4.28. Now we shall give a second approach to verifying the positive definiteness (uniformly with respect to $h \leq h_0$) of symmetric operators A_h via a priori estimates with the C-norm, which are based on the monotonicity condition $M(A_h, \overline{\omega})$ from (4.2.1). The difference operators A_h of the type (PB) or (MD) so far taken into account are symmetric and positive definite at least for each fixed h ($h \leq h_0$), which was proved by (4.4.27) and (4.4.26), respectively. We now consider the eigenvalue problem

$$A_h y = \lambda y \quad \text{for } y \in D_0, \; \|y\|_0 \neq 0. \tag{4.4.37}$$

Clearly, all eigenvalues λ are real and positive for $h \leq h_0$. If λ_0 denotes the smallest eigenvalue of A_h on D_0, then we have $(A_h y, y) \geq \lambda_0 (y, y)$

for each $y \in D_o$, since the Rayleigh-quotient has the real positive number λ_o as a lower bound. To prove the positive definiteness of A_h, uniformly with respect to $h \leq h_o$, it is sufficient to find a lower bound of λ_o, which does not depend on h, $h \leq h_o$. In Section 4.5.3., a priori estimates of the type

$$\|y\|_{C(\bar{\omega})} \leq C\|A_h y\|_{C(\bar{\omega})} \quad \text{for } y \in D,\ D \text{ from (2.2.8)} \qquad (4.4.38)$$

and $\|\cdot\|_{C(\bar{\omega})}$ from (4.3.9), are derived, where $0 < C = O(1)$ holds, cf. Theorems 4.39., 4.40. and Corollary 4.42. From (4.4.38) we may conclude that A_h^{-1} exists and is uniformly bounded with respect to h, $h \leq h_o$. Consequently, from $A_h^{-1} y_o = \lambda_o^{-1} y_o$, y_o eigenfunction to the smallest eigenvalue λ_o, we get

$$\|\lambda_o^{-1} y_o\|_{C(\bar{\omega})} = \lambda_o^{-1}\|y_o\|_{C(\bar{\omega})} = \|A_h^{-1} y_o\|_{C(\bar{\omega})} \leq \|A_h^{-1}\|\|y_o\|_{C(\bar{\omega})} \leq C\|y_o\|_{C(\bar{\omega})}.$$

Since $\|y_o\|_{C(\bar{\omega})} \neq 0$ holds, the relation $\lambda_o^{-1} \leq C$ is obvious and, therefore,

$$C^{-1}(y,y) \leq \lambda_o(y,y) \leq (A_h y,y) \quad \text{for } y \in D_o,\ h \leq h_o,\ 0 < C = O(1), \qquad (4.4.39)$$

holds, where C is taken from (4.4.38). We summarize this result in

Lemma 4.31. Let the assumptions of Theorems 4.39., 4.40. or Corollary 4.42. be satisfied, where a priori estimates of the type $\|y\|_{C(\bar{\omega})} \leq C\|A_h y\|_{C(\bar{\omega})}$ for $y \in D$ are derived for difference operators A_h. Furthermore, let A_h be symmetric and positive definite for each fixed h, $h \leq h_o$. Then, A_h is also "uniformly with respect to $h \leq h_o$ positive definite" in the sense of (4.4.39), i.e.,

$$\|y\|_o^2 \leq C(A_h y,y) \quad \text{for } y \in D_o. \qquad (4.4.40)$$

Proof: In fact, the proof was given in the text immediately preceding Lemma 4.31.

It must be borne in mind that some sufficient conditions for A_h to be "of monotonic type" and for the technique of deriving estimate (4.4.38) are required to be fulfilled. For many boundary value problems of interest and the networks used, these assumptions are satisfied, and we are able to give explicitly lower bounds of $\lambda_o(A_h)$. On the other hand, the method of proving the positive definiteness of A_h via Friedrichs-Poincaré-type inequalities works in the general case and without additional restrictions on the network or on the coefficients k_{ij} (i,j=1,2). But the constants are not explicitly available in the general case. Nevertheless, they can be calculated approximately, cf. LANGER/JUNG [1], or estimated for some special cases, cf. Remark 4.29.

We conclude this section with a remark concerning the properties of the operators $\tilde{A}_h$, B_h and the matrices $\tilde{\mathcal{A}}_h$, $\mathcal{B}_h$, which are related with A_h.

Remark 4.32. The operator $\tilde{A}_h$ associated with A_h by (4.1.4) acts on grid functions $y \in \tilde{D}$, $\tilde{D}$ from (2.2.8), and is symmetric and positive definite on $\tilde{D}$, since A_h has these properties on D_o. The operator B_h defined on $\tilde{D}$ differs from $\tilde{A}_h$ in the factors $H(x)$, $h(x)$ only, cf. (4.1.5). The matrix $\mathcal{B}_h$ from (4.1.6) is symmetric and positive definite in the sense of

$$\mathcal{B}_h = \mathcal{B}_h^T, \ (\mathcal{B}_h y,y)_{R^{\tilde{n}}} \geq \bar{C}h^2(y,y)_{R^{\tilde{n}}} \quad \text{for } y \in R^{\tilde{n}},\ h \leq h_o, \tag{4.4.41}$$

where $(.,.)_{R^{\tilde{n}}}$ denotes the usual scalar product in $R^{\tilde{n}}$, $\tilde{n} := |\omega + \gamma_{23}|$ from (2.2.8), and $\bar{C}$ does not depend on h ($h \leq h_o$) and $y \in R^{\tilde{n}}$. Due to (4.1.6), the properties of $\tilde{\mathcal{A}}_h$ can be derived from $\mathcal{B}_h$, cf. Theorem 4.3.

4.5. A priori estimates for A_h using the W_2^1- and C-norm

4.5.1. A priori estimates using the W_2^1-norm.

A priori estimates using the L_2-, W_2^1- and W_2^{-1}-norms are the basis of stability theorems for A_h with respect to various pairs of such "discrete spaces", and, together with the corresponding approximation theorems, they lead to estimates of the convergence $y \to u$ as $h \to 0$ (see Section 5.5.), where u is the solution of the BVP $Au = F$, with $u \in C^k(\bar{\Omega})$ or $u \in W_2^k(\Omega)$, $k \geq 2$.
The fundamental inequality for such a priori estimates is

$$\|y\|_1^2 \leq C(A_h y,y) \quad \text{for } y \in D_o \text{ and } h \leq h_o,\ 0 < C = O(1), \tag{4.5.1}$$

which was proved in Theorem 4.28. by means of Friedrichs-Poincaré-type inequalities (4.4.15), (4.4.20). But the positive definiteness of A_h suffices to derive the inequality (4.5.1), which will be clear by

Remark 4.33. Let A_h be a difference operator from Theorem 3.1. (PB-operators A_h^j, $j=0,1$) or Theorem 3.4. (MD-operators A_h^j, $j=0,1$) on irregular grids $\bar{\omega}$. Assume that the positive definiteness of A_h was proved with a constant C_o, i.e.,

$$\|y\|_o^2 \leq C_o(A_h y,y) \quad \text{for } y \in D_o \text{ and } h \leq h_o,\ 0 < C_o = O(1). \tag{4.5.2}$$

Then, inequality (4.5.1) holds, and C can be specified by $C := (1+k_oC_o)/k_o$. ∎

Proof: By means of Green's formula, cf. Section 4.4.1., estimate (4.4.26) and $(dy,y) \geq 0$, we get $(A_h y,y) \geq k_o|y|_1^2$ for $y \in D_o$. Combining this estimate with (4.5.2), and taking into account $\|y\|_1^2 := \|y\|_o^2 + |y|_1^2$, we obtain (4.5.1), with the constant C given in Remark 4.33.

Due to the inequalities $|(y,v)| \leq \|y\|_o \|v\|_o$, $|(y,v)| \leq \|y\|_{-1} \|v\|_1$ for $y,v \in D_o$ and (4.5.2), (4.5.1), we get the estimates

$$\|y\|_0 \le C_0\|A_h y\|_0 \quad \text{and} \quad \|y\|_1 \le C\|A_h y\|_{-1} \quad \text{for } y \in D_0 \text{ and } h \le h_0, \tag{4.5.3}$$

with positive constants C_0 and C which are independent of y and h. The inequalities in (4.5.3) express the L_2-L_2- and W_2^{-1}-W_2^1-stability of A_h. Since the relations $\|v\|_{-1} \le \|v\|_0 \le \|v\|_1$ hold for $v \in D_0$, we might derive the L_2-L_2-stability from the W_2^{-1}-W_2^1-stability alone. The norm $\|.\|_{-1}$ from (4.3.16) is not so convenient for explicit calculations, e.g. for $\|A_h y\|_{-1}$. It is often sufficient to take such majorants of the norm $\|\cdot\|_{-1}$, which can be easily calculated and which are in a certain sense norms between $\|.\|_{-1}$ and $\|.\|_0$. Since a priori estimates with such "majorant norms" of $\|A_h y\|_{-1}$ play an essential role in the field of error estimation and convergence proofs for the FDM on irregular networks, we shall be concerned with this subsequently.
Let us consider the difference operator A_h of second order, which was constructed in Chapter 3. (Theorem 3.1. and 3.4.). Then A_h may be represented as the sum of the principal part A_h^H ("Hauptteil") and the secondary part A_h^N ("Nebenteil"), where A_h^H consists of the discrete analogues of the operator L_0 from (2.1.3) on ω, the first order operator $(K\nabla.,n)$ from (2.1.3) on γ_{23} and the identity operator I on γ_1. The j-dependence, $j \in \{0,1\}$, is involved in A_h^N, i.e. $A_h^j = A_h^H + A_h^{N_j}$. More exactly, we suppose

$$A_h y = A_h^H y + A_h^N y \text{ for } y \in D_0, \text{ with } A_h^H y := A_h y \text{ by setting tempo-} \tag{4.5.4}$$
$$\text{rarily } c = 0 \text{ on } \bar{\Omega} \text{ and } \alpha = 0 \text{ on } \Gamma_{23},\ A_h^N y := A_h y - A_h^H y.$$

The operators A_h^H are given explicitly by

<u>Definition 4.34.</u> Let A_h^H be defined as follows,
case (PB): $k := k_{11} = k_{22}$, $k_{12} = 0$, cf. (3.2.18),

$$(A_h^H y)(x) := \begin{cases} y(x) & x \in \gamma_1, \\ -\frac{1}{h(x)} \sum_{\xi \in S'(x)} s^0(x'_\xi) k_h(x'_\xi) y_{\bar{h}}(x'_\xi) & \text{for } x \in \gamma_{23}, \\ -\frac{1}{H(x)} \sum_{\xi \in S'(x)} s(x'_\xi)\, k_h(x'_\xi) y_{\bar{h}}(x'_\xi) & x \in \omega, \end{cases} \tag{4.5.5}$$

case (MD): cf. (3.3.19),

$$(A_h^H y)(x) := \begin{cases} y(x) & x \in \gamma_1, \\ -\frac{1}{h(x)} \sum_{x'' \in S''(x)} (K_h \nabla_{\bar{h}} y, \frac{\bar{h}^{\bullet}-\bar{h}^{o}}{2})(x'') & \text{for } x \in \gamma_{23}, \\ -\frac{1}{H(x)} \sum_{x'' \in S''(x)} (K_h \nabla_{\bar{h}} y, \frac{\bar{h}^{\bullet}-\bar{h}^{o}}{2})(x'') & x \in \omega. \equiv \end{cases} \tag{4.5.6}$$

We now consider the scalar product $(A_h^H y, v)$ for $y, v \in D_o$ and apply Green's formula from Section 4.4.1., cf. (4.4.2) and (4.4.5). Thus, we get the identities

$$\begin{aligned} &\text{case (PB):} \quad (A_h^H y, v)_p = (k_h y_{\bar{x}}, v_{\bar{x}})', \\ &\text{case (MD):} \quad (A_h^H y, v)_m = \sum_{i,j=1}^{2} (k_{ij}^{(h)} \frac{\partial y}{\partial x_i}, \frac{\partial v}{\partial x_j})'', \end{aligned} \qquad \text{for } y, v \in D_o, \qquad (4.5.7)$$

with $K_h = (k_{ij}^{(h)})_{2x2}$, scalar products $(.,.)_r$, $r \in \{p,m\}$, $(.,.)'$, $(.,.)''$ from (4.3.1), (4.3.3), (4.3.4), respectively. Using (4.4.21) and (4.5.7) for $v := y$, the estimates

$$\begin{aligned} &\text{case (PB):} \quad k_o \|y_{\bar{x}}\|_o'^2 \leq (A_h^H y, y)_p \leq \| k_h y_{\bar{x}} \|_o' \; \| y_{\bar{x}} \|_o', \qquad (4.5.8) \\ &\text{case (MD):} \quad k_o \| |\nabla_{\bar{x}} y| \|_o''^2 \leq (A_h^H y, y)_m \leq \| |K_h \nabla_{\bar{x}} y| \|_o'' \; \| |\nabla_{\bar{x}} y| \|_o'', \end{aligned} \qquad \text{for } y \in D_o,$$

can be derived; for $\|\cdot\|_o'$, $\|\cdot\|_o''$ see (4.3.11), (4.3.12). By means of (4.5.1), (4.5.4), (4.5.8), Cauchy-Bunjakowski's inequality and by further estimation, the inequalities

$$\begin{aligned} \|y\|_{1,p}^2 &\leq C\{(A_h^H y, y)_p + (A_h^N y, y)_p\} \leq C\{\|k_h y_{\bar{x}}\|_o' \|y_{\bar{x}}\|_o' + \|A_h^N y\|_{o,p} \|y\|_{o,p}\} \\ &\leq C\{\|k_h y_{\bar{x}}\|_o'^2 + \|A_h^N y\|_{o,p}^2\}^{1/2} \{\|y_{\bar{x}}\|_o'^2 + \|y\|_{o,p}^2\}^{1/2} \quad \text{for } y \in D_o \end{aligned} \qquad (4.5.9)$$

will be obtained and, therefore,

$$\text{case (PB):} \quad \|y\|_{1,p} \leq C\{\|k_h y_{\bar{x}}\|_o'^2 + \|A_h^N y\|_{o,p}^2\}^{1/2} =: C\|A_h y\|_{-,p} \quad \text{for } y \in D_o. \qquad (4.5.10)$$

The norm $\|A_h y\|_-$ from (4.5.10) is an upper bound of $\|A_h y\|_{-1}$ such that estimate (4.5.10) can be interpreted as a cruder form of the second inequality in (4.5.3). This will be clear from

$$\|A_h y\|_{-1} \leq \|A_h^H y\|_{-1} + \|A_h^N y\|_o \leq \|k_h y_{\bar{x}}\|_o' + \|A_h^N y\|_o =: \|A_h y\|_- \qquad (4.5.11)$$

for $y \in D_o$, which is partially based on

$$\|A_h^H y\|_{-1} := \sup_{\substack{v \in D_o \\ v \neq 0}} \frac{|(A_h^H y, v)|}{\|v\|_1} = \sup_{\substack{v \in D_o \\ v \neq 0}} \frac{|(k_h y_{\bar{x}}, v_{\bar{x}})|}{\|v\|_1} \leq \|k_h y_{\bar{x}}\|_o' \quad \text{for } y \in D_o. \qquad (4.5.12)$$

Relations quite similar to (4.5.9), (4.5.10), (4.5.11) and (4.5.12) can be easily proved in the

$$\text{case (MD):} \quad \|y\|_{1,m} \leq C\{\| |K_h \nabla_{\bar{x}} y| \|_o''^2 + \|A_h^N y\|_{o,m}^2\}^{1/2} =: C\|A_h y\|_{-,m} \quad \text{for } y \in D_o, \qquad (4.5.13)$$

with $\|A_h^H y\|_{-1} \le \|\|K_h \nabla_h y\|\|_o''$. Summarizing these foregoing considerations, the following theorem can be formulated.

Theorem 4.35. - Stability of A_h in $W_2^1(\bar{\omega})$. For the difference operators A_h constructed in Theorem 3.1. (PB) and 3.4. (MD), the following a priori estimates hold:

$$\|y\|_1 \le C\|A_h y\|_o, \ \|y\|_1 \le C\|A_h y\|_{-1} \le C\|A_h y\|_- \quad \text{for } y \in D_o \qquad (4.5.14)$$

and $h \le h_o$, with sufficiently small h_o. Since $\|y\|_o \le \|y\|_1$ is valid, the inequalities in (4.5.14) are also true for $\|y\|_o$ instead of $\|y\|_1$.

Here, $\|\cdot\|_o$, $\|\cdot\|_1$ and $\|\cdot\|_{-1}$ denote discrete L_2-, W_2^1- and W_2^{-1}-norms from (4.3.5), (4.3.16), and $\|\cdot\|_-$ is taken from (4.5.10), (4.5.13). The constant C is independent of y and h, $h \le h_o$. ≡

Proof: In Theorem 4.28., cf. (4.4.29), inequality (4.5.1) was shown, which was the starting point for the derivation of (4.5.3), (4.5.10) and (4.5.13) proved immediately before Theorem 4.35.

4.5.2. "Weak imbedding" of $W_2^1(\bar{\omega})$ in $C(\bar{\omega})$. Estimates of the discrete W_2^1-norm $\|y\|_1$ are often available, e.g. $\|y\|_1 \le C\|A_h y\|_{-1}$, and we want to get an a priori estimate for the C-norm $\|y\|_{C(\bar{\omega})}$ by means of an inequality $\|y\|_{C(\bar{\omega})} \le E_h\|y\|_1$. For $\|\cdot\|_C$, $\|\cdot\|_1$ and $\|\cdot\|_{-1}$, see (4.3.9), (4.3.16). Indeed, for one-dimensional grids $\omega, \bar{\omega} \subset \bar{\Omega} \subset R^1$, the constant E_h is bounded for $h \le h_o$. But for plane grids $\bar{\omega} \subset \bar{\Omega} \subset R^2$, we have $E_h \to \infty$ as $h \to 0$, since the imbedding operator $W_2^1(\Omega) \hookrightarrow C(\bar{\Omega})$ for two-dimensional domains Ω is not bounded. For instance, in OGANESYAN/RUKHOVETS [2, pp. 74-77] E_h was estimated as a function of h, and the inequality

$$\|\hat{u}\|_{C(\Omega_{ex}^h)} \le E|\ln h|^{1/2}\|\hat{u}\|_{W_2^1(\Omega_{ex}^h)} \qquad \text{for } h \le h_o,\ 0 < E = O(1), \quad (4.5.15)$$

was proved for continuous functions $\hat{u}$ linear on each triangle Δ, where the constant E does not depend on u and h, $h \le h_o$. Here, Ω_{ex}^h is a polygonal domain which contains Ω and does not coincide, in general, with Ω_h used here. It will be turned out that the method of proving such inequalities can be transferred to the polygonal domains Ω_h and to the discrete norms $\|y\|_{C(\bar{\omega})}$ and $\|y\|_1$, which are here employed.

Theorem 4.36. - Weak imbedding of $W_2^1(\bar{\omega})$ in $C(\bar{\omega})$. Let $\Omega \subset R^2$ satisfy Assumption 2.1. - $V(\bar{\Omega})$, i.e., $\Gamma := \partial\Omega \in C^{0,1} \cap PC^2$, and Γ has at most a finite number of corners with angles $\beta_k \in (0, 2\pi)$. Assume that the grid $\bar{\omega}$ which is based on a network of triangles fulfils the grid regularity condition VO(MD, $\bar{\omega}$) (Condition 2.24.). Then, the following inequalities hold:

$$\|y\|_{C(\bar{\omega})} \leq C|\ln h|^{1/2}\|y\|_{1,\hat{m}} \leq C|\ln h|^{1/2}\|y\|_{1,m} \quad \text{for } y \in D_o \qquad (4.5.16)$$

and $h \leq h_o$, where C does not depend on y and h. Here, $\|\cdot\|_{C(\bar{\omega})}$ and $\|\cdot\|_{1,\hat{m}}$, $\|\cdot\|_{1,m}$ are taken from (4.3.9) and (4.3.14), respectively, and h_o is assumed to be sufficiently small. ■

Proof: Obviously, we may define some spherical sector $\mathcal{K}_o$ (cone) with the angle $\theta_o > 0$ and the radius $r_o > 0$ such that any $x \in \bar{\Omega}$ is the vertex of a finite cone $\mathcal{K}_x$ congruent to $\mathcal{K}_o$, with $\mathcal{K}_x \setminus \{x\} \subset \Omega$. The same can be asserted for the family of polygonal domains $\{\Omega_h\}$ instead of Ω (possibly after a small modification of θ_o and r_o), if $h \leq h_o$ holds and h_o is chosen sufficiently small. This is due to the fact that Ω_h approximates Ω and all corner points P_k of Γ are contained in Γ_h (and γ), and, the angle β_k of Γ at $P_k \in \Gamma$ will be approximated by the angle β_{kh} of Γ_h at the same corner point P_k. Now we consider an arbitrary grid function $y \in D_o$ and the piecewise linear function $\hat{y} \in PP_1(\bar{\Omega}_h)$ assigned to y by (4.3.20). The grid function $|y|$ takes on the value $\|\hat{y}\|_{C(\bar{\Omega}_h)} := \max_{\bar{x} \in \bar{\Omega}_h} |\hat{y}(\bar{x})|$ on the grid $\bar{\omega}$, i.e. on $\omega + \gamma_{23}$, since $y|_{\gamma_1} = 0$ holds. Let $x_o \in \omega + \gamma_{23}$ denote the point where $\|\hat{y}\|_{C(\bar{\Omega}_h)} = \|y\|_{C(\bar{\omega})} = |y(x_o)|$ is satisfied. Then, there is a fixed cone $\mathcal{K}_{x_o}$ congruent to $\mathcal{K}_o$ and equipped with the following properties: x_o is the vertex of $\mathcal{K}_{x_o}$ and $\mathcal{K}_{x_o} \setminus \{x_o\} \subset \Omega_h$ for $h \leq h_o$. Now, the ideas and the proof from OGANESYAN/RUKHOVETS [2, pp. 75, 76] or OGANESYAN/RIVKIND/RUKHOVETS [1, part 1, pp. 110-113] can be transferred word by word to our polygonal domain Ω_h. Thus, we get the basic inequality

$$\|\hat{y}\|_{C(\bar{\Omega}_h)} \leq C|\ln h|^{1/2}\|\hat{y}\|_{W_2^1(\Omega_h)} \quad \text{for } h \leq h_o,\ 0 < C = O(1). \qquad (4.5.17)$$

Due to the norm relations (4.3.41), (4.3.51) and (4.5.17), we immediately get (4.5.16).

In the following Corollary 4.37., we shall add inequalities of the type (4.5.16) for networks of triangles and (or) rectangles and for secondary networks of the type (PB).

Corollary 4.37. Let VO(PB, $\bar{\omega}$) from (2.3.8) be satisfied for networks of triangles and (or) rectangles. Then, the inequalities

$$\|y\|_{C(\bar{\omega})} \leq C_1|\ln h|^{1/2}\|y\|_{1,\hat{p}} \leq C_1|\ln h|^{1/2}\|y\|_{1,p} \quad \text{for } y \in D_o \qquad (4.5.18)$$

and $h \leq h_o$ hold, where C_1 does not depend on y and h. Here, $\|\cdot\|_{C(\bar{\omega})}$ and $\|\cdot\|_{1,\hat{p}}$, $\|\cdot\|_{1,p}$ are taken from (4.3.9) and (4.3.13), respectively. ■

Proof: In each rectangle, we fit in a diagonal such that $\bar{\Omega}_h$ is triangulated by triangles only ("complete triangulation"). Now, the continuous function $\hat{y} \in PP_1(\bar{\Omega}_h)$ can be associated with the grid function $y \in D_0$, cf. (4.3.20). Since the two triangles in each rectangle have right angles opposite to the diagonal inserted in the rectangle, the secondary network (PB) does not change after the "complete triangulation". Therefore, the norms $\|\cdot\|_{i,p}$, $\|\cdot\|_{i,\hat{p}}$ (i=0,1) on the original network and on the new network of triangles coincide. Due to the equivalence of the norms $\|\cdot\|_{1,\hat{m}}$ and $\|\cdot\|_{1,\hat{p}}$ as well as $\|y\|_{1,\beta} \leq \|y\|_{1,p}$, cf. (4.3.50), (4.3.52), we get inequalities (4.5.18) as a consequence of (4.5.16).

Especially, domains Ω which are rectangles or are composed of rectangles equipped with regular or irregular networks of rectangles □ are included in Corollary 4.37., too.

4.5.3. A priori estimates using the C-norm. A priori estimates using the C-norm are needed to verify the C-C-stability of A_h and may be utilized for proving the convergence $y \to u$, cf. Section 5.1.4., where u is a classical solution of the BVP Au = F, with $u \in C^k(\bar{\Omega})$, $k \geq 2$. It is well known that a priori estimates with the C-norm can be derived for difference operators A_h satisfying Condition 4.1. - $M(A_h, \bar{\omega})$, see e.g. BRAMBLE/HUBBARD [1], CIARLET [1], MICHLIN/SMOLIZKI [1], SAMARSKIĬ/ANDREEV [1] and HEINRICH [2,3]; and for operators A_h which arise by FEM-approximations and fulfil $M(A_h, \bar{\omega})$, see e.g. CIARLET/RAVIART [1], IKEDA [1]. For mixed boundary conditions and networks of triangles occuring at the same time, such C-estimates do not seem to have been reported in the literature on the FDM up to now, but they will be discussed subsequently. Due to $M(A_h, \bar{\omega})$, some restrictions on the angles of the triangles Δ and the coefficients k_{ij} (i,j=1,2) will be made. The technique of deriving the C-estimates is based on certain modifications of Gerschgorin-Batschelet's method, cf. HEINRICH [3].

In the following, locally irregular networks will be considered preferentially. We classify the points $x \in \bar{\omega}$ of the locally irregular grids $\bar{\omega}$ by defining six subsets ω_i (i=1,2,...,6) of $\bar{\omega}$. Therefore, a corresponding subdivision of the difference operator A_h into six components A_{hi} (i=1,2,...,6) is implied, with $A_h = (A_{h1}, A_{h2}, \ldots, A_{h6})^T$. The components L_h, l_{h1}, l_{h2}, l_{h3} of A_h, which are in accordance with the partition $\bar{\omega} = \omega + \gamma_1 + \gamma_2 + \gamma_3$, will be arranged, for the sake of convenient notation, in the form $A_h = (l_h, L_h)^T$, with $l_h = (l_{h1}, l_{h2}, l_{h3})^T$. The j-dependence of A_h^j occurs only in the third component $l_{h3} = l_{h3}^j$ of l_h. The partition of A_h^j is defined by the following scheme (4.5.19), where the grid components ω_i are noted, and the corresponding local approximation errors $\tilde{\psi}(x)$ for $x \in \omega_i$, $\tilde{\psi}(x) := (Au - A_h u)(x)$, are given for functions $u \in C^2(\bar{\Omega})$, $k \in C^1(\bar{\Omega})$, $c \in C(\bar{\Omega})$, $\alpha \in PC^1(\Gamma_3)$, cf. HEINRICH[6].

Sometimes, $\tilde{\psi}(x)$ is called the "local truncation error", cf. MITCHELL/GRIFFITHS [1].

Assumption 4.38. - Partition of $A_h, \bar{\omega}$. Let $\bar{\omega}$ be locally irregular. Then the partition of $A_h, \bar{\omega}$ will be given by the following scheme.

A_h-components		$\bar{\omega}$- components	$\tilde{\psi}:=Au-A_hu$ for $u\in C^2(\bar{\Omega})$
$l_h^j y:=$	$A_{h1}y:=l_{h1}y$	$x\in\omega_1:=\gamma_1$	0
	$A_{h2}y:=l_{h2}y$	$x\in\omega_2:=\gamma_2$	$O(h)$
	$A_{h3}^j y:=l_{h3}^j y$	$x\in\omega_3:=\gamma_3$	$O(h)$
$L_h y:=$	$A_{h4}y$	$x\in\omega_4:=\overset{\circ}{\omega}$	$o(1)$
	$A_{h5}y$	$x\in\omega_5:=\overset{*}{\omega}_1$	$O(1)$
	$A_{h6}y$	$x\in\omega_6:=\overset{*}{\omega}_{23}$	$O(1)$

(4.5.19)

The asymptotic behaviour of $\tilde{\psi}(x)$ given in the third column of (4.5.19) is no assumption, but a result obtained by Taylor's expansion. The sets $\overset{*}{\omega}_1$ and $\overset{*}{\omega}_{23}$ are defined by $\overset{*}{\omega}_1:=\{x\in\overset{*}{\omega}:S'(x)\cap\gamma_1\neq\emptyset\}$, $\overset{*}{\omega}_{23}:=\overset{*}{\omega}\setminus\overset{*}{\omega}_1$. The set $\overset{*}{\omega}_{23}$ is divided into $\overset{*}{\omega}_2$, $\overset{*}{\omega}_3$ by $\overset{*}{\omega}_{23}=\overset{*}{\omega}_2+\overset{*}{\omega}_3$, where $\overset{*}{\omega}_i$ $(i=2,3)$ is defined such that $x\in\overset{*}{\omega}_i$ has at least one neighbouring point $\xi\in S'(x)$ on γ_i ($i=2,3$; $\overset{*}{\omega}_2\cap\overset{*}{\omega}_3\neq\emptyset$ does not affect the results). The remaining subsets γ_i $(i=1,2,3)$, $\overset{*}{\omega}$ and $\overset{\circ}{\omega}$ are already given by (2.2.6). ■

Since the results obtained for the difference operators A_h^p(PB) and A_h^m(MD) differ slightly, they will be formulated separately.

Theorem 4.39. - Stability of A_h(PB) in $C(\bar{\omega})$, $\bar{\omega}$ locally irregular

(i) Let $A_h:=A_h^j$ $(j=0,1)$ from Theorem 3.1. be given, with the assumptions made there and, especially, with (2.1.8).

(ii) Let the grids $\bar{\omega}$ be locally irregular, with $\bar{\omega}$ from Definition 2.10, 2.11. or 2.12. Furthermore, let Condition 2.16. - VO(PB,$\bar{\omega}$) and the following "coupling condition" for $x\in\overset{*}{\omega}_i$ ($i=1,2,3$, cf. (4.5.19)) be satisfied: The inequalities

$$s^j(x_\xi^!)\geq\varepsilon h>0,\ \varepsilon \text{ fixed, or } \theta_-+\theta_+\leq\pi-\theta_0,\quad \theta_0>0, \qquad (4.5.20)$$

hold at least for one point $\xi\in S'(x)\cap\gamma_i$ and $h\leq h_0$, with $s^j(x_\xi^!)$ from (3.2.18c), $j\in\{0,1\}$ and angles $\theta_\pm:=\theta_\pm(x,\xi)$ opposite to $h(x,\xi)$.

(iii) Assume that there is a so-called "majorant function" v with the following properties:

$$v\in C^2(\bar{\Omega}),\ Lv\geq\tfrac{3}{2} \text{ in } \Omega,\ lv\geq\tfrac{3}{2} \text{ on } \Gamma_2,\ L \text{ and } l \text{ from (3.2.1)}. \quad (4.5.21)$$

Then, the following a priori estimate holds:

$$\|y\|_{C(\bar\omega)} \leq \sum_{i=1}^{6} N_{hi} \|A_{hi}y\|_{C(\omega_i)} \leq C\|A_h y\|_{C(\bar\omega)} \text{ for } y \in D, \qquad (4.5.22)$$

D from (2.2.8), and $h \leq h_o$, h_o sufficiently small. The positive constants N_{hi}, C show the asymptotic behaviour

$$N_{hi} = O(1) \text{ for } i = 1,2,3,4,\ N_{h5} = O(h^2),\ N_{h6} = O(h),\ C = O(1), \qquad (4.5.23)$$

and this assertion is valid for $j = 0$ as well as for $j = 1$. ■

Proof: The essential ideas of the proof, i.e., the Gerschgorin-Batschelet-technique and its extension as well as more technical details of determining the constants f_5, f_6 and lower bounds of $A_{hi}w_j$ used in the following, can be found in HEINRICH [3,6]. In HEINRICH [3, p. 78, Theorem 4] , cf. also MICHLIN/SMOLIZKI [1, p. 73], the following theorem is formulated: Let $A_h = (A_{h1},A_{h2},\dots,A_{hn})^T$ be a difference operator equipped with the property $M(A_h,\bar\omega)$ from (4.2.1) on the grid $\bar\omega = \bigcup_{i=1}^{n} \omega_i$. Assume that there are n functions w_i $(i=1,2,\dots,n)$ defined on $\bar\omega$ and satisfying the inequalities

$$A_{hi}w_j \geq \delta_{ij} := \begin{cases} 1 & \text{for } i = j, \\ 0 & \text{for } i \neq j, \end{cases} \quad i,j = 1,2,\dots,n. \qquad (4.5.24)$$

Then, the estimate

$$\|y\|_{C(\bar\omega)} \leq \sum_{i=1}^{n} N_{hi} \|A_{hi}y\|_{C(\omega_i)},\ N_{hi} := \|w_i\|_{C(\bar\omega)}, \qquad (4.5.25)$$

holds for any $y \in D$, D from (2.2.8), and $h \leq h_o$, if h_o is sufficiently small.

Since $M(A_h,\bar\omega)$ was already proved in Theorem 4.3., we may restrict ourselves to the study of the majorant functions w_i. By means of v from (4.5.21), we may find a function w with the property

$$w \in C^2(\bar\Omega),\ Lw \geq \tfrac{3}{2} \text{ in } \Omega,\ lw \geq \tfrac{3}{2} \text{ on } \Gamma_2 \cup \Gamma_3,\ w \geq 1 \text{ on } \bar\Omega, \qquad (4.5.26)$$

cf. HEINRICH [3, p. 78], and the number 3/2 will be used here without loss of generality. Taking w from (4.5.26), we define functions w_i $(i=1,2,\dots,6)$ on $\bar\omega$ as follows:

$$w_1 := 1,\ w_i := w \text{ for } i = 2,3,4,\ w_5 := f_5 \begin{cases} 0 & \text{on } \gamma_1, \\ h^2 & \text{on } \omega + \gamma_{23}, \end{cases} \qquad (4.5.27)$$

$$w_6 := f_6 wh + f_5 \begin{cases} 0 & \text{on } \gamma, \\ h^2 & \text{on } \omega, \end{cases}$$

with appropriate, positive constants $f_i = O(1)$ $(i=5,6)$.

To prove (4.5.25), i.e. (4.5.22), we must verify the inequalities in (4.5.24) for w_j ($j=1,2,\ldots,6$) from (4.5.27) and A_{hi} from (4.5.19). Estimating $A_{hi}w_j$ we can show, for sufficiently small h_o, that the inequalities

$$A_{hi}w_j \geq \delta_{ij} \text{ for } \begin{cases} i = 1,2,\ldots,6 & \text{and} \quad j = 1,5,6, \\ i = 1,2,3,4 & \quad\quad\;\; j = 2,3,4, \end{cases} \tag{4.5.28}$$

$$A_{hi}w_j \geq - C_w \text{ for } i = 5,6 \text{ and } j = 2,3,4,$$

hold, where C_w is a positive constant independent of h, $h \leq h_o$. The inequalities $A_{hi}w_j \geq \delta_{ij}$ in (4.5.28) are verified in the same way as in Theorem 6 stated in HEINRICH [3], but here the local approximation error $\tilde{\psi}(x)$ from (4.5.19) is taken into account. The inequalities $A_{hi}w_j \geq -C_w$ in (4.5.28) result from $\tilde{\psi}(x) = O(1)$ for $x \in \omega_5$, $x \in \omega_6$ according to (4.5.19). The coupling condition (4.5.20) is specifically used for the estimation of f_5, f_6 as positive constants of the type f_5, $f_6 = O(1)$. First, the smallest value of the factor f_5 will be taken such that $A_{h5}w_5 \geq 1$ is fulfilled. Then, the same can be carried out for the factor f_6 with respect to $A_{h2}w_6 \geq 0$, i.e., $A_{h2}w_6 \geq 0$ determines f_6. Unfortunately, due to $A_{hi}w_j \geq -C_w$ for some i and j, the inequalities (4.5.24) for getting (4.5.25) are not satisfied completely. Therefore, a new difference operator $R_h = (R_{h1},R_{h2},\ldots,R_{h6})^T$ will be defined by means of A_h and the constant C_w, viz.

$$R_{hi} := \begin{cases} A_{hi} & \quad i = 1,2,3,4, \\ & \text{for} \\ A_{hi} + C_w I_{hi} & \quad i = 5,6,\ I_{hi}\text{: identity operator.} \end{cases} \tag{4.5.29}$$

This operator R_h clearly fulfils the inequalities $R_{hi}w_i \geq \delta_{ij}$ ($i,j=1,2,\ldots,6$) and the Condition 4.1. - $M(R_h, \bar{\omega})$, since R_h is a "partial diagonal modification" of A_h with a positive constant C_w. Clearly, inequality (4.5.25) can be noted for R_h instead of A_h, i.e.

$$\|y\|_{C(\bar{\omega})} \leq \sum_{i=1}^{6} M_{hi} \|R_{hi}y\|_{C(\omega_i)},\ M_{hi} := \|w_i\|_{C(\bar{\omega})},\ \text{for } y \in D \tag{4.5.30}$$

and $h \leq h_o$, w_i from (4.5.27). Estimate (4.5.30) obviously yields

$$\|y\|_{C(\bar{\omega})} \leq \sum_{i=1}^{6} M_{hi}\|A_{hi}y\|_{C(\omega_i)} + C_w \sum_{i=5,6} M_{hi}\|y\|_{C(\omega_i)},$$

and this directly leads to inequalities (4.5.22), where the constants M_{hi}, N_{hi} and their asymptotic behaviour are specified by

$$N_{hi} := M_{hi}/(1 - \sum_{j=5,6} M_{hj}),\ M_{h5} = O(h^2),\ M_{h6} = O(h), \tag{4.5.31}$$

$M_{hi} = O(1)$ for $i = 1,2,3,4$, $0 < N_{hi} = O(M_{hi})$ for $i = 1,2,\ldots,6$, if $h \leq h_0$ holds, with sufficiently small h_0. This completes the proof.

The constants N_{hi} depend on the majorant functions, on the parameters of the BVP $Au = F$, inclusively $\bar{\Omega} = \Omega \cup \Gamma$, and on the geometry of the given family of networks. They can be calculated in many cases, since the function v from (4.5.21) and, therefore, w from (4.5.26) can be given explicitly for some essential classes of BVPs $Au = F$. The asymptotic behaviour of the constants N_{h5}, N_{h6}, viz. $N_{h5} = O(h^2)$ and $N_{h6} = O(h)$, is crucial for error estimates $\|y-u\|_{C(\bar{\omega})} \leq M\,\varphi(h)$ using (4.5.22) on locally irregular networks ($u \in C^l(\bar{\Omega})$, $l \geq 2$).

To prove the positive definiteness of A_h via (4.4.39), (4.4.40), it is sufficient to have a lower bound of the smallest eigenvalue λ_0 of the operator M_h from (4.4.25), since (4.4.26) holds. The coefficient k in the principal part Λ_h from M_h is constant, i.e. $k = 1$. For such operators A_h with $k = \text{const}$, the following Theorem 4.40. can be formulated for networks which are irregular even in the whole domain $\bar{\Omega}$ and, additionally, the constant C_0 can be determined explicitly.

<u>Theorem 4.40. - Stability of A_h(PB) in $C(\bar{\omega})$, $\bar{\omega}$ irregular</u>

(i) Let the operator $A = (L,l)^T$ from (3.2.1) be given, with $k = \text{const} \geq k_0 > 0$ and assumption (2.1.8), and for the following cases of Γ:

a) $\Gamma = \Gamma_1 \cup \Gamma_3$, i.e. $\Gamma_2 = \emptyset$, (4.5.32a)

or

b) $\Gamma = \Gamma_1 \cup \Gamma_2 \cup \Gamma_3$, where Γ_2 is "monotonic" (see the text preceeding the proof below). (4.5.32b)

(ii) Let $A_h := A_h^j$ ($j=0,1$) be the difference operators from Theorem 3.1. constructed via method (PB) on irregular networks and under the Condition 2.16. - VO(PB, $\bar{\omega}$).

Then, the a priori estimate

$$\|y\|_{C(\bar{\omega})} \leq C_0 \|A_h y\|_{C(\bar{\omega})}, \text{ for } y \in D \text{ and } h \leq h_0,\ 0 < C_0 = O(1), \qquad (4.5.33)$$

holds, where C_0 can be specified explicitly. ≡

The notion "monotonic" used in (4.5.32b) here means that Γ_2 is a smooth arc of the boundary Γ and has a representation $x_2 = \varphi(x_1)$, where the derivative $\varphi'(x_1)$ does not change sign, say $-\infty < \varphi'(x_1) \leq 0$ for $(x_1,x_2) \in \Gamma_2$. Clearly, the cases $\Gamma = \Gamma_1$ and $\Gamma = \Gamma_3$ are included in (4.5.32).

<u>Proof:</u> We consider the function

$$w(x_1,x_2) := C_0 - C_1 \sum_{i=1}^{2} (x_i - a_i)^2, \text{ with } C_0,\ C_1 > 0 \text{ and } a_1, a_2 \in R^1. \qquad (4.5.34)$$

Then, there are fixed constants C_0, C_1, a_1, a_2 depending on Ω, Γ_2, k_0 and α_0 such that the inequalities

$$Lw \geq \frac{3}{2} \text{ in } \Omega,\ lw \geq \frac{3}{2} \text{ on } \Gamma_{23},\ w \geq 1 \text{ on } \Gamma_1 \qquad (4.5.35)$$

are satisfied, and w is strictly positive on $\bar{\Omega}$. This was proved in HEINRICH [3] (Lemma 1 and 3) for $\Gamma_2 = \emptyset$, and in HEINRICH [1, p. 13] for Γ_2 from (4.5.32b). In the case $\Gamma_2 = \emptyset$, we may freely choose a_1, a_2. Although the local approximation error $\tilde{\Psi}(x) := (Au-A_hu)(x)$, $x \in \omega$, has the asymptotic behaviour $O(1)$ for irregular networks even in the case $u \in C^\infty(\bar{\Omega})$, the error $\tilde{\Psi}(x)$ for the special polynomial function w from (4.5.34) obeys at least the relations

$$|\tilde{\Psi}(x)| = |(Aw-A_hw)(x)| = |O(h)| \leq \frac{1}{2} \quad \text{for } x \in \bar{\omega}, \text{ and } h \leq h_o, \qquad (4.5.36)$$

if h_o is chosen sufficiently small. Especially for $x \in \gamma_{23}$, $\tilde{\Psi}(x) = O(h)$ is obvious, since $w \in C^2(\bar{\Omega})$ holds, cf. (4.5.19). But in the case $x \in \omega$, we are able to confirm $Lw - L_hw = 0$ by direct calculation (for PB-operators only). For $x \in \gamma_1$, $lw - l_hw = 0$ is clear by definition. Relations (4.5.35) and (4.5.36) lead evidently to $(A_hw)(x) \geq 1$ for $x \in \bar{\omega}$. Moreover, Condition 4.1. - $M(A_h, \bar{\omega})$ is satisfied, which was proved in Theorem 4.3. Due to Theorem 4 in HEINRICH [3, p. 78], which was already explained in the proof of Theorem 4.39., we immediately get estimate (4.5.33) with the constant C_o from (4.5.34), $C_o \geq \|w\|_{C(\bar{\omega})}$. The constant C_o is given explicitly in HEINRICH [3] (pointer (12), $\Gamma_2 = \emptyset$) or in HEINRICH [1] (p. 12; the constants a_1, a_2 are fixed by Γ_2).

<u>Remark 4.41.</u> Inequality (4.5.33) can be refined and written in the form (4.5.22) with constants N_{hi} from (4.5.23), if e.g. Condition 2.17. - VO"(PB, $\bar{\omega}$) is satisfied, and $\bar{\omega}$ is, in general, not regular. ■

We now proceed with the study of difference operators A_h obtained by method (MD). Already Theorem 4.7. indicates that the monotonicity condition $M(A_h, \bar{\omega})$ can be satisfied only for certain special cases of the coefficients k_{il} $(i,l=1,2)$ and for angle restrictions of the type VO(PB, $\bar{\omega}$) or VO"(PB, $\bar{\omega}$).

<u>Corollary 4.42. - Stability of A_h(MD) in $C(\bar{\omega})$.</u> Let the difference operators $A_h := A_h^j$ $(j=0,1)$ from Theorem 3.4. be given, cf. (3.3.19), where A_h is a discrete analogue of $A = (L,l)^T$ from (3.2.1), i.e. $k := k_{11} = k_{22}$, $k_{12} = 0$. Furthermore, let VO(PB, $\bar{\omega}$) or VO"(PB, $\bar{\omega}$) be satisfied for $k = \text{const}$ or $k \neq \text{const}$, respectively, and finally, (iii) from Theorem 4.39. Then, the assertion of Theorem 4.39. holds literally for the difference operators A_h of the type (MD), i.e. (4.5.22), (4.5.23). Additionally, under the assumptions of Theorem 4.40., the assertion in Theorem 4.40. is also valid for MD-operators A_h, i.e., (4.5.33) holds. ■

The proof of Theorem 4.39. and 4.40. for MD-operators A_h is quite similar to that of PB-operators A_h. But a modification must be made in the proof of $(A_hw) \geq 1$ on $\bar{\omega}$, with w from (4.5.34). Due to $Lw - L_hw = O(1)$, the relation $L_hw \geq 1$ on ω must be proved by direct calculation, and (4.5.36) can be used for $x \in \gamma$ only.

5. ERROR ESTIMATES AND CONVERGENCE

5.1. Error splitting and approaches to the error estimation

5.1.1. Introductory remarks and error splitting for PB-schemes. In the classical FDM for the BVP $Au = F$ considered here, error estimates and convergence are usually studied for classical solutions $u \in C^k(\bar{\Omega})$, $k \geq 2$, and for regular or, at most, locally irregular networks of rectangles. Estimating the local truncation error $Au - A_h u + F_h - F$ at $x \in \bar{\omega}$ and utilizing a priori estimates with the C-norm, e.g. $\|y\|_{C(\bar{\omega})} \leq C\|A_h y\|_{C(\bar{\omega})}$, the uniform convergence $y \to u$ is proved in the following sense: $\lim_{h\to 0} \|y-u\|_{C(\bar{\omega})} = 0$. Here, y and u denote respectively the exact solutions of the FDS $A_h y = F_h$ and the BVP $Au = F$. This method was used e.g. by BRAMBLE/HUBBARD [1], COLLATZ [2], HEINRICH [2], MICHLIN/SMOLIZKI [1], SAMARSKIJ [1], and others. It has the disadvantage of being applicable only to classical solutions and regular or locally irregular networks. Nevertheless, there are also other approaches for proving convergence, which take into account discrete analogues of the L_2- and W_2^1-norms and classical solutions, see e.g. FRYASINOV [1,4], SAMARSKIĬ/ANDREEV [1], as well as networks of rectangles and generalized solutions, cf. e.g. LAZAROV [1,2], LAZAROV et al. [1], TEMAM [1], WEINELT [1,2]. Irregular networks of quadrangles are considered in GIRAULT [1,2], where the error estimates are proved for $u \in C^3(\bar{\Omega})$.

The main purpose of this chapter is to develop an approach to error estimates and convergence proofs with discrete analogues of the W_2^1- and C-norm, for generalized solutions $u \in W_2^k(\Omega)$, $k \geq 2$, and networks which are irregular in the whole domain $\bar{\Omega}$. The special features of this approach are the following: the errors are studied via the balance equations, which were used for the derivation of the FDSs $A_h y = F_h$, and, the errors are split in such a way that linear functionals on Sobolev spaces and the well-known Bramble-Hilbert Lemma, cf. BRAMBLE/HILBERT [1], can be utilized. Since the methods for investigating the elementary errors within the approaches (PB) and (MD) are sometimes quite different due to the choice of quadrature formulas and the boxes $\mathcal{K}(x)$, we are sometimes forced to divide the studies into PB- and MD-parts.
We now define certain errors and error functionals needed for the evaluation of the approximations $A_h \approx A$, $F_h \approx F$ and $y \approx u$, where the FDS $A_h y = F_h$ is a discrete analogue of the BVP $Au = F$. Let y and u denote respectively the unique solutions of $A_h y = F_h$ and $Au = F$, under the assumptions $V(\bar{\Omega})$, $V(A,F)$ and $V(u)$ from Section 2.1., but always with $u \in W_2^2(\Omega)$ at least, i.e. u from (2.1.9).
Then, the so-called <u>local approximation error z of the approximate solution y</u> with respect to the solution u (in short: error z) and the <u>local approximation error ψ of the FDS $A_h y = F_h$</u> with respect to the BVP $Au = F$ will be defined by

$$z(x):= y(x) - u(x), \quad \psi(x):= (A_h y)(x) - (A_h u)(x), \; x \in \bar{\omega}. \qquad (5.1.1)$$

Obviously, the relations

$$\psi(x) = (A_h z)(x) \text{ for } x \in \bar{\omega}, \; \psi(x) = z(x) = 0 \text{ for } x \in \gamma_1, \text{ i.e. } \psi, z \in D_o, \qquad (5.1.2)$$

hold, and therefore, ψ and z must be studied for $x \in \omega + \gamma_{23}$ only. By means of the functional $\tilde{\varphi}$ defined by

$$\tilde{\varphi}(x):= \int_{\mathcal{K}(x)} \varphi(\bar{x})d\bar{x} \cdot \begin{cases} \frac{1}{H(x)} & x \in \omega, \\ & \text{for} \\ \frac{1}{h(x)} & x \in \gamma_{23}, \end{cases} \qquad (5.1.3)$$

with $H(x)$ and $h(x)$ from (2.3.13), and due to $\widetilde{Lu} = \tilde{f}$, cf. (2.1.1a) and (5.1.3), as well as to $A_h y = F_h$, the approximation error ψ can be represented by

$$\psi(x) = (\widetilde{Lu})(x) - (A_h u)(x) + F_h(x) - \tilde{f}(x), \; x \in \omega + \gamma_{23}. \qquad (5.1.4)$$

Although y, ψ, z and certain constants associated with the FDSs $A_h^j y = F_h^j$, $j \in \{0,1\}$, depend on the index j, we shall omit j or write j only in such cases where the distinction between $j = 0$ or $j = 1$ is either essential or convenient. First of all, we consider error splittings with respect to the principal and secondary parts of the operator A and the right-hand side F. In ψ_H ("Hauptteil"), the error of the principal parts L_o and $(K\nabla\, .,n)$ is incorporated and, in ψ_N ("Nebenteil"), the error of the remaining secondary parts of L, l and the error of the right-hand side F are gathered. We now take into account FDSs $A_h^j y = F_h^j$ ($j=0,1$) from Theorem 3.1.

<u>Error splitting for PB-schemes $A_h y = F_h$:</u>

<u>case $x \in \omega$:</u> We insert $\widetilde{Lu}$, $\tilde{f}$, $A_h u$ and F_h in (5.1.4), apply Gauss' formula, cf. (3.1.3), and obtain the following representation of ψ, with $\psi(x) = \psi^j(x)$ ($j=0,1$):

$$\psi^j(x) = \psi_H(x) + \psi_N^j(x) \text{ for } j = 0,1 \text{ and } x \in \omega. \qquad (5.1.5)$$

Here, ψ_H is defined by

$$\psi_H(x):= -\frac{1}{H(x)} \sum_{\xi \in S'(x)} s^o(x'_\xi)\, \mathcal{K}(x'_\xi), \; x \in \omega, \qquad (5.1.6)$$

and expresses the local approximation error of the principal part of L_h, where the local error functional

$$\varkappa(x'_{\xi}) := \frac{1}{s^{\circ}(x'_{\xi})} \int_{\mathcal{S}(x,\xi)} k \frac{\partial u}{\partial n}\, ds - k_h(x'_{\xi}) u_{\bar{x}}(x'_{\xi}), \quad \xi \in S'(x), \tag{5.1.7}$$

will be used later for estimating ψ_H. It is obvious that $\varkappa$ describes the approximation error of the averaged flux of the diffusing quantity through the "wall" $\mathcal{S}(x,\xi)$. The notations $s^{\circ}(x'_{\xi}) := s(x'_{\xi}) :=$ mes $\mathcal{S}(x,\xi)$, $k_h(x'_{\xi})$ and $u_{\bar{x}}(x'_{\xi})$ are taken from (3.2.18), with $u_{\bar{x}}(x') := (u(\xi)-u(x))/h(x,\xi)$. In the term ψ_N^j which is defined by the relation

$$\psi_N^j(x) := \psi_{cu}(x) - \psi_f(x), \; x \in \omega, \tag{5.1.8}$$

the local approximation errors ψ_{cu} of the secondary part of L_h and ψ_f of the right-hand side f_h are collected, with

$$\psi_{cu}(x) := (\widetilde{cu})(x) - (cu)(x), \quad \psi_f(x) := \tilde{f}(x) - f_h(x), \; x \in \omega, \tag{5.1.9}$$

and $\widetilde{cu}$, $\tilde{f}$ are taken in the sense of (5.1.3), $f_h(x) := F_h(x)$ for $x \in \omega$. Although the errors ψ^j and ψ_N^j do not actually depend on j in the case $x \in \omega$, the index j will frequently be written with regard to ψ^j and ψ_N^j for $x \in \gamma_{23}$.

case $x \in \gamma_{23}$: In analogy to the case $x \in \omega$, but with a proper dependence on $j \in \{0,1\}$, we have now

$$\psi^j(x) = \psi_H(x) + \psi_N^j(x) \text{ for } j = 0,1 \text{ and } x \in \gamma_{23}. \tag{5.1.10}$$

The error ψ_H is defined by

$$\psi_H(x) := -\frac{1}{h(x)} \sum_{\xi \in S'(x)} s^{\circ}(x'_{\xi})\, \varkappa(x'_{\xi}), \; x \in \gamma_{23}, \tag{5.1.11}$$

$$\varkappa(x'_{\xi}) := \frac{1}{s^{\circ}(x'_{\xi})} \int_{\mathcal{S}(x,\xi)} k \frac{\partial u}{\partial n}\, ds - k_h(x'_{\xi}) u_{\bar{x}}(x'_{\xi}), \quad \xi \in S'(x). \tag{5.1.12}$$

The functional $\varkappa$ is quite similar to that in (5.1.7), with the only difference that now $s^{\circ}(x'_{\xi}) = s(x'_{\xi})$ for $x'_{\xi} \in \omega'$ and $s^{\circ}(x'_{\xi}) \neq s(x'_{\xi})$ (in general) for $x'_{\xi} \in \gamma_{23}$ hold, cf. (2.3.5), (2.3.7) and (2.3.13). Furthermore, ψ_N^j is given by

$$\psi_N^j(x) := -\frac{1}{h(x)} \int_{\mathcal{S}(x)} g\, ds + g_h^j(x) + \frac{1}{h(x)} \int_{\mathcal{S}(x)} \alpha u\, ds - (\alpha^j u)(x) + (\widetilde{cu})(x)$$
$$- \tilde{f}(x) - \frac{1}{8h(x)} \sum_{\xi \in S'(x) \cap \gamma} h^2(x'_{\xi})\, \alpha_h(x'_{\xi}) u_{\bar{x}}(x'_{\xi}) = \psi_{\alpha u}^j(x) - \psi_g^j(x),$$
$$x \in \gamma_{23}, \tag{5.1.13}$$

with the partial errors

$$\psi_g^j(x) := \frac{1}{h(x)} \int_{\mathcal{S}(x)} g\, ds - g_h^j(x) - \tilde{f}(x), \; x \in \gamma_{23},$$

$$\psi^j_{\alpha u}(x) := \frac{1}{h(x)} \int\limits_{\mathfrak{S}(x)} \alpha u ds - (\alpha^j u)(x) + (\widetilde{cu})(x)$$

$$- \frac{j}{8h(x)} \sum_{\mathfrak{s} \in S'(x) \cap \gamma} h^2(x'_\mathfrak{s}) \alpha_h(x'_\mathfrak{s}) u_\hbar(x'_\mathfrak{s}), \quad x \in \gamma_{23}, \qquad (5.1.14)$$

and with h(x) from (2.3.13), $\mathfrak{S}(x)$ from (2.3.4); g^j_h, α^j, α_h and $u_\hbar$ are taken from (3.2.18c), $\tilde{f}$ and $\widetilde{cu}$ according to (5.1.3).

<u>5.1.2. Two kinds of error splitting for MD-schemes.</u> Although the general concept of error estimation for PB- and MD-schemes is identical, there are some differences due to

a) the geometry of the boxes $\mathcal{H}(x)$, especially for $x \in \gamma_{23}$,

b) the position of quadrature points for the approximation of the line integral $\int(K\nabla u,n)ds$ along $\partial\mathcal{H}(x)$,

c) the general coefficients k_{ij} (i,j=1,2).

For instance, a) has the effect that the exponent k in the inequality $|s^\pm(x) - \frac{1}{2}h(x,\mathfrak{s}^\pm_\Gamma)| \le K_2 h^k$, with $x, \mathfrak{s}^\pm_\Gamma \in \gamma$, always has the value k = 3 for PB-boxes and $k \in \{2,3\}$ for MD-boxes, cf. (3.2.12) and (3.3.8). The facts b) and c) influence the splitting of the approximation error $\psi_H(x)$ associated with the main part of L_h, l_h. So this error can be split according to the quadrature points, i.e., $x' \in \omega' + \gamma'_{23}$ (midpoints of mesh segments) for PB-schemes and $x'' \in \bar{\omega}''$ (centres of gravity) for MD-schemes. It should be pointed out that error splitting with respect to the midpoints x' of mesh segments generally yields smaller error bounds, since the symmetry of regular difference stars causes the elimination of certain error terms Therefore, we shall also take into account the x'-splitting for MD-schemes.

We now use the errors z, ψ and the functional $\tilde{\varphi}$ in the same sense as in (5.1.1) - (5.1.4), with H(x) according to method (MD), the BVP Au = F from (2.1.1), $u \in W^2_2(\Omega)$ and $A^j_h y = F^j_h$ from Theorem 3.4., cf. (3.3.19). Starting from (5.1.4), we get the

<u>(i) x"-error splitting for MD-schemes $A_h y = F_h$:</u>

<u>case $x \in \omega$:</u>

$$\psi^j(x) = -\frac{1}{H(x)} \int\limits_{\partial\mathcal{H}(x)} (K\nabla u,n)ds + (\widetilde{cu})(x) - (L_h u)(x) + f_h(x) - \tilde{f}(x)$$

$$= \psi_H(x) + \psi^j_N(x), \quad x \in \omega, \qquad (5.1.15)$$

with

$$\psi_H(x) := -\frac{1}{H(x)} \sum_{x'' \in S''(x)} \Big\{ \int\limits_{\mathfrak{S}(x,x'')} (K\nabla u,n)ds - (K_h \nabla_\hbar u, \frac{\hbar^\bullet - \hbar^\circ}{2})(x,x'') \Big\}, \qquad (5.1.16)$$

$\psi^j_N(x)$, $\psi_{cu}(x)$, $\psi_f(x)$ from (5.1.8), (5.1.9), but here for MD-boxes $\mathcal{H}(x)$.

case $x \in \gamma_{23}$:

$$\psi^j(x) = -\frac{1}{h(x)} \int_{\partial\mathcal{X}(x)} (K\nabla u, n)ds + (\tilde{\sigma u})(x) - (l_h^j u)(x) + g_h^j(x) - \tilde{f}(x)$$

$$= \psi_H(x) + \psi_N^j(x), \quad x \in \gamma_{23}, \tag{5.1.17}$$

with

$$\psi_H(x) := -\frac{1}{h(x)} \sum_{x'' \in S''(x)} \left\{ \int_{S(x,x'')} (K\nabla u, n)ds - (K_h \nabla_{h} u, \frac{h^{\bullet} - h^{o}}{2})(x,x'') \right\}, \tag{5.1.18}$$

$\psi_N^j(x)$, $\psi_{\sigma u}^j(x)$, $\psi_g^j(x)$ from (5.1.13), (5.1.14), but here for boxes $\mathcal{X}(x)$ and arcs $S(x)$ of the type (MD).

From (5.1.16) and (5.1.18), it will be clear that $\psi_H(x)$ describes, as in the case (PB), the approximation error of the flux, which is scaled by $1/H(x)$ or $1/h(x)$, through the "wall" $\partial\mathcal{X}(x) \cap \Omega$ of the box $\mathcal{X}(x)$. However, if we use $x'' \in \overline{\omega}''$ as a reference point of error splitting, then it is more convenient to assemble the contributions of $\psi_H(x)$, $\psi_H(\xi)$ and $\psi_H(\xi^+)$ to the triangle $\Delta(x'') := \Delta(x,\xi,\xi^+)$. Thus, the error ψ_H is, properly speaking, considered on the triangles Δ only. Since ψ_H is estimated in a weak norm via the L_2-norm of a vector functional $\varkappa$, we shall analyze $\varkappa$ only. To define the vector $\varkappa(x'') \in R^2$, $x'' \in \overline{\omega}''$, we introduce three subsets ω_i'' $(i=0,1,2)$ of $\overline{\omega}''$ and flux error functionals $r(.,x'')$, where the symbol $\partial^2\Delta$ denotes the set of the vertices x,ξ,ξ^+ of $\Delta(x'')$:

$\omega_0'' := \{x'' \in \overline{\omega}'' : \partial^2\Delta(x'') \cap \gamma_1 = \emptyset\}$, $\partial^2\Delta(x'')$ is not in contact with γ_1

$\omega_1'' := \{x'' \in \overline{\omega}'' : \partial^2\Delta(x'') \cap \gamma_1 = \{\xi^+\}\}$, $\partial^2\Delta(x'')$ is in contact with one point from γ_1, say ξ^+, (5.1.19)

$\omega_2'' := \{x'' \in \overline{\omega}'' : \partial^2\Delta(x'') \cap \gamma_1 = \{\xi,\xi^+\}\}$, $\partial^2\Delta(x'')$ is in contact with two points from γ_1, say ξ, ξ^+,

cf. Figs. 8a, d, e, f,

$$\begin{aligned} r(x,x'') &:= \int_{S(x,x'')} (K\nabla u,n)ds - (K_h \nabla_{h} u, \frac{h^{\bullet}-h^{o}}{2})(x''), \quad x \in \partial^2\Delta(x''), \\ r(\xi,x'') &:= \int_{S(\xi,x'')} (K\nabla u,n)ds - (K_h \nabla_{h} u, \frac{-h^{\bullet}}{2})(x''), \quad \xi \in \partial^2\Delta(x''), \\ r(\xi^+,x'') &:= \int_{S(\xi^+,x'')} (K\nabla u,n)ds - (K_h \nabla_{h} u, \frac{h^{o}}{2})(x''), \quad \xi^+ \in \partial^2\Delta(x''). \end{aligned} \tag{5.1.20}$$

The vector functional $\varkappa$ is now defined as follows (cf. Fig. 8):

$$x'' \in \omega_0'':\quad \varkappa(x'') := \frac{1}{\operatorname{mes} \Delta(x'')} \Big\{ \Big[\int_{\mathcal{S}^-(x,x'')} (K\nabla u, n)\,ds \Big] h + \Big[\int_{\mathcal{S}^-(\xi,x'')} (K\nabla u, n)\,ds \Big] (h^+ - h)$$

$$- \Big[\int_{\mathcal{S}^-(\xi^+,x'')} (K\nabla u, n)\,ds \Big] h^+ \Big\} - (K_h \nabla_h u)(x''), \qquad (5.1.21)$$

$$x'' \in \omega_1'':\quad \varkappa(x'') := \frac{1}{\operatorname{mes} \Delta(x'')} \{ r(x,x'')h^+ + r(\xi,x'')(h^+ - h) \},$$

$$x'' \in \omega_2'':\quad \varkappa(x'') := \frac{1}{\operatorname{mes} \Delta(x'')}\, r(x,x'')h,$$

with $r(\cdot,x'')$ from (5.1.20). The vector $\varkappa$ originates from ψ_H via Green's formula (5.1.26).

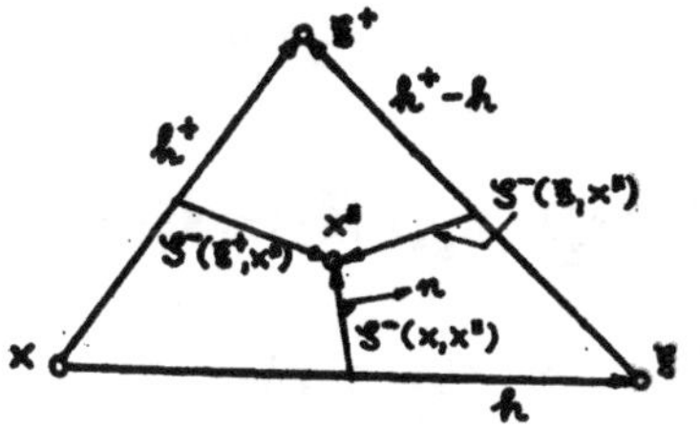

Fig. 8a

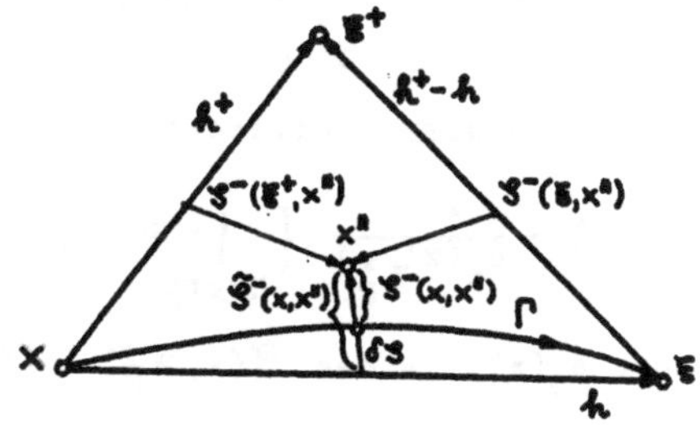

Fig. 8b

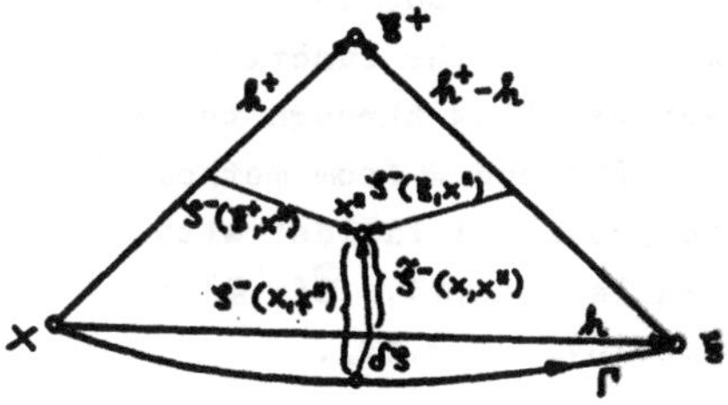

Fig. 8c

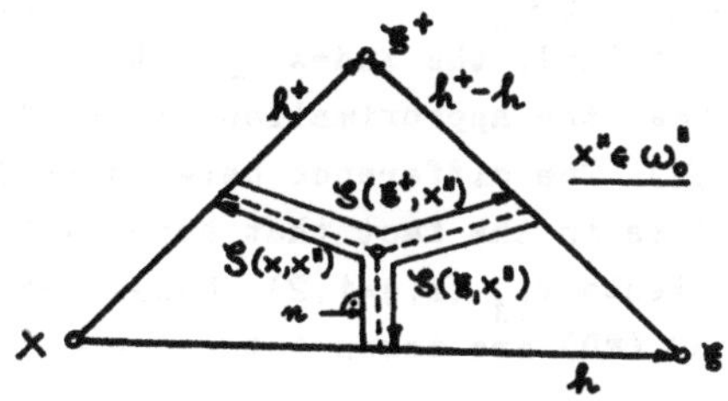

Fig. 8d

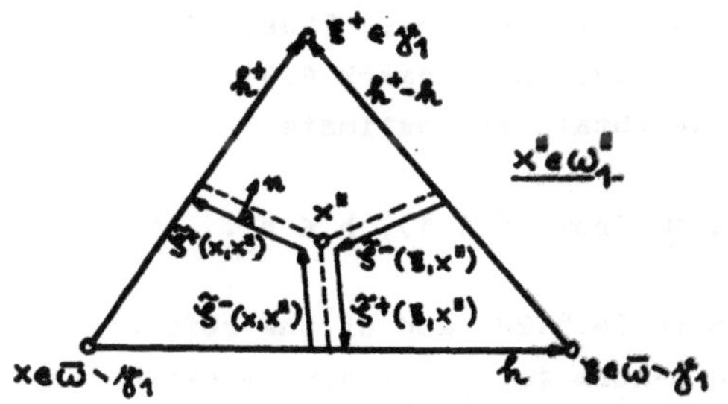

Fig. 8e

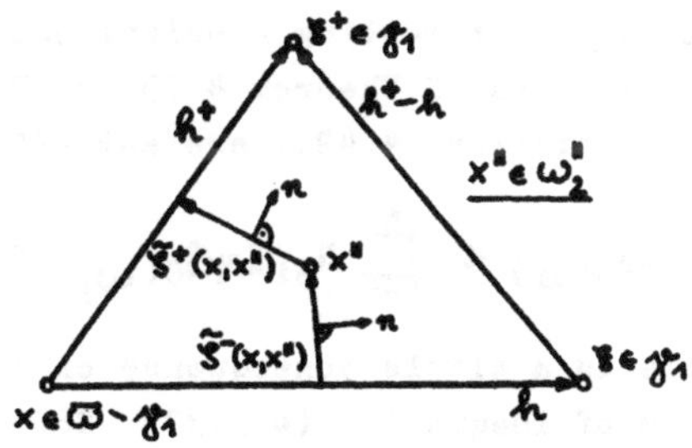

Fig. 8f

Now we shall deal with the x'-splitting of $\psi_H(x)$ and take the summation index $\xi \in S'(x)$ and $\mathcal{S}(x,\xi)$, $\mathcal{S}^{\pm}(x,\xi)$ instead of $x'' \in S''(x)$ and $\mathcal{S}(x,x'')$, cf. (2.3.15), (2.3.16) and (2.3.17). After some calculations, we get

(ii) x'-error splitting for MD-schemes $A_h y = F_h$

$$\psi_H(x) := - \sum_{\xi \in S'(x)} s_0(x'_\xi)\,\varkappa(x'_\xi) \begin{cases} \frac{1}{H(x)} & x \in \omega, \\ & \text{for} \\ \frac{1}{h(x)} & x \in \gamma_{23}, \end{cases} \tag{5.1.22a}$$

with

$$\varkappa(x'_\xi) := \varkappa^-(x'_\xi) + \varkappa^+(x'_\xi), \quad s_0(x'_\xi) = \text{mes}\, S(x,\xi) = |S^-(x,\xi)| + |S^+(x,\xi)|,$$

$$\varkappa^{\pm}(x'_\xi) := \frac{1}{s_0(x'_\xi)} \left\{ \int_{S^{\pm}(x,\xi)} (K\nabla u, n)\,ds - (K_h \nabla_h u(x''_{\pm}), n_{S\pm})\, |S^{\pm}(x,\xi)| \right\} \tag{5.1.22b}$$

for $x'_\xi \in \omega'$

and, with boundary modifications $\tilde{S}^{\pm}(x, \xi^{\pm}_{\Gamma}) := \tilde{S}(x, \xi^{\pm}_{\Gamma})$,

$$\varkappa(x'_{\xi^{\pm}_{\Gamma}}) := \frac{1}{s_0(x'_{\xi^{\pm}_{\Gamma}})} \left\{ \int_{S(x,\xi^{\pm}_{\Gamma})} (K\nabla u, n)\,ds - (K_h \nabla_h u(x''_{\pm}), n_{\tilde{S}\pm})\, |\tilde{S}^{\pm}(x, \xi^{\pm}_{\Gamma})| \right\}, \tag{5.1.22c}$$

where $s_0(x'_{\xi^{\pm}_{\Gamma}}) := \text{mes}\, \tilde{S}(x, \xi^{\pm}_{\Gamma}) = |\tilde{S}^{\pm}(x, \xi^{\pm}_{\Gamma})|$ for $x'_{\xi^{\pm}_{\Gamma}} \in \gamma'_{23}$.

Here, $\tilde{S}(x, \xi^{\pm}_{\Gamma})$ denote the boundary modifications of $S(x, \xi^{\pm}_{\Gamma})$ for $x \in \gamma_{23}$, $\xi^{\pm}_{\Gamma} \in \gamma$, cf. (2.3.16), and $|\,.\,|$ symbolizes the Euclidean norm. In (5.1.22c), the index "$\pm$" at $\tilde{S}^{\pm}$ conforms to $\xi^{\pm}_{\Gamma}$. The functional $\varkappa$ describes the approximation error of the averaged flux through the wall $S(x,\xi)$. The difference between $\varkappa$ from (5.1.22) and $\varkappa$ from method (PB) consists in the fact that here "anisotropic potential fields" with coefficients k_{ij} $(i,j=1,2)$, boxes $\mathcal{H}(x)$ and walls $S(x,\xi)$, $\tilde{S}(x,\xi)$ of method (MD) are taken into account, cf. (2.3.16), (2.3.17).

5.1.3. A priori estimates of the error. Since $z \in D_0$ and $\psi = A_h z \in D_0$ hold, z from (5.1.1), all a priori estimates of Section 4.5. can be utilized for getting a priori estimates of the error z. Thus, if the assumptions of Theorem 4.39. or Theorem 4.40. and Remark 4.41., cf. also Corollary 4.42., are satisfied, we obtain the estimate

$$\|z\|_{C(\bar{\omega})} \leq \sum_{i=1}^{6} N_{hi} \|\psi\|_{C(\omega_i)} \text{ for } z, \psi \text{ from (5.1.1), } h \leq h_0. \tag{5.1.23}$$

This is a simple consequence of estimate (4.5.22) and of the refined form of inequality (4.5.33). Now, the C-norm $\|\psi\|_{C(\omega_i)}$ can be evaluated for the six groups of grid points ω_i, $i = 1,2,\ldots,6$. By virtue of $A_h y - F_h = 0$, $Au - F = 0$ and under the assumption $u \in C^2(\bar{\Omega})$, the error ψ can be taken in the form (5.1.4) or $\psi(x) = \{Lu - L_h u + f_h - f\}(x)$ e.g. for regular grid points x. If the grid $\bar{\omega}$ is locally irregular, then the convergence $\lim_{h\to 0} \|z\|_{C(\bar{\omega})} = 0$ can be proved via estimate (5.1.23).

Our main interest here is in exploring error estimates with the discrete W_2^1-norm $\|z\|_1$, since the Stability Theorem 4.35. allows us to exploit weak norms of ψ, e.g. $\|\psi\|_-$, which are suitable for generalized solutions u and irregular networks. First of all, we prove analogues of Green's formula which are similar to (4.4.2) and (4.4.5), but for ψ_H and z instead of $L_h y$ and v, respectively. Recalling the definition of the scalar products $(.,.)'$, $(.,.)''$ from (4.3.3), (4.3.4) and introducing a third, similar scalar product $'(.,.)$ defined by

$$'(y,v) := \sum_{x' \in \omega' + \gamma'_{23}} y(x')v(x')H_o(x') \quad \text{for } y,v \in \tilde{D}' \text{ from (2.2.8)}, \tag{5.1.24}$$
$$\text{with } H_o(x') := s_o(x')h(x'),\ s_o(x') \text{ from (5.1.22)},$$

which is associated with the x'-splitting of ψ_H(MD), we are in a position to formulate

<u>Lemma 5.1. - Green's formulas for ψ_H.</u> Let $(.,.)_r$ $r \in \{p,m\}$, $(.,.)'$, $(.,.)''$ and $'(.,.)$ denote the scalar products given in (4.3.1), (4.3.3), (4.3.4) and (5.1.24), respectively. Then, for the errors ψ_H and $z := y-u \in D_o$, the following identities hold:

case (PB): $(\psi_H, z)_p = (\varkappa, z_{\bar{h}})'$, $\quad \psi_H, \varkappa$ from (5.1.6), and (5.1.11), $\quad$ (5.1.25)

case (MD) and x"-splitting:

$$(\psi_H, z)_m = \sum_{i=1}^{2} \left(\varkappa_i, \frac{\delta z}{\delta x_i}\right)'', \quad \psi_H, \varkappa = (\varkappa_1, \varkappa_2)^T \in R^2 \text{ from (5.1.16), (5.1.18) and (5.1.21)}, \tag{5.1.26}$$

case (MD) and x'-splitting:

$$(\psi_H, z)_m = {}'(\varkappa, z_{\bar{h}}), \quad \psi_H, \varkappa \text{ from (5.1.22)}, \tag{5.1.27}$$

where $z_{\bar{h}}$ and $\frac{\delta z}{\delta x_i} := (\nabla_{\bar{h}} z, e_i)$ $(i=1,2)$ are the difference quotients defined in (3.3.16) and (3.3.15). ■

<u>Proof:</u> To verify Green's formulas with ψ_H, we apply the summation by parts and take into account the skew symmetry $\varkappa(x,\xi) = -\varkappa(\xi,x)$ as well as the symmetry $s^o(x,\xi) = s^o(\xi,x)$, i.e.,

$$(\psi_H, z)_p = -\sum_{x \in \omega + \gamma_{23}} \Big\{ \sum_{\xi \in S'(x)} s^o(x,\xi)\, \varkappa(x,\xi) \Big\} z(x)$$
$$= -\sum_{x'_\xi \in \omega' + \gamma'_{23}} \Big\{ s^o(x,\xi)\, \varkappa(x,\xi) z(x) + s^o(\xi,x)\, \varkappa(\xi,x) z(\xi) \Big\}$$
$$= \sum_{x'_\xi \in \omega' + \gamma'_{23}} \varkappa(x'_\xi) \frac{z(\xi)-z(x)}{h(x'_\xi)} s^o(x'_\xi) h(x'_\xi) = (\varkappa, z_{\bar{h}})'$$

Hence, (5.1.25) is confirmed. The proof of (5.1.27) is quite similar. In identity (5.1.26), we transist from the summation with respect to $\omega + \gamma_{23}$ to the summation on $\bar{\omega}''$, i.e.,

$$(\psi_H, z)_m = -\sum_{x \in \omega+\gamma_{23}} \sum_{x'' \in S''(x)} r(x,x'')z(x) = -\sum_{x'' \in \omega_0''} \{z(x)r(x,x'') + z(\xi)r(\xi,x'')$$
$$+ z(\xi^+)r(\xi^+,x'')\} - \sum_{x'' \in \omega_1''} \{z(x)r(x,x'') + z(\xi)r(\xi,x'')\} - \sum_{x'' \in \omega_2''} z(x)r(x,x'').$$

After some calculations, taking into account $\nabla_h z = (\frac{\delta z}{\delta x_1}, \frac{\delta z}{\delta x_2})^T$ from (3.3.14) and $z \in D_0$, i.e. $z|_{\gamma_1} = 0$, we get (5.1.26).

In the following Lemma 5.2., a new norm $'\| \cdot \|_{1,m}$ will be used, which is based on the scalar product $'(.,.)$ defined in (5.1.24) and adapted to the x'-splitting of errors for MD-schemes.

Lemma 5.2. Let Condition 2.17. - VO"(PB, $\bar{\omega}$) be satisfied. Then, the norm $'\| \cdot \|_{1,m}$ defined by

$$'\|y\|^2_{1,m} := \|y\|^2_{0,m} + '\|y_h\|^2_0, \text{ with } '\|y_h\|^2_0 := '(y_h, y_h),\ '(.,.) \text{ from (5.1.24)}, \tag{5.1.28}$$

is equivalent to the norm $\|\cdot\|_{1,m}$ defined in (4.3.14). Especially, the inequality

$$'\|y\|_{1,m} \le C_0 \|y\|_{1,m}, \text{ for } y \in D_0 \text{ and } C_0 \text{ from (5.1.30)}, \tag{5.1.29}$$

holds. ≡

Proof: We shall verify (5.1.29) only. Due to Lemma 4.25., cf. (4.4.12), and by definition (5.1.28), we can note

$$\|y\|^2_{1,m} = \|y\|^2_{0,m} + \sum_{x' \in \omega'+\gamma'_{23}} y_h^2(x')s^0(x')h(x'),$$

$$'\|y\|^2_{1,m} = \|y\|^2_{0,m} + \sum_{x' \in \omega'+\gamma'_{23}} y_h^2(x')s_0(x')h(x'),$$

for $y \in D_0$. By virtue of VO"(PB, $\bar{\omega}$), the behaviour of s^0 and s_0 can be described by $s_0(x') \le \varepsilon^{-1}h$, $s^0(x') \ge \varepsilon h > 0$ for all $x' \in \omega' + \gamma'_{23}$, with a fixed ε, $0 < \varepsilon \le 1$, which depends on θ_0, ε_0, h_0. Taking this ε, we are obviously led to the inequality

$$'\|y\|^2_{1,m} \ \|y\|^{-2}_{1,m} \le \varepsilon^{-2} =: C_0^2 \ (0 < \varepsilon \le 1) \text{ for } y \in D_0. \tag{5.1.30}$$

We now summarize the results about error splittings and a priori estimates in

Theorem 5.3. Suppose that the assumptions V($\bar{\Omega}$), V(A,F) and V(u) are satisfied and that the solution u of the BVP Au = F belongs to $C^2(\bar{\Omega})$

or $u \in W_2^2(\Omega)$. Let y be the solution of the FDSs $A_h y = F_h$ constructed in Theorem 3.1. (PB) and 3.4.(MD), where VO(PB, $\bar{\omega}$) and VO(MD, $\bar{\omega}$) are fulfilled, and h_o is assumed to be sufficiently small. Then, the error $z := y - u$, with $z \in D_o$, and the local approximation error ψ as well as the splittings and functionals employed in Section 5.1.1. and 5.1.2. are well-defined. Moreover, the following a priori estimate of the error z holds:

$$\|z\|_1 \leq C\|\psi\|_- \text{ for } h \leq h_o,\ z \text{ and } \psi \text{ from (5.1.1), (5.1.4).} \tag{5.1.31}$$

The weak norm $\|\psi\|_-$ will be always expressed by discrete L_2-norms of the flux approximation error $\varkappa$ and the error ψ_N of the secondary part of A_h and of the right-hand side F_h, i.e.,

<u>case (PB)</u>, x'-splitting:

$$\|z\|_{1,p} \leq C_p\{\|\varkappa\|_o'^2 + \|\psi_N\|_{o,p}^2\}^{1/2},\ \|\varkappa\|_o'^2 := \sum_{x' \in \omega' + \gamma_{23}'} \varkappa^2(x')s^o(x')h(x'), \tag{5.1.32}$$

with $\varkappa$, ψ_N from Section 5.1.1., cf. (5.1.5) - (5.1.14), and PB-norms from (4.3.5), (4.3.13);

<u>case (MD)</u>, x"-splitting:

$$\|z\|_{1,m} \leq C_m\{\||\varkappa|\|_o''^2 + \|\psi_N\|_{o,m}^2\}^{1/2},\ \||\varkappa|\|_o''^2 := \sum_{x'' \in \bar{\omega}''} |\varkappa(x'')|^2 \operatorname{mes} \Delta(x''), \tag{5.1.33}$$

with $\varkappa$, ψ_N from Section 5.1.2., cf. (5.1.15) - (5.1.21), and MD-norms from (4.3.5), (4.3.14);

<u>case (MD)</u>, with VO"(PB, $\bar{\omega}$) and x'-splitting:

$$\|z\|_{1,m} \leq C_m'\{'\|\varkappa\|_o^2 + \|\psi_N\|_{o,m}^2\}^{1/2},\ '\|\varkappa\|_o^2 := \sum_{x' \in \omega' + \gamma_{23}'} \varkappa^2(x')s_o(x')h(x'), \tag{5.1.34}$$

with $\varkappa$, ψ_N from Section 5.1.2., cf. (5.1.22), and MD-norms from (4.3.5), (4.3.14).

The constants C_p, C_m, C_m' are independent of $h (h \leq h_o)$, z, $\varkappa$ and ψ_N. $\equiv$

<u>Proof:</u> The proof is partially carried out in the foregoing Sections 5.1.1., 5.1.2. and in this section. Due to the smoothness assumptions on u and the functions involved in A and F, using the imbedding and trace theorems from Appendix IM and checking the arithmetic of the splittings, we see that all error functionals and their parts are well-defined, cf. the errors and their splitting in Section 5.1.1., 5.1.2. Taking the rough splitting $A_h z = \psi = \psi_H + \psi_N$, using Green's formulas from Lemma 5.1. for ψ_H, e.g. $(\psi_H, z)_p = (\varkappa, z_{\bar{x}})'$, the a priori estimate $\|z\|_{1,p}^2 \leq C(A_h z, z)_p$ from (4.4.29), and an estimation quite similar to that of (4.5.9), we get the estimates

$$\|z\|^2_{1,p} \leq C(A_h z,z)_p = C\{(\varkappa,z_{\bar{x}})' + (\psi_N,z)_p\} \leq C\{\|\varkappa\|'_o \|z_{\bar{x}}\|'_o \tag{5.1.35}$$

$$+ \|\psi_N\|_{o,p} \|z\|_{o,p}\} \leq C\{\|\varkappa\|'^2_o + \|\psi_N\|^2_{o,p}\}^{1/2} \|z\|_{1,p} \text{ for } z \in D_o.$$

Hence, inequality (5.1.32) is proved, and the estimates (5.1.33), (5.1.34) can be confirmed analogously by means of $\|z\|^2_{1,m} \leq C(\psi,z)_m$ and (5.1.26), (5.1.27). Additionally, for the verification of inequality (5.1.34), we must regard the relation $'\|z\|_{1,m} \leq C_o \|z\|_{1,m}$ proved in Lemma 5.2., cf. (5.1.29).

Corollary 5.4. Under the assumptions of Theorem 5.3., the following estimate is valid:

$$\|z\|_{C(\bar{\omega})} \leq M|\ln h|^{1/2} \|\psi\|_- \quad \text{for } z \in D_o,\ h \leq h_o,\ \psi = A_h z, \tag{5.1.36}$$

if h_o is chosen sufficiently small. The estimates (5.1.32), (5.1.33) and (5.1.34) hold with the C-norm $\|z\|_{C(\bar{\omega})}$ instead of $\|z\|_1$, if the constants C_p, C_m and C'_m are replaced by $M_p|\ln h|^{1/2}$, $M_m|\ln h|^{1/2}$ and $M'_m|\ln h|^{1/2}$, respectively.∎

Proof: Combining the inequalities $\|z\|_1 \leq C\|\psi\|_-$ from Theorem 5.3. and $\|z\|_{C(\bar{\omega})} \leq C|\ln h|^{1/2}\|z\|_1$ from Theorem 4.36., cf. also Corollary 4.37., we get (5.1.36) and, by the same way, the remaining inequalities mentioned in Corollary 5.4.

5.1.4. Convergence for classical solutions of $C^l(\bar{\Omega})$-type ($l \geq 2$)

A solution u of the BVP Au = F is called classical, if u belongs to $C^2(\Omega) \cap C^1(\Omega \cup \Gamma_{23}) \cap C(\bar{\Omega})$. For the sake of simplicity, we shall be concerned with solutions u of the type $C^l(\bar{\Omega})$, $l \geq 2$. Under this assumption and for PB-schemes $A_h y = F_h$, error estimates and convergence proofs are extensively studied by the author, cf. HEINRICH [6, 7]. Extracting the results from these papers, we are in position to formulate

Theorem 5.5. - Convergence for PB-schemes. Let u be the solution of Au = F from (3.2.1), y the solution of $A_h y = F_h$ from Theorem 3.1. Moreover, let the assumptions of Theorem 4.39. be satisfied, i.e., the grid $\bar{\omega}$ is locally irregular and taken from Definition 2.10., 2.11. or 2.12., the "coupling condition" at the boundary γ is fulfilled, and the "majorant function" v from (4.5.21) exists. Then, the inequality

$$\|y-u\|_{C(\bar{\omega})} \leq \sum_{i=1}^{6} N_{hi} \|\psi\|_{C(\omega_i)} \text{ for } h \leq h_o \tag{5.1.37}$$

holds. Especially, we have

(i) $\|y-u\|_{C(\bar{\omega})} \to 0$ as $h \to 0$ for $u \in C^2(\bar{\Omega})$, $k \in C^1(\bar{\Omega})$,

(ii) $\|y-u\|_{C(\bar{\omega})} \leq Mh$ for $u \in C^3(\bar{\Omega})$, $k \in C^2(\bar{\Omega})$, (5.1.38)

(iii) $\|y-u\|_{C(\bar\omega)} \leq Mh^2$ for $u \in C^4(\bar\Omega)$, $k \in C^3(\bar\Omega)$, $\Gamma = \Gamma_1$, (5.1.38)

$h \leq h_0$, $0 < M = O(1)$, and if further smoothness conditions concerning c, f, α, g are satisfied, e.g. $c, f \in C(\bar\Omega)$ and $\alpha, g \in PC^1(\Gamma_{23})$, $g \in C(\Gamma_1)$. ■

Proof: Inequality (5.1.37) was already checked in (5.1.23), where the constants N_{hi} are to be taken from (4.5.23). It remains to evaluate $\psi(x)$ for $x \in \omega_i$ $(i=1,2,\ldots,6)$, which can be made easily by Taylor's expansion. The error representation ψ from (5.1.4) may be replaced by $\psi = Au - A_hu + F_h - F$, which is the "classical splitting" of the local approximation error for FDSs $A_hy = F_h$, cf. CIARLET [1], MITCHELL/GRIFFITHS [1], SAMARSKIJ [1], HEINRICH [2,3]. Thus, we are led to the following approximation errors $\psi(x)$ for $x \in \bar\omega$:

$x \in \mathring{\omega}$: $\psi(x) = o(1)$ for $u \in C^2(\bar\Omega)$, $k \in C^1(\bar\Omega)$,

$\psi(x) = O(h^i)$ $\quad u \in C^{2+i}(\bar\Omega)$, $k \in C^{1+i}(\bar\Omega)$, $i = 1,2$,

$x \in \omega$: $\psi(x) = O(1)$ for $u \in C^2(\bar\Omega)$, $k \in C^1(\bar\Omega)$, (5.1.39)

and for c, $f \in C(\bar\Omega)$ in all cases previously noted,

$x \in \gamma_{23}$: $\psi^j(x) = O(h)$ for $u \in C^2(\bar\Omega)$, $k \in C^1(\bar\Omega)$; $\alpha, g \in PC^1(\Gamma_{23})$ and $c, f \in C(\bar\Omega)$ for $j = 0,1$,

$x \in \gamma_1$: $\psi(x) = 0$ for $u = g \in C(\Gamma_1)$.

Inserting ψ from (5.1.39) and N_{hi} from (4.5.23) into (5.1.37), we immediately get the inequalities in (5.1.38) with constants $M = O(1)$, which depend especially on the "majorant function" v and on bounds of the derivatives of u and of the other parameters of $Au = F$. Moreover, $h \leq h_0$ must be satisfied, with sufficiently small h_0.

Clearly, a quite similar theorem could be formulated for MD-schemes. But the main disadvantages of using a priori error estimates of type (5.1.37) are the following two. Firstly, grids $\bar\omega$ which are irregular in the whole domain lead to errors $\psi = O(1)$ and, therefore, they cannot be considered within this theory. Secondly, this concept is not appropriate for generalized solutions. Fortunately, we have a proper way out of this situation, since the C-norm of the error z is bounded by $\|z\|_{C(\bar\omega)} \leq M|\ln h|^{1/2}\|\psi\|_-$, cf. (5.1.36). This inequality can be also exploited for classical solutions $u \in C^l(\bar\Omega)$, $l > 2$, and leads to more efficient error estimates with the following features: The order of convergence is generally higher and the requirements with respect to the smoothness of u and the grid regularity are weaker than in the classical concept of estimating the error z.

In the following Section 5.1.5., we shall be concerned with generalized solutions $u \in W_2^l(\Omega)$, $l > 2$, and with estimates $\|z\|_1 \leq C\|\psi\|_-$ and $\|z\|_C \leq M|\ln h|^{1/2}\|\psi\|_-$. Clearly, since $u \in C^l(\bar\Omega)$ implies $u \in W_2^l(\Omega)$, we can omit the further investigation of $u \in C^l(\bar\Omega)$ and refer to the methods and results developed for $u \in W_2^l(\Omega)$.

5.1.5. Convergence for generalized solutions of $W_2^l(\Omega)$-type ($l \geq 2$)

The approach to estimating the error z via $\|z\|_1 \leq C\,\|\psi\|_-$ is a very natural one, since $\|\psi\|_-$ measures directly the errors made in the finite difference approximation of the system of balance equations. The general procedure of deriving estimates of the error z for generalized solutions $u \in W_2^l(\Omega)$, $l \in \{2,3\}$, will be developed in the next few sections and generally comprises the following ideas and steps:

a) The inequality $\|z\|_1 \leq C\,\|\psi\|_-$ is taken into account, with $\|\psi\|_- = \{\|\varkappa\|_o'^2 + \|\psi_N\|_o^2\}^{1/2}$ for PB-schemes and analogous representations for MD-schemes.

b) The functional $\varkappa$ is defined at least for $u \in W_2^2(\Omega)$ and sufficiently smooth parameters of the BVP Au = F, which is proved via imbedding and trace theorems. Moreover, $\varkappa(x')$ is of a local nature and consists, in general, of three partial errors: the error of the quadrature formula, the error of the finite difference approximation of first order derivatives and the "geometric error" which occurs at the boundary Γ only, i.e. for $x \in \gamma_{23}$, if $\partial\mathcal{X}(x)$ is replaced by $\partial\tilde{\mathcal{X}}(x)$. Similar statements can be made about ψ_N, but here the error of the cubature formula used plays an essential role.

c) The functionals $\varkappa$ and ψ_N are split into a sum of other functionals which can be estimated by Taylor's expansion or they are linear functionals on Sobolev spaces and will be estimated via the Bramble-Hilbert Lemma, cf. BRAMBLE/HILBERT [1]. Since these linear functionals are defined on Sobolev spaces $W_2^l(G_h)$, where G_h depends on h, and since the constants in the imbedding and trace theorems, but also in the Bramble-Hilbert Lemma, depend on G_h, we are forced to transform the so-called "estimation regions" G_h onto a fixed "reference domain" $\hat{G}$ which does not depend on h. We shall consider only affine transformations which are invertible, uniformly with respect to h, $h \leq h_o$; cf. Appendix TR.

d) After this transformation, the linear functionals are now expressed in terms of the "$\wedge$"-variables, the so-called "reference variables", and can be estimated by means of the Bramble-Hilbert Lemma, where certain seminorms of the functions of interest appears in the bounds of the estimates. After the estimation of the linear functionals, the inverse mapping $\hat{G} \to G_h$ is applied, and the bounds of all partial error functionals are added. Thus, local bounds of $|\varkappa|$ and $|\psi_N|$ are obtained.

e) Bounds of the discrete L_2-norms of $\varkappa$ and ψ_N can now be calculated and inserted in the estimate (5.1.32), which leads to bounds of $\|z\|_1$ in terms of powers of h and constants which depend on the solution u, on the parameters A, F, Ω and on fixed parameters of the geometry of the network. But these constants are independent of h, for $h \leq h_o$ and sufficiently small h_o.

f) If $\Omega_h \not\subset \Omega$ occurs, then $u \in W_2^l(\Omega)$ is extended via Calderon's Theorem to $u \in W_2^l(\tilde{\Omega})$, where $\tilde{\Omega} \supset \Omega \cup \Omega_h$ holds for $h \leq h_o$, and $\tilde{\Omega}$ is fixed. Due to assumption $V(\tilde{\Omega})$, this is always possible, cf. Appendix EX, Theorem 1. Moreover, there is a constant C_E such that the inequalities

$$\|u\|_{l,\Omega_h} \leq \|u\|_{l,\tilde{\Omega}} \leq C_E \|u\|_{l,\Omega} \quad \text{for } \Omega_h \cup \Omega \subset \tilde{\Omega}, h \leq h_0, \tag{5.1.40}$$

and $l \in \{0,1,2,3\}$ hold. If the extension by Geymonat is used, cf. Theorem 2 in Appendix EX, then in (5.1.40) certain seminorms $|.|$ can be used instead of the full norms $\|.\|$.

g) The estimates of the error functionals will often be given in more detail, where a great number of constants must be taken into account. These constants always exist and most of them can be easily calculated. Several others are significant constants of analysis as e.g. C_I, C_E, the constants in the imbedding and extension theorems. The determination or estimation of such constants would merit further study, but the reader is instead referred to the literature, cf. ADAMS [1], MICHLIN [1,2]. In our monograph, the asymptotic behaviour of the constants with respect to h and the way of proving error estimates are of particular interest and, therefore, a detailed representation of the constants will not be given. Unless otherwise specified, the constants M_i, C_i, K_i etc. do not depend on h for $h \leq h_0$.

The estimation of the error norm $\|z\|_1 := \|y-u\|_1$ for generalized solutions $u \in W_2^l(\Omega)$, $l \geq 2$, is the main subject of Sections 5.2., 5.3., 5.4. and 5.5.

5.2. The error $\varkappa$ of the principal part of PB-operators

5.2.1. The splitting of $\varkappa$ and first estimates.

To apply inequality (5.1.32), we consider $\varkappa$ from (5.1.7), (5.1.12) and split $\varkappa$ into three partial errors $\varkappa_k$, $\varkappa_u$ and $\varkappa_s$, which denote respectively the quadrature error, the error of the difference quotient and the error caused by the geometric approximation of the curved boundary Γ:

$$\varkappa(x') := \varkappa(x',u,k) = \varkappa_k(x') + k_h(x')\,\varkappa_u(x') + k_h(x')\,\varkappa_s(x'), \tag{5.2.1a}$$

$$\varkappa_k(x'_\xi) := \varkappa_k(x'_\xi,u,k) := \frac{1}{s^o(x'_\xi)} \int_{S(x,\xi)} \{k(s)-k_h(x'_\xi)\} \frac{\partial u}{\partial n}\, ds, \tag{5.2.1b}$$

$$\varkappa_u(x'_\xi) := \varkappa_u(x'_\xi,u,k) := \frac{1}{s(x'_\xi)} \int_{S(x,\xi)} \frac{\partial u}{\partial n}\, ds - \frac{u(\xi)-u(x)}{h(x,\xi)}, \tag{5.2.1c}$$

$$\varkappa_s(x'_\xi) := \varkappa_s(x'_\xi,u) := \left\{\frac{1}{s^o(x'_\xi)} - \frac{1}{s(x'_\xi)}\right\} \int_{S(x,\xi)} \frac{\partial u}{\partial n}\, ds. \tag{5.2.1d}$$

All functionals $\varkappa.$ are well-defined, since $u \in W_2^2(\Omega)$ implies $u \in C(\bar{\Omega})$ and $\frac{\partial u}{\partial x_i}$, $\frac{\partial u}{\partial n} \in L_2(S)$ for sufficiently smooth one-dimensional manifolds S, with $S \subset \Omega$ or $S \subset \partial\Omega$, e.g. the straight segments $S(x') := S(x'_\xi) := S(x,\xi)$ of the perpendicular bisectors and the continuously curved arcs

$\mathfrak{S}^{\pm}(x)$. $\widehat{x\xi_\Gamma^+}$ on $\partial\Omega$, which will be considered later. Beside (5.2.1), we could give other splittings, which are convenient for further estimates, cf. (5.2.46) and HEINRICH [7, p. 109]. For the presentation of a basic approach to the estimation of $\varkappa$, we assume that the grid regularity condition V1"(PB, $\bar{\omega}$) is satisfied,

$$\text{V1"(PB}, \bar{\omega}\text{): VO(PB}, \bar{\omega}\text{) and } 0 < \theta_0 \leq \theta \leq \frac{\pi}{2} - \theta_0 \text{ for the angles } \theta \text{ of all triangles, and } \theta = \frac{\pi}{2} \text{ is allowed for some triangles } \Delta(\Delta \cap \gamma_{23} = \emptyset). \quad (5.2.2)$$

Then, the circumcentre $p(\Delta)$ is contained in Δ and mes $\mathfrak{S}^{\pm}(x,\xi) \geq \varepsilon h > 0$ holds, cf. Remark 2.20, or we have $\mathfrak{S}^{+}(x,\xi) = \emptyset$ or (and) $\mathfrak{S}^{-}(x,\xi) = \emptyset$. Some weakening of (5.2.2) will be given in Section 5.2.5.

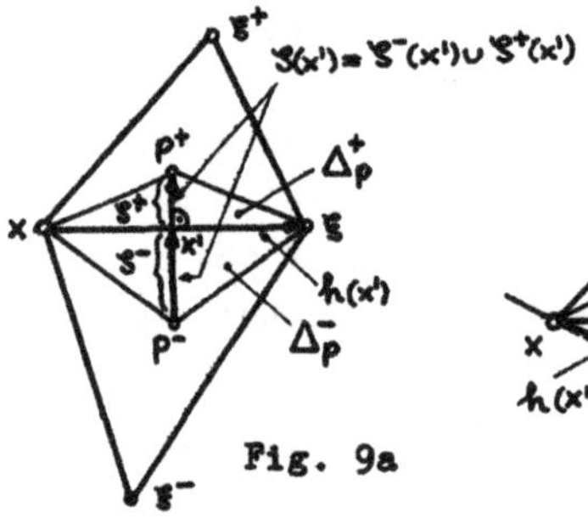

Fig. 9a

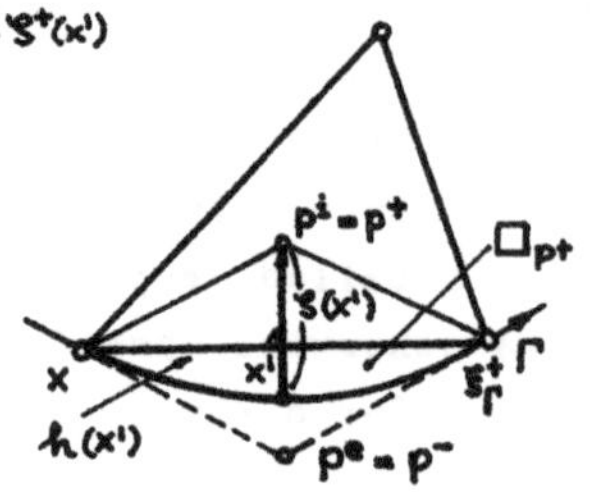

Fig. 9b

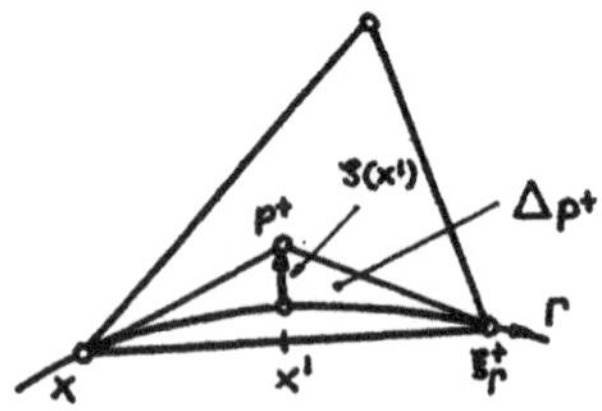

Fig. 9c

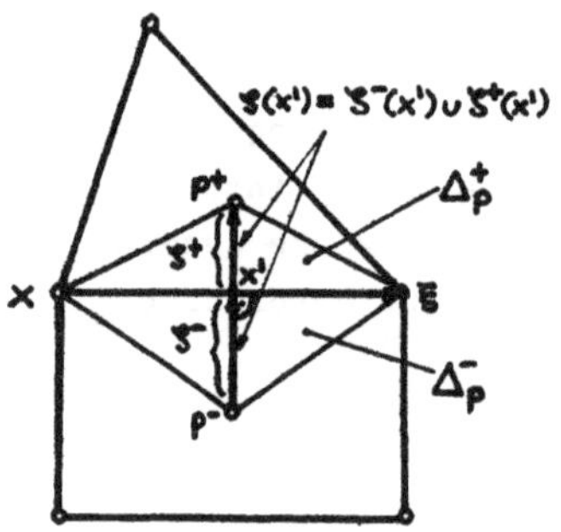

Fig. 9d

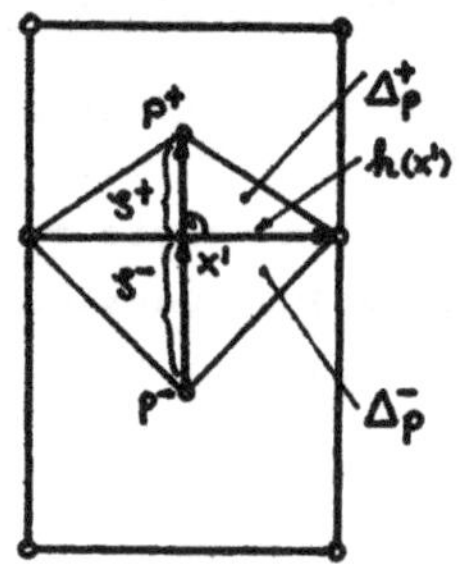

Fig. 9e

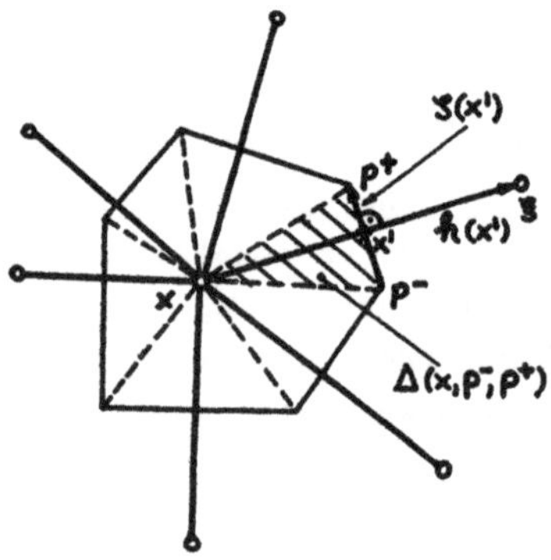

Fig. 9f

The functional $\varkappa_k$ from (5.2.1b) is a bilinear functional with respect to $k - k_h$ and $\frac{\partial u}{\partial n}$, and $\varkappa_k$ is well-defined for $k \in C^1(\bar{\Omega})$ and $u \in W_2^2(\Omega)$, cf. Theorem 2 from Appendix IM. The error $k - k_h$ can be easily isolated by an L_∞- L_1- or L_2-L_2-estimate of $\varkappa_k$ with respect to $k - k_h$ and $\frac{\partial u}{\partial n}$. For this, we note Taylor's expansion for $k(s)$ (s: the arc length) at the point $x' := x'_\xi$ for segments $\mathfrak{S}$ with $x' \in \mathfrak{S}(x') \subset \bar{\Omega}$.

$$k(s) = k(x') + \frac{\partial k(\check{s})}{\partial s}(s-s') \text{ for } k \in C^1(\bar{\Omega}),\ s' = s(x'). \quad (5.2.3a)$$

$$k(s) = k(x') + \frac{\partial k(x')}{\partial s}(s-s') + \frac{\partial^2 k(\hat{s})}{\partial s^2}\frac{(s-s')^2}{2} \text{ for } k \in C^2(\bar{\Omega}). \quad (5.2.3b)$$

In (5.2.3), $\check{s}$ and $\hat{s}$ denote intermediate values of s on $\mathfrak{S}(x')$, cf. also Figs. 9a, b. For simplicity, we do not modify the symbol k if we change from (x_1,x_2)-coordinates to the s-coordinate and vice versa, and this is also maintained for other functions as α, g etc. For the case $x' \notin \mathfrak{S}(x')$, which can arise at $x' \in \gamma'$ (cf. Fig. 9c), we choose that boundary point $\tilde{x}' \in \partial\mathfrak{S}(x')$ as the point of Taylor's expansion which is nearest to x'. Then, for $k_h(x') := k(x')$ or $k_h(x') := \{k(x) + k(\xi)\}/2$, cf. (3.2.5), we get the estimate

$$|k(s) - k_h(x')| \leq M_1 h \text{ for } k \in C^1(\bar{\Omega}),\ M_1 = O(|\partial^1 k|). \tag{5.2.4}$$

Here, the difference $k(x') - k_h(x')$ is estimated by Taylor's expansion along $\hbar(x,\xi)$ or, in the case $x' \notin \mathfrak{S}(x')$, along $\widehat{x\xi} \subset \partial\Omega$. For $\varkappa_k$ from (5.2.1b), we get a first estimate, viz.

$$|\varkappa_k(x')| \leq \frac{M_1 h}{s_0(x')} \int_{\mathfrak{S}(x')} \left|\frac{\partial u}{\partial n}\right| ds \leq M_2 h^{1/2} \left\|\frac{\partial u}{\partial n}\right\|_{0,\mathfrak{S}} \quad \text{for } k \in C^1(\bar{\Omega}),\ u \in W_2^2(\Omega), \tag{5.2.5}$$

where M_2 does not depend on u and h, $h \leq h_0$. We now consider the special case

$$\hbar(x') \subset \bar{\Omega},\ x' \text{ is the midpoint of } \mathfrak{S}(x'),\ k \in C^2(\bar{\Omega}),\ u \in W_2^3(\Omega). \tag{5.2.6}$$

That x' is the midpoint of $\mathfrak{S}(x')$ is important for regular or locally irregular grids. For instance, we encounter a situation like this, if two equilateral triangles or two congruent rectangles have the mesh segment $\hbar(x')$ as a common side. Using (5.2.3b) and some simple estimates, we obtain

$$|\varkappa_k(x')| \leq \frac{1}{s_0(x')} \left\{ |k(x')-k_h(x')| \int_{\mathfrak{S}(x')} \left|\frac{\partial u}{\partial n}\right| ds + \left|\frac{\partial k(x')}{\partial s}\right| \left|\int_{\mathfrak{S}(x')} (s-s') \frac{\partial u}{\partial n}\, ds\right| + \left\|\frac{\partial^2 k}{\partial s^2}\right\|_{C(\mathfrak{S})} \int_{\mathfrak{S}(x')} \frac{(s-s')^2}{2} \left|\frac{\partial u}{\partial n}\right| ds \right\}. \tag{5.2.7}$$

Since x' is also the midpoint of $\hbar(x')$, $k(x') - k_h(x') = O(h^2)$ holds. Applying integration by parts in the second term of the right-hand side of (5.2.7) and some transformations and estimations, we get

$$\left|\int_{\mathfrak{S}(x')} (s-s') \frac{\partial u}{\partial n}\, ds\right| = \left|\left[\frac{(s-s')^2}{2} \frac{\partial u}{\partial n}\right]_{s_A}^{s_B} - \int_{\mathfrak{S}(x')} \frac{(s-s')^2}{2} \frac{\partial^2 u}{\partial s \partial n}\, ds\right| \leq M_3 h^{5/2} \left\|\frac{\partial^2 u}{\partial s \partial n}\right\|_{0,\mathfrak{S}}, \quad s' = s(x'), \tag{5.2.8}$$

where s_A, s_B and s' denote respectively the arc length of the boundary points of $\mathfrak{S}(x')$ and of the midpoint $x' \in \mathfrak{S}(x')$. Due to $u \in W_2^3(\Omega)$ and Theorem 2 from Appendix IM, we see that $\frac{\partial^2 u}{\partial s \partial n} \in L_2(\mathfrak{S})$. Hence, owing to (5.2.7), (5.2.8) and by further estimation

$$|\varkappa_k(x')| \leq M_4\, h^{3/2} \left\{ \left\| \frac{\partial u}{\partial n} \right\|_{0,\mathfrak{S}} + \left\| \frac{\partial^2 u}{\partial s \partial n} \right\|_{0,\mathfrak{S}} \right\} \quad \text{for } k \in C^2(\bar{\Omega}),\ u \in W_2^3(\Omega), \tag{5.2.9}$$

can be verified, with $M_4 = O(\,|\partial^2 k| + |\partial^1 k|)$, and M_4 is independent of u and h, $h \leq h_0$. Moreover, if x' differs from the midpoint of $\mathfrak{S}(x')$ by a distance of $O(h^2)$, the term $h^2 \| \frac{\partial u}{\partial n} \|_{C(\mathfrak{S})}$ must be added on the right-hand side of (5.2.9). Owing to $u \in W_2^3(\Omega)$ and Theorem 1 from Appendix IM, we have $\frac{\partial u}{\partial x_i} \in C(\bar{\Omega})$ and, therefore, $\frac{\partial u}{\partial n} \in C(\mathfrak{S})$.

<u>5.2.2. The choice of local estimation regions.</u> For further estimations in (5.2.5), (5.2.9) and for $|\varkappa_u|$, we must find bounds of the seminorms of u on $\mathfrak{S}$ in terms of seminorms of u on R^2-subsets, which will be called "local estimation regions" and which are characterized by containing $\mathfrak{S}$ as a subset and having a positive R^2-measure with the asymptotic behaviour $O(h^2)$. In HEINRICH [7, Section 5.4.2] a short review of various choices of such local estimation regions is given. For instance, we might use the straight or curved triangles of the primary network. But a transformation which maps a curved triangle Δ onto a straight reference triangle, say $\hat{\Delta}$, is a nonlinear one and is not affine, cf. KORNEEV [1], ODEN/REDDY [1], ZLAMAL [1,2]. Moreover, the property of beeing a "perpendicular bisector segment" is not preserved, if an affine mapping is applied. But the triangles of the primary network could be also used for estimations within the method (PB), cf. HEINRICH [7].

For PB-schemes considered here, we shall use only one class of local estimation regions which are defined by certain triangles of the secondary network. In advance it should be pointed out that, conversely, for MD-schemes the straight triangles of the primary network will be preferred. For the motivation of a choice of estimation regions, we proceed with some remarks.

<u>Remark 5.6.</u> According to assumption (5.2.2), for each flux point $x' \in \omega'$, there are straight triangles $\Delta_p^{\pm}(x')$ which are defined by the vertices $x, \bar{x}$ of the primary triangles $\Delta(x''_{\pm})$ and the circumcentres $p_{\pm}$ of these triangles $\Delta(x''_{\pm})$, or $p_{\pm}$ are the circumcentres of primary rectangles $\square(x''_{\pm})$. The combination of these cases is also possible. These situations are illustrated in Figs. 9a, d, e. The triangles $\Delta_p^{\pm}(x')$ have $\mathfrak{h}(x')$ as a common side and are contained in the primary elements Δ or $\square$. For the whole segment $\mathfrak{S}(x') := \overline{p_- p_+}$ or the partial segments $\mathfrak{S}^{\pm}(x') := \overline{x' p_{\pm}}$ of the perpendicular bisectors to $\mathfrak{h}(x')$, the following relations hold:

$$\mathfrak{S}(x') \subset \Box_p(x') := \Delta_p^-(x') \cup \Delta_p^+(x'),\ \Box_p := \Box_p(p_-, \mathfrak{z}, p_+, x):\ \text{deltoid}, \tag{5.2.10}$$

$\mathfrak{S}^{\pm}(x') \subset \Delta_p^{\pm}(x')$, $\mathfrak{S}(x')$ from (2.3.6), for $x' \in \omega'$.

The partial segments $\mathfrak{S}^{\pm}(x')$ are medians in $\Delta_p^{\pm}(x')$, but $\Delta_p^+ = \emptyset$ or (and) $\Delta_p^- = \emptyset$ $(\theta = \frac{\pi}{2})$ may also occur. ≡

Remark 5.7. For $x' \in \gamma'_{23}$, the segment $\mathfrak{S}(x')$ may be regarded as a subset of the following sets, if $V(\bar{\Omega})$, (5.2.2) and $h \leq h_o$ hold, and h_o is sufficiently small.

a) case int $\mathfrak{h}(x') \cap \bar{\Omega} = \emptyset$, cf. (2.2.1) and Fig. 9c: Here, $\mathfrak{S}(x')$ is a subset of the straight triangle $\Delta_p(x')$, with the side $\mathfrak{h}(x')$ lying on Γ_h. The triangle $\Delta_p(x')$ is defined by $x, \mathfrak{z}_\Gamma \in \gamma$ and the circumcentre p, and $\mathfrak{S}(x')$ is portion of a median in $\Delta_p(x')$. Consequently, we have

$$\mathfrak{S}(x') \subset \Delta_p(x'),\ \mathfrak{S}(x') \text{ from (2.3.4), for } x' \in \gamma'_{23} \text{ and int } \mathfrak{h}(x') \cap \bar{\Omega} = \emptyset$$

$$\text{formally: } \mathfrak{S}(x') \subset \Box_p(x') := \Delta_p^-(x') \cup \Delta_p^+(x'), \text{ where either } \Delta_p^- = \emptyset$$

$$\text{or } \Delta_p^+ = \emptyset \text{ holds, i.e., } \Box_p(x') \text{ is actually a triangle.} \tag{5.2.11a}$$

b) case $\mathfrak{h}(x') \subset \bar{\Omega}$, cf. (2.2.1) and Fig. 9b: Here, $\mathfrak{S}(x')$ is a subset of $\Delta_p^i(x') \cup \Delta_p^e(x')$, where Δ_p^i and Δ_p^e denote the "interior" and "exterior" triangle, respectively. The triangle Δ_p^i is generated by $x, \mathfrak{z}_\Gamma$ and the circumcentre p of the interior primary triangle. Thus, we always have $\Delta_p^i \subset \bar{\Omega}$. The triangle Δ_p^e is the "mirror image" of Δ_p^i obtained by reflection of Δ_p^i in the side $\mathfrak{h}(x')$. That $\mathfrak{S} \subset \Delta_p^i \cup \Delta_p^e$ is valid will become clear from $\delta(x') := s(x') - \tilde{s}(x') = O(h^2)$ and, consequently, $\delta(x') \leq \tilde{s}(x')$, if $h \leq h_o$ holds and h_o is sufficiently small, cf. (3.2.12) and Fig. 9b. Thus, we have

$$\mathfrak{S}(x') \subset \Box_p(x') := \Delta_p^-(x') \cup \Delta_p^+(x'),\ \Box_p \text{ is a rhomboid, for } x' \in \gamma'_{23},\ \mathfrak{h}(x') \subset \bar{\Omega}. \tag{5.2.11b}$$

Here, $\Delta_p^{\pm}(x')$ are congruent triangles taken from $\{\Delta_p^i, \Delta_p^e\}$, and $\mathfrak{S}(x')$ is a subset of a diagonal of $\Box_p$. ≡

It should be noted that the special choice $\Box_p$ from (5.2.11b) is not the only one available and that Δ_p^e is not needed in the case $\mathfrak{h}(x') \subset \Gamma$. Finally, Remark 5.8. is devoted to the problem of finding a so-called "reference domain" of a given "estimation region".

Remark 5.8. The assumption (5.2.2) ensures that the triangles Δ_p and the deltoids $\Box_p$ from Remark 5.6. and 5.7. are regular, uniformly with respect to h $(h \leq h_o)$, i.e., they do not degenerate as $h \to 0$. Moreover, (5.2.2) implies the relations mes $\mathfrak{S}^+(x') = 0$ or (and) mes $\mathfrak{S}^-(x') = 0$ $(\theta = \frac{\pi}{2})$, or

$$\text{mes } \mathfrak{S}(x') \geq \varepsilon h > 0 \text{ for } x' \in \omega' + \gamma'_{23},\ \text{mes } \mathfrak{S}^{\pm}(x') \geq \varepsilon h > 0 \text{ for } x' \in \omega', \tag{5.2.12}$$

cf. Remarks 2.18., 2.20. and 2.21. Each triangle $\Delta := \Delta_p$ can be represented as the image of a fixed straight reference triangle $\hat{\Delta}$ (cf. Appendix TR) by a unique invertible mapping F which depends on Δ, viz.

$$\Delta = F(\hat{\Delta}), \ x = F(\hat{x}) = B\hat{x} + b, \ B \in L(R^2, R^2), \ b \in R^2. \tag{5.2.13a}$$

Clearly, B and b here depend on Δ. But throughout this paper, we shall omit the index Δ at F, B and b. The matrix B is regular iff Δ is a regular triangle. Moreover, the reference triangle $\hat{\Delta}$ can be represented as the image of Δ by the inverse mapping F^{-1}:

$$\hat{\Delta} = F^{-1}(\Delta), \ \hat{x} = F^{-1}(x) = B^{-1}x - B^{-1}b, \ \text{B and b from (5.2.13a)}, \equiv (5.2.13b)$$

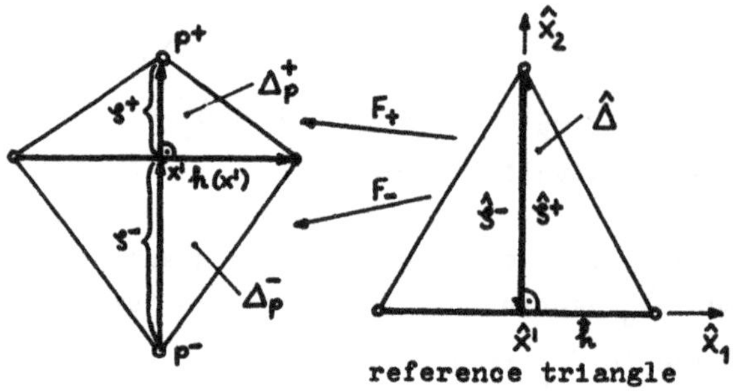

Fig. 10a

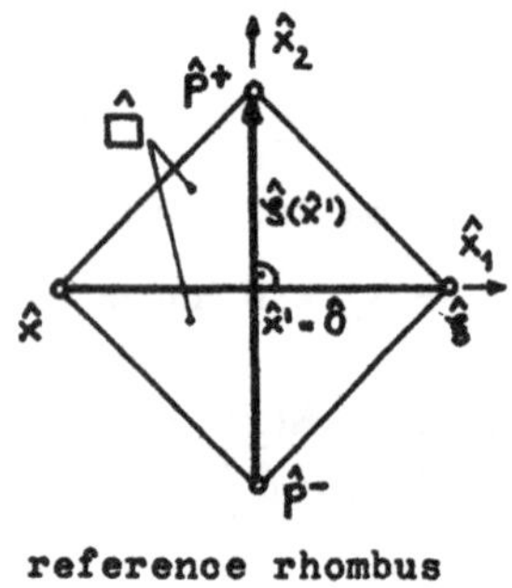

Fig. 10b

It is well known that the properties " x' is the midpoint of a triangle side $h(x')$" or " x" is the centre of gravity of a triangle" are preserved by an affine mapping. Likewise, the medians of the triangle Δ are transformed into the medians of the reference triangle $\hat{\Delta}$. The partial segments $\mathfrak{s}^{\pm}(x')$ of the perpendicular bisectors of $\Delta_p^{\pm}(x')$ are simultaneously located on a median of $\Delta_p^{\pm}(x')$ and, therefore, the image $F_{\pm}^{-1}(\mathfrak{s}^{\pm})$ always lies on a median of the reference triangle $\hat{\Delta}$, since F^{-1} is affine. If $\Box := \Box_p$ is a rhombus, then $\Box$ can be mapped on a fixed reference rhombus $\hat{\Box}$, say a square, by means of an affine transformation F^{-1} from (5.2.13b).

After this discussion about one proposal of defining local estimation regions and mappings, which put these regions in an affine one-to-one correspondence to a fixed reference domain, we proceed with further estimations of the $L_2(\mathfrak{s})$-norms in (5.2.5) and (5.2.9).

<u>5.2.3. The estimation of $\varkappa_k$ and $\varkappa_s$.</u> We now take the regions $\Delta_p^{\pm}$ and $\Box_p$ from (5.2.10), (5.2.11) and define the closed set $\bar{\Omega}_h^*$ by

$$\bar{\Omega}_h^* := \bigcup_{x' \in \omega' + \gamma'_{23}} \Box_p(x'), \quad \Box_p(x') \text{ from (5.2.10), (5.2.11)}, \tag{5.2.14}$$

i.e., $\bar{\Omega}_h^*$ contains all estimation regions. If a corner point $P_k \in \Gamma$ has an reentrant angle, then an overlap of the triangles Δ_p^e from (5.2.11b) with some other triangles Δ_p cannot be excluded. But this overlap takes place only in an O(h)-neighbourhood of P_k and will not restrict the generality of the following considerations. We extend the function u by Calderon or Geymonat to a fixed, bounded domain $\tilde{\Omega}$, with $\tilde{\Omega} \supset \Omega \cup \Omega_h^*$ for $h \leq h_o$. In order to visualize a more concret situation, we could assume that $\partial\tilde{\Omega}$ has the same smoothness properties as $\partial\Omega$ and lies in an $O(h_o)$-neighbourhood of $\partial\Omega$.
We now go on to estimate the $L_2(\mathfrak{s})$-norms in (5.2.9),(5.2.10) under the assumption $u \in W_2^2(\Omega)$. Taking into account the affine transformation $F^{-1}: \Delta_p \longrightarrow \hat{\Delta}$ according to (5.2.13b), we write the functionals to be estimated in terms of the "∧"-variables of the reference plane. For this, cf. (5.2.13) and Appendix TR as well as Fig. 10a. Writing temporarily $\mathfrak{s}, \Delta, \mathfrak{s} \subset \Delta$ instead of $\mathfrak{s}^{\pm}, \Delta_p^{\pm}, \mathfrak{s}^{\pm} \subset \Delta_p^{\pm}$, we obtain, after transformation $\hat{\Delta} = F^{-1}(\Delta)$ according to (5.2.13b), the following relations for the further estimation of $\left\|\frac{\partial u}{\partial n}\right\|_{o,\mathfrak{s}}$:

$$\begin{aligned} \left\|\frac{\partial u}{\partial n}\right\|_{o,\mathfrak{s}}^2 &:= \int_{\mathfrak{s}} (\nabla u, n)^2 ds = \int_{\hat{\mathfrak{s}}} (B^{T-1}\hat{\nabla}\hat{u}, B\hat{n})^2 \frac{|B\hat{\mathfrak{s}}|}{|\hat{\mathfrak{s}}|} d\hat{s} \\ &\leq \|B^{-1}\|^2 \|B\| \int_{\hat{\mathfrak{s}}} |\hat{\nabla}\hat{u}|^2 d\hat{s} \leq \|B^{-1}\|^2 \|B\| C_I^2 \, \||\hat{\nabla}\hat{u}|\|_{1,\hat{\Delta}}^2, \end{aligned} \tag{5.2.15}$$

where the Trace Theorem 2 (Corollary) from Appendix IM was additionally used. The constant C_I does not depend on $\hat{u}$ or u, and, what is essential, C_I is also independent of the individual segment $\mathfrak{s}$, since each $\mathfrak{s}$-transformation $\hat{\mathfrak{s}}$ lies on a median of the reference triangle $\hat{\Delta}$. Using the relations between the seminorms $|\hat{u}|_{1,\hat{\Delta}}$ and $|u|_{1,\Delta}$ for the transformation $F: \hat{\Delta} \longrightarrow \Delta$, cf. Appendix TR, we get, first of all,

$$\||\hat{\nabla}\hat{u}|\|_{1,\hat{\Delta}}^2 \leq |\hat{u}|_{1,\hat{\Delta}}^2 + |\hat{u}|_{2,\hat{\Delta}}^2 \leq \|B\|^2 |\det B|^{-1} \{|u|_{1,\Delta}^2 + \|B\|^2 |u|_{2,\Delta}^2\}. \tag{5.2.16}$$

In accordance with Appendix TR and for B from (5.2.13), the estimates

$$\|B\| \leq C_1 h, \quad \|B^{-1}\| \leq C_2 h^{-1}, \quad |\det B|^{-1} \leq C_3 h^{-2} \tag{5.2.17}$$

hold, since all triangles Δ are regular, uniformly with respect to h, $h \leq h_o$. Combining (5.2.14), (5.2.15), (5.2.16) and interpreting $\mathfrak{s} \subset \Delta$ by $\mathfrak{s}^{\pm} \subset \Delta_p^{\pm}$, we are led to the estimate

$$\left\|\frac{\partial u}{\partial n}\right\|_{o,\mathfrak{s}^{\pm}}^2 \leq C_4 h^{-1} \{ |u|_{1,\Delta_p^{\pm}}^2 + h^2 |u|_{2,\Delta_p^{\pm}}^2 \} \quad \text{for } \mathfrak{s}^{\pm}(x') \subset \Delta_p^{\pm}(x') \subset \bar{\Omega}_h^*, \quad x' \in \omega' + \gamma_{23}', \tag{5.2.18}$$

where $\Delta_p^- = \emptyset$ or $\Delta_p^+ = \emptyset$ is possible in some cases, cf. (5.2.11a).

We now turn to the important special case that $u \in W_2^3(\Omega)$ and $\mathcal{S}(x') \subset \subset \square_p(x') \subset \bar{\Omega}$ hold for $x' \in \omega'$, cf. (5.2.6), where $\square := \square_p(x')$ is a rhombus, and proceed to estimate $\left\| \frac{\partial^2 u}{\partial s \partial n} \right\|_{0,\mathcal{S}}$ from (5.2.9). After the transformation $\hat{\square} = F^{-1}(\square)$, with F^{-1} from (5.2.13b), we get the following inequalities, which can be apparently derived from relations presented in Appendix DI, TR and IM (in the following, the summation over the repeated subscripts is understood):

$$\left\| \frac{\partial^2 u}{\partial s \partial n} \right\|_{0,\mathcal{S}}^2 := \int_{\mathcal{S}} \left(\frac{\partial^2 u}{\partial s \partial n}\right)^2 ds \leq \int_{\mathcal{S}} \left(\frac{\partial^2 u}{\partial x_i \partial x_j}\right)^2 ds \leq \|B^{-1}\|^4 \, \|B\| \left\| \frac{\partial^2 \hat{u}}{\partial \hat{x}_i \partial \hat{x}_j} \right\|_{0,\hat{\mathcal{S}}}^2 ,$$

$$\left\| \frac{\partial^2 \hat{u}}{\partial \hat{x}_i \partial \hat{x}_j} \right\|_{0,\hat{\mathcal{S}}}^2 \leq C_I^2 \left\| \frac{\partial^2 \hat{u}}{\partial \hat{x}_i \partial \hat{x}_j} \right\|_{1,\hat{\square}}^2 \leq C_I^2 \left\{ |\hat{u}|_{2,\hat{\square}}^2 + |\hat{u}|_{3,\hat{\square}}^2 \right\} \tag{5.2.19}$$

$$\leq C_I^2 \, \|B\|^4 \, |\det B|^{-1} \left\{ |u|_{2,\square}^2 + \|B\|^2 \, |u|_{3,\square}^2 \right\}.$$

Hence, (5.2.19) and (5.2.17) yield

$$\left\| \frac{\partial^2 u}{\partial s \partial n} \right\|_{0,\mathcal{S}}^2 \leq C_5 h^{-1} \left\{ |u|_{2,\square}^2 + h^2 |u|_{3,\square}^2 \right\} \text{ for } \mathcal{S}(x') \subset \square := \square_p(x'),\ x' \in \omega',$$
$$\square\text{: rhombus, } u \in W_2^3(\Omega). \tag{5.2.20}$$

On the other hand, we obtain the inequality

$$\left\| \frac{\partial u}{\partial n} \right\|_{0,\mathcal{S}}^2 \leq C_6 h^{-1} \left\{ |u|_{1,\square}^2 + h^2 |u|_{2,\square}^2 \right\} \text{ for } \mathcal{S}(x') \subset \square := \square_p(x'),$$
$$x' \in \omega' + \gamma'_{23},$$
$$\square\text{: deltoid, } u \in W_2^2(\Omega_h^*). \tag{5.2.21}$$

If $\square$ is a rhombus, we get (5.2.21) in the same way as (5.2.18). But if $\square := \square_p(x')$ is not a rhombus, then, by splitting $\mathcal{S}(x') = \mathcal{S}^-(x') \cup \mathcal{S}^+(x')$, $\mathcal{S}^{\pm}(x') \subset \Delta_p^{\pm}(x')$, and by addition of two inequalities of the type (5.2.18) for $\Delta_p^{\pm}$, we arrive at (5.2.21).

Now we are in position to estimate $\varkappa_k$ from (5.2.1b). Inequalities (5.2.5) and (5.2.9), together with (5.2.18), (5.2.20) and (5.2.21), imply the estimates (5.2.22), viz.

$$|\varkappa_k(x')|^2 \leq C_7 \left\{ |u|_{1,\square_p}^2 + h^2 |u|_{2,\square_p}^2 \right\} \quad \text{for } x' \in \omega' + \gamma'_{23},\ \mathcal{S}(x') \subset \square_p(x'), \tag{5.2.22a}$$

$k \in C^1(\bar{\Omega})$, $u \in W_2^2(\Omega_h^*)$, $\square_p$ stands as a symbol for a deltoid, a rhombus or a triangle (case (5.2.11a); for Ω_h^*, see (5.2.14)),

$$|\varkappa_k(x')|^2 \leq C_8 h^2 \left\{ |u|^2_{2,\square_p} + h^2 |u|^2_{3,\square_p} \right\} \text{ for } x' \in \omega' : \tag{5.2.22b}$$

x' is the midpoint of $\mathfrak{S}(x')$, i.e. $\square_p$ is a rhombus, $k \in C^2(\bar{\Omega})$, $u \in W_2^3(\Omega)$.

The functional $\varkappa_k$ is globally evaluated by the norm $\|\varkappa_k\|_o'$ given in (5.2.23), which will be inserted via (5.2.1a) into (5.1.32). Starting from (5.2.22a), we get

$$\begin{aligned} \|\varkappa_k\|_o'^2 &:= \sum_{x' \in \omega' + \gamma_{23}'} \varkappa_k^2(x')\; s^o(x')h(x') \leq C_9 h^2 \sum_{x' \in \omega' + \gamma_{23}'} \{ |u|^2_{1,\square_p}(x') \\ &+ h^2 |u|^2_{2,\square_p}(x') \} \leq C_9 h^2 \{ |u|^2_{1,\Omega_h^*} + h^2 |u|^2_{2,\Omega_h^*} \\ &+ C_9' [|u|^2_{1,\omega_h^*} + h^2 |u|^2_{2,\omega_h^*}] \}. \end{aligned} \tag{5.2.23}$$

The symbol ω_h^* denotes a "convergent boundary strip" along Γ and of width $O(h)$ (cf. Theorem 4 in Appendix IM), which contains the small areas possibly arising by the overlap of estimation regions caused by the use of exterior triangles Δ_p^e introduced in (5.2.11b). The constant C_9' counts the number of overlaps, where $0 \leq C_9' = O(1)$ holds for $h \leq h_o$. Due to Theorem 4 from Appendix IM and Theorem 1 from Appendix EX, the inequality $|u|^2_{1,\omega_h^*} + h^2 |u|^2_{2,\omega_h^*} \leq C_9'' h \|u\|^2_{2,\Omega}$ holds, which implies that the effect of overlap in (5.2.23) is only of the order $O(h^3)$. Thus, by means of (5.1.40) noted for Ω_h^* instead of Ω_h and by (5.2.23), the following estimate can be derived

$$\|\varkappa_k\|_o' \leq C_{10} h \|u\|_{2,\Omega} \quad \text{for } k \in C^1(\bar{\Omega}),\ u \in W_2^2(\Omega),\ \bar{\omega} \text{ irregular.} \tag{5.2.24}$$

Next we try to find bounds of $\|\varkappa_k\|_o'$ under the assumption that the grid $\bar{\omega}$ and the functions k, u are more regular than in (5.2.24). More precisely, we assume that $\bar{\omega}$ is regular or locally irregular, and $k \in C^2(\bar{\Omega})$, $u \in W_2^3(\Omega)$. First of all, we divide $\bar{\omega}'$ into subsets $\overset{\circ}{\omega}'$, $\overset{*}{\omega}'$ and γ' related by

$$\bar{\omega}' = \overset{\circ}{\omega}' + \overset{*}{\omega}' + \gamma', \text{ with } \overset{\circ}{\omega}' := \{x' \in \omega' : x' \text{ is the midpoint of } \mathfrak{S}(x')\}$$
$$\text{and } \overset{*}{\omega}' := \omega' \setminus \overset{\circ}{\omega}', \text{ and } \bar{\omega}', \omega', \gamma' \text{ are taken from (2.3.14).} \tag{5.2.25}$$

Apparently, the points $x' \in \overset{*}{\omega}' + \gamma_{23}'$ assigned to the locally irregular grids $\bar{\omega}$ from Definition 2.10., 2.11. or 2.12. are located in an $O(h)$-neighbourhood of Γ. Moreover, some mesh segments $\mathfrak{h}(x')$ with $\mathfrak{S}(x') = \{x'\}$, i.e. mes $\mathfrak{S}(x') = 0$, may occur. But the error $\varkappa(x')$ vanishes for $x' \in \overset{\circ\circ}{\omega}' := \{x' \in \omega' : \text{mes } \mathfrak{S}(x') = 0\}$ and, therefore, the discussion of $\overset{\circ\circ}{\omega}'$ is superfluous. Using (5.2.22) we can derive bounds of $\|\varkappa_k\|_o'$ as follows:

$$\|\varkappa_k\|_0'^2 := \Big\{\sum_{x'\in\overset{*}{\omega}'} + \sum_{x'\in\overset{*}{\omega}'+\gamma_{23}'}\Big\}\varkappa_k^2(x')s^o(x')h(x') \le \Big\{\max_{x'\in\omega'+\gamma_{23}'} s^o(x')h(x')\Big\}\, N(u),$$

with $N(u) := C_8h^2\sum_{x'\in\overset{*}{\omega}'}\|u\|^2_{3,\square_p(x')} + C_7\sum_{x'\in\overset{*}{\omega}'+\gamma_{23}'}\Big\{|u|^2_{1,\square_p(x')} + h^2|u|^2_{2,\square_p(x')}\Big\}$,

that is,

$$\|\varkappa_k\|_0'^2 \le C_{11}\Big\{h^4\|u\|^2_{3,\Omega} + h^2C_{11}'\Big[|u|^2_{1,\omega_h} + h^2|u|^2_{2,\omega_h}\Big]\Big\}. \qquad (5.2.26)$$

Here, ω_h is a "convergent boundary strip" along Γ and of width $O(h)$, which contains the estimation regions $\square_p(x')$ for the flux points $x'\in\overset{*}{\omega}'+\gamma_{23}'$ near and on the boundary Γ_h. Moreover, ω_h also includes the small "overlap areas" of the estimation regions, which was already discussed in (5.2.23). Due to $|u|^2_{1,\omega_h} + h^2|u|^2_{2,\omega_h} \le C_{11}''h\|u\|^2_{2,\Omega}$ and after further estimations in (5.2.26), we get the inequality

$$\|\varkappa_k\|_0' \le C_{12}h^{3/2}\|u\|_{3,\Omega} \text{ for } k\in C^2(\bar\Omega),\ u\in W_2^3(\Omega);\ \bar\omega \text{ is locally irregular}$$

$$\text{and taken from Definition 2.10., 2.11. or 2.12.,} \qquad (5.2.27a)$$

and, especially,

$$\|\varkappa_k\|_0' \le C_{13}h^2\|u\|_{3,\Omega} \quad \text{for } \Gamma = \Gamma_1,\ \bar\omega \text{ is regular } (\overset{*}{\omega}'+\gamma_{23}'=\emptyset). \qquad (5.2.27b)$$

We now turn to the functional $\varkappa_s(x')$ from (5.2.1d), the estimation of which is analogous to that of $\varkappa_k(x')$ and even simpler. Indeed, due to $s^o(x') = s(x')$ for $x'\in\omega'$, we have $\varkappa_s(x') = 0$ for $x'\in\omega'$, and for $x'\in\gamma_{23}'$ the estimate $|s^o(x') - s(x')|\le K_1h^2$ (cf. (3.2.12)) is known and gives rise to

$$|\varkappa_s(x')| \le \Big|\frac{1}{s^o(x')} - \frac{1}{s(x')}\Big| \int_{S(x')}\Big|\frac{\partial u}{\partial n}\Big|ds \le C_{14}h^{1/2}\Big\|\frac{\partial u}{\partial n}\Big\|_{o,S} \text{ for } u\in W_2^2(\Omega). \qquad (5.2.28)$$

This means that bounds such as those found for $\varkappa_k$ in (5.2.5) are now obtained for $\varkappa_s(x')$, $x'\in\gamma_{23}'$. All estimation regions for $\varkappa_s(x')$, where $\varkappa_s(x')\ne 0$, are located in the convergent boundary strip ω_h from (5.2.26). Therefore, $\|\varkappa_s\|_0'^2 \le C_{14}h^2\Big\{|u|^2_{1,\omega_h} + h^2|u|^2_{2,\omega_h}\Big\}$ holds also for grids which are irregular. Applying Theorem 4 from Appendix IM and Theorem 1 from Appendix EX, we get the estimate

$$\|\varkappa_s\|_0' \le C_{15}h^{3/2}\|u\|_{2,\Omega} \quad \text{for } u\in W_2^2(\Omega),\ \bar\omega \text{ irregular}, \qquad (5.2.29)$$

especially we have

$\|\varkappa_s\|_0' = 0$ for $\gamma_{23}' = \emptyset$ ($\Gamma_{23}=\emptyset$), or $s^0(x') = s(x')$ for all $x' \in \gamma_{23}'$.

<u>5.2.4. The estimation of $\varkappa_u$.</u> We now consider the functional $\varkappa_u(x')$ from (5.2.1c) and see clearly that $\varkappa_u$ is well-defined, linear and bounded on $W_2^2(\Omega)$, cf. Appendix IM. First of all, $\varkappa_u$ will be decomposed into functionals $\varkappa^{\pm}$ which can be treated on the triangles $\Delta_p^{\pm}$, i.e.,

$$\varkappa_u = \varkappa_u^- + \varkappa_u^+, \quad \varkappa_u^{\pm}(x') := \frac{s^{\pm}(x')}{s(x')} \left\{ \frac{1}{s^{\pm}(x')} \int_{\mathcal{S}^{\pm}(x')} \frac{\partial u}{\partial n} \, ds - \frac{u(\xi)-u(x)}{h(x,\xi)} \right\}, \tag{5.2.30}$$

with $\mathcal{S}^{\pm}(x') = \mathcal{S}(x') \cap \Delta_p^{\pm}(x')$; $\Delta_p^{\pm}(x')$ are the triangles defined in (5.2.10), (5.2.11), under the assumption (5.2.2). The case $\mathcal{S}(x') \subset \Delta_p(x')$, cf. (5.2.11a), is contained in (5.2.30), namely for $\mathcal{S}^- = \emptyset$ or $\mathcal{S}^+ = \emptyset$, i.e. $\varkappa_u^- = 0$ or $\varkappa_u^+ = 0$.
In order to estimate $\varkappa_u^{\pm}$ and $\varkappa_u$ from (5.2.30), we apply the Bramble-Hilbert Lemma, i.e. Theorem 1 from Appendix ES (in this lemma and in the sequel, let k be an integer which should not be confused with the coefficient k from L in (3.2.1)), and perform the following steps:

1° The affine transformation of $\Delta_p^{\pm}(x')$ onto a fixed reference triangle $\hat{\Delta}$ and of the rhombus $\Box_p(x')$ onto a fixed reference rhombus $\hat{\Box}$. If $\Box_p(x')$ is a deltoid, then the splitting $\Box_p = \Delta_p^- \cup \Delta_p^+$ will be applied.

2° The representation of $\varkappa_u^{\pm}$, $\varkappa_u$ by the "$\wedge$"-variables of the reference plane, i.e. $\hat{\varkappa}_{\hat{u}}^{\pm}$, $\hat{\varkappa}_{\hat{u}}$.

3° The verification of the boundedness of $\hat{\varkappa}_{\hat{u}}^{\pm}$, $\hat{\varkappa}_{\hat{u}}$ on $W_2^k(\hat{\Delta})$, $W_2^k(\hat{\Box})$, respectively, for $k \in \{2,3\}$.

4° The verification of the vanishing of the functionals $\hat{\varkappa}_{\hat{u}}^{\pm}$, $\hat{\varkappa}_{\hat{u}}$ on certain classes of polynomials.

5° The application of the Bramble-Hilbert Lemma (Theorem 1 from Appendix ES) and the local estimation of $\hat{\varkappa}_{\hat{u}}^{\pm}$ and $\hat{\varkappa}_{\hat{u}}$ by certain Sobolev seminorms of $\hat{u}$ on $\hat{\Delta}$ and $\hat{\Box}$, respectively.

6° The backward transformation and the local estimation of the error functionals $\varkappa_u^{\pm}$, $\varkappa_u$ by Sobolev seminorms from $W_2^k(\Delta_p^{\pm})$, $W_2^k(\Box_p)$, respectively.

7° The global estimation of $\varkappa_u$ over the whole grid by the norm $\|\varkappa_u\|_0'$, i.e., bounds of $\|\varkappa_u\|_0'$ in terms of powers of h and Sobolev (semi-)norms from $W_2^k(\Omega)$, $k \in \{2,3\}$, will be given.

First of all, we consider the case $\mathcal{S}(x') \subset \Delta_p(x') =: \Delta(x')$, transform Δ onto $\hat{\Delta}$ by means of F^{-1} from (5.2.13b) and represent the functional $\varkappa_u$ from (5.2.1c) by the "$\wedge$"-variables of the reference plane. Thus, we get

$$\varkappa_u(x') = \hat{\varkappa}_{\hat{u}}(\hat{x}') = \frac{1}{|B\hat{\mathfrak{s}}|} \int_{\hat{\mathfrak{s}}} (B^{T-1}\hat{\nabla}\hat{u}, B\hat{n}) \frac{|B\hat{\mathfrak{s}}|}{|\hat{\mathfrak{s}}|} d\hat{s} - \frac{\hat{u}(\hat{\xi})-\hat{u}(\hat{x})}{|B\hat{h}|} . \tag{5.2.31}$$

Here in step 2°, some rules of affine transformations from Appendix TR are used, especially $s(x') := |\mathfrak{s}| = |B\hat{\mathfrak{s}}|$, $h(x') := |h| = |B\hat{h}|$, $\nabla u = B^{T-1}\hat{\nabla}\hat{u}$, $ds = |B\hat{\mathfrak{s}}|\,|\hat{\mathfrak{s}}|^{-1} d\hat{s}$, $u(x) = \hat{u}(\hat{x})$. Obviously, $\hat{\varkappa}_{\hat{u}}$ is well-defined for $\hat{u} \in W_2^2(\hat{\Delta})$, linear and bounded on $W_2^2(\hat{\Delta})$, since $W_2^2(\hat{\Delta})$ is continuously imbedded in $C(\hat{\Delta})$, and the $L_2(\hat{\mathfrak{s}})$-norm of the traces of the first order derivatives of $\hat{u}$ on $\hat{\mathfrak{s}}$ can also be bounded from above by the $W_2^2(\hat{\Delta})$-norm, cf. Theorems 1, 2 from Appendix IM. Hence, estimating $|\hat{\varkappa}_{\hat{u}}|$ from (5.2.31), we get the inequalities

$$\begin{aligned} |\hat{\varkappa}_{\hat{u}}| &\le |B\hat{\mathfrak{s}}|^{-1}\|B^{-1}\|\|B\| \int_{\hat{\mathfrak{s}}} |\hat{\nabla}\hat{u}|\, d\hat{s} + |B\hat{h}|^{-1}\left\{|\hat{u}(\hat{\xi})| + |\hat{u}(\hat{x})|\right\} \\ &\le M_1 h^{-1}\left\{\||\hat{\nabla}\hat{u}|\|_{0,\hat{\mathfrak{s}}} + \|\hat{u}\|_{C(\hat{\Delta})}\right\} \le M_1(C_{I1}+C_{I2})h^{-1}\|\hat{u}\|_{2,\hat{\Delta}}, \end{aligned} \tag{5.2.32}$$

where bounds of $\|B\|$, $\|B^{-1}\|$ were taken from (5.2.17). Thus, the first assumption of the Bramble-Hilbert-Lemma is verified, and the step 3° is finished. We proceed with step 4° and show that $\hat{\varkappa}_{\hat{u}}$ vanishes on the space of all polynomials of degree ≤ 1, i.e.

$$\hat{\varkappa}_{\hat{p}_1}(\hat{x}') = 0 \quad \text{for each } \hat{p}_1(\hat{x}) := \hat{a}_o + (\hat{a},\hat{x}),\ \hat{a}_o \in R^1 \text{ and } \hat{a},\ \hat{x} \in R^2. \tag{5.2.33}$$

Evidently, because of the relations $\hat{\xi} - \hat{x} = \hat{h} = B^{-1}h$, $\hat{n} = B^{-1}h/h = B^{-1}n$ and $\hat{\nabla}\hat{p}_1 = \hat{a}$, we get the identities

$$\hat{\varkappa}_{\hat{p}_1}(\hat{x}') = |\hat{\mathfrak{s}}|^{-1} \int_{\hat{\mathfrak{s}}} (B^{T-1}\hat{a}, B\hat{n})d\hat{s} - |B\hat{h}|^{-1}(\hat{a}, \hat{\xi}-\hat{x}) = (\hat{a},\hat{n}) - (\hat{a},\frac{\hat{h}}{h}) = 0.$$

Consequently, the second assumption of the Bramble-Hilbert Lemma is confirmed, and relations (5.2.32), (5.2.33) admit the application of Theorem 1 from Appendix ES for $p = k = 2$. Thus, step 5° can be performed:

$$|\hat{\varkappa}_{\hat{u}}(\hat{x}')| \le M_2 C_{BH} h^{-1}|\hat{u}|_{2,\hat{\Delta}} \quad \text{for } \hat{u} \in W_2^2(\hat{\Delta}),\ M_2 := M_1(C_{I1}+C_{I2}), \tag{5.2.34}$$

where C_{BH} is the constant from the Bramble-Hilbert Lemma. According to step 6°, the backward transformation $\Delta = F(\hat{\Delta})$ from (5.2.13a) and further estimation yield

$$|\varkappa_u(x')| = |\hat{\varkappa}_{\hat{u}}(\hat{x}')| \le M_2 C_{BH} h^{-1}\|B\|^2\,|\det B|^{-1/2}|u|_{2,\Delta(x')} \le M_3|u|_{2,\Delta(x')} \tag{5.2.35}$$

for $u \in W_2^2(\Delta)$, where $|\hat{u}|_{2,\hat{\Delta}}$ was estimated in terms of $|u|_{2,\Delta}$, cf. (5.2.17) and Appendix TR. The constant M_3 does not depend on u, x' and h, $h \le h_o$. Estimate (5.2.35) was proved under the assumption $\mathfrak{s}(x') \subset \Delta(x')$.

Otherwise, if $\mathfrak{S}(x') \subset \square(x')$ holds for $x' \in \omega'$, then $\mathfrak{S}(x')$ will be divided into two portions $\mathfrak{S}^{\pm}(x')$, with $\mathfrak{S}^{\pm}(x') \subset \Delta^{\pm}(x')$, which implies the splitting of $\varkappa_u$ into $\varkappa_u^{\pm}$, cf. (5.2.30). Now, the same estimate (5.2.35) will be obtained for $\varkappa_u^{\pm}$ instead of $\varkappa_u$. By addition of two such inequalities, the result is

$$|\varkappa_u| \leq |\varkappa_u^-| + |\varkappa_u^+| \leq M_3\{|u|_{2,\Delta^-} + |u|_{2,\Delta^+}\} \leq M_4|u|_{2,\square} \tag{5.2.36}$$

for $x' \in \omega'$ and the deltoid $\square := \square(x')$, $u \in W_2^2(\square)$, $h \leq h_o$.

If $\square$ is a rhombus, then $\hat{\square} = F^{-1}(\square)$ can be performed and, in this case, $\hat{\varkappa}_{\hat{u}}$ can be estimated on the reference rhombus $\hat{\square}$ without splitting (5.2.30), but with the same results as given in (5.2.36). Thus, all cases are checked and, collecting the results, we get

$$|\varkappa_u(x')| \leq M_5|u|_{2,\square(x')} \quad \text{for } x' \in \omega' + \gamma'_{23},\ u \in W_2^2(\square),\ h \leq h_o, \tag{5.2.37}$$

where $\square$ is a deltoid, a rhombus or a triangle (case (5.2.11a)).

The next problem consists in finding smaller local bounds of $|\varkappa_u|$ for more regular solutions u and grids $\bar{\omega}$. More precisely, we assume $u \in W_2^3(\Omega)$ and that the grid $\bar{\omega}$ is regular or locally irregular. For flux points $x' \in \mathring{\omega}'$, $\mathring{\omega}'$ from (5.2.25), the deltoid $\square(x') := \square_p(x')$ is always a rhombus which may be transformed onto a fixed reference rhombus $\hat{\square}$ by means of F^{-1} from (5.2.13b). Without loss of generality, let the diagonal $\hat{\mathfrak{h}}(\hat{x},\hat{\mathfrak{z}})$ of $\hat{\square}$ be a subset of the $\hat{x}_1$-axis such that $\hat{x}' = \hat{0}$ (origin) and $\hat{x} = -\hat{\mathfrak{z}}$ hold, cf. Fig. 10b. The functional $\varkappa_u$ from (5.2.30) expressed by the "$\wedge$"-variables is also of the form (5.2.31). Furthermore, $\hat{\varkappa}_{\hat{u}}$ is bounded on $W_2^2(\hat{\square})$ as well as on $W_2^3(\hat{\square})$, and we may write

$$|\hat{\varkappa}_{\hat{u}}(\hat{x}')| \leq M_6 h^{-1}\|\hat{u}\|_{2,\hat{\square}} \leq M_6 h^{-1}\|\hat{u}\|_{3,\hat{\square}} \quad \text{for } x' \in \mathring{\omega}', \text{ cf. (5.2.25).} \tag{5.2.38}$$

In the special case $x' \in \mathring{\omega}'$, the functional $\hat{\varkappa}_{\hat{u}}$ vanishes on the space of polynomials of degree ≤ 2, i.e.

$$\hat{\varkappa}_{\hat{p}_2}(\hat{x}') = 0 \text{ for } \hat{p}_2(\hat{\underline{x}}) := \hat{a}_o + (\hat{a},\hat{\underline{x}}) + \tfrac{1}{2}(\hat{A}\hat{\underline{x}},\hat{\underline{x}}), \tag{5.2.39}$$

with $\hat{a}_o \in R^1$ and $\hat{a},\hat{\underline{x}} \in R^2$, $\hat{A} = \hat{A}^T \in L(R^2,R^2)$.

In order to verify (5.2.39), we insert $\hat{p}_2$ into $\hat{\varkappa}_{\hat{u}}$ and utilize the partial result (5.2.33) as well as the relations (cf. Fig. 10b) $\hat{x} = -\hat{\mathfrak{z}}$ $(\hat{A}\hat{\mathfrak{z}},\hat{\mathfrak{z}}) = (\hat{A}\hat{x},\hat{x})$, $d\hat{s} = d\hat{\underline{x}}_2$, $\hat{\underline{x}}|_{\hat{\mathfrak{S}}} = (0,\hat{\underline{x}}_2)^T$ and $e_2 := (0,1)^T$ for the proof of the following identities:

$$\hat{\varkappa}_{\hat{p}_2}(\hat{x}') = |\hat{S}|^{-1} \int_{\hat{S}} (B^{T-1}\hat{A}\hat{\hat{x}}, B\hat{n})d\hat{s} - (2|B\hat{A}|)^{-1} \{(\hat{A}\hat{s},\hat{s}) - (\hat{A}\hat{x},\hat{x})\}$$
$$= |\hat{S}|^{-1} \int_{\hat{S}} (\hat{A}\hat{\hat{x}},\hat{n})d\hat{s} = |\hat{S}|^{-1} (\hat{A}e_2,\hat{n}) \int_{\hat{S}} \hat{\hat{x}}_2 \, d\hat{\hat{x}}_2 = 0.$$

Thus, (5.2.38) and (5.2.39) verify that $\hat{\varkappa}_{\hat{u}}$ satisfies the assumptions of the Bramble-Hilbert Lemma (Theorem 1 from Appendix ES) for $p = 2$, $k = 3$. Consequently, we get

$$|\varkappa_u(x')| = |\hat{\varkappa}_{\hat{u}}(\hat{x}')| \leq M_6 C_{BH} h^{-1} |\hat{u}|_{3,\hat{\Box}(\hat{x}')} \text{ for } \hat{u} \in W_2^3(\hat{\Box}), \quad (5.2.40)$$

and, after backward transformation and further estimations (cf. Appendix TR), this leads to

$$|\varkappa_u(x')| \leq M_6 C_{BH} h^{-1} \|B\|^3 |\det B|^{-\frac{1}{2}} |u|_{3,\Box(x')} \leq M_7 h |u|_{3,\Box(x')}$$
$$\text{for } u \in W_2^3(\Box(x')), \ h \leq h_0, \ x' \in \mathring{\omega}', \ \mathring{\omega}' \text{ from (5.2.25).} \quad (5.2.41)$$

At this point we notice that the local estimates of the error functional $\varkappa_u$ are complete and given in (5.2.37) and (5.2.41).

In accordance with step 7°, we proceed with the global estimation of $\varkappa_u$, which can be performed just as for $\varkappa_k$. Reasoning the same way as in (5.2.23), (5.2.24) and starting from (5.2.37), we get the inequalities

$$\|\varkappa_u\|_0'^2 := \sum_{x' \in \mathring{\omega}' + \gamma_{23}'} \varkappa_u^2(x') s^{\circ}(x') h(x') \leq M_8 h^2 \{|u|^2_{2,\Omega_h^*} + M_8' |u|^2_{2,\omega_h^*}\} \quad (5.2.42)$$
$$\leq M_9 h^2 \|u\|^2_{2,\Omega} .$$

Especially, if $\bar{\omega}$ is locally irregular, we obtain the estimates

$$\|\varkappa_u\|_0'^2 \leq M_{10} \{h^4 \sum_{x' \in \mathring{\omega}'} |u|^2_{3,\Box(x')} + h^2 M_{10}' |u|^2_{2,\omega_h}\} \leq M_{11} h^3 \|u\|^2_{3,\Omega}, \quad (5.2.43)$$

which follows from (5.2.37), (5.2.41); cf. the technique of deriving (5.2.26), (5.2.27a). Particularly, for $\Gamma = \Gamma_1 (\Gamma_{23} = \emptyset)$ and for regular grids $\bar{\omega}$, we have $\overset{*}{\omega}' + \gamma_{23}' = \emptyset$ and, consequently, $\omega_h = \emptyset$ in (5.2.43). Thus, the inequality

$$\|\varkappa_u\|_0'^2 \leq M_{10} h^4 |u|^2_{3,\Omega} \text{ for } u \in W_2^3(\Omega), \ \overset{*}{\omega}' + \gamma_{23}' = \emptyset, \quad (5.2.44)$$

holds, which follows from (5.2.43).

<u>5.2.5. The weakening of the assumptions and conclusions.</u> For the sake of simplicity, some assumptions were employed for estimating the error functionals $\varkappa_k$, $\varkappa_s$ and $\varkappa_u$. In particular, the restriction (5.2.2)

ensured the regularity of the transformation F from (5.2.13) and (for $\mathfrak{S}^{\pm} \neq \emptyset$) mes $\mathfrak{S}^{\pm}(x') \geq \varepsilon h > 0$. Nevertheless, some weakening of the assumptions may be given.

<u>Remark 5.9.</u> Instead of V1"(PB,ϖ) from (5.2.2), with $\theta \sim \frac{\pi}{2}$ in some cases, let the following more general condition be fulfilled:
Condition 2.17. - V0"(PB,ω^1) and $0 < \theta_o \leq \theta \leq \frac{\pi}{2}$ for the angles θ of all triangles. (5.2.45)

Then, the global estimates of $\varkappa_k$, $\varkappa_s$ and $\varkappa_u$ are maintained, but now, in general, with constants greater than those obtained under assumption (5.2.2). ≡

Assumption (5.2.45) only guarantees mes $\mathfrak{S}(x') \geq \varepsilon h > 0$ and mes $\mathfrak{S}^{\pm}(x') \geq 0$ instead of (5.2.12). Hence, the triangles $\Delta_p^{\pm}(x')$ may, in general, degenerate and are not regular (uniformly with respect to h, $h \leq h_o$). In HEINRICH [7], a technique of "geometric majorant of the estimation region" is introduced, which offers an outlet in such cases and may described as follows. The deltoid $\square_p(x')$ from (5.2.10) is not split into two triangles $\Delta_p^{\pm}(x')$, but it is extended to a rhombus $\tilde{\square}_p(x')$ which contains $\square_p(x')$ and is regular, uniformly with respect to h, $h \leq h_o$. The rhombus is affine-equivalent to a fixed reference rhombus $\hat{\square}$, and the local estimates of $\varkappa_k$, $\varkappa_s$ and $\varkappa_u$ which were obtained in the previous sections are preserved. But the extended estimation region $\tilde{\square}_p$, which is a geometric majorant of $\square_p$, causes some overlap of several regions of the type $\tilde{\square}_p$. Nevertheless, this overlap does not disturb the asymptotic behaviour of the bounds of $\|\varkappa_k\|_o'$, $\|\varkappa_s\|_o'$ and $\|\varkappa_u\|_o'$. Due to the fact that the number N of grid points which are contained in an O(h)-neighbourhood of each point $x \in \tilde{\Omega}$ is finite, i.e. $N = O(1)$, we get, in general, greater constants. This technique is studied in HEINRICH [7]. Moreover, (5.2.45) is not the weakest condition for getting estimate (5.2.47). For instance, it is already sufficient to have segments $\mathfrak{S}(x')$ with mes $\mathfrak{S}(x') \geq \varepsilon h > 0$ or mes $\mathfrak{S}(x') = 0$, i.e. V0"(PB,ω^1) alone could be used.

<u>Remark 5.10.</u> The splitting of the error functional $\varkappa$, the choice of the estimation regions and the extension of $u \in W_2^2(\Omega)$ can also be performed in a different manner. For instance, instead of (5.2.1a) we could use

$$\varkappa(x_s') := \frac{1}{s^o(x_s')} \int_{\mathfrak{S}^o(x_s')} (k - k_h(x_s')) \frac{\partial u}{\partial n}\, ds + k_h(x_s') \left\{ \frac{1}{s^o(x_s')} \int_{\mathfrak{S}^o(x_s')} \frac{\partial u}{\partial n}\, ds - u_h(x_s') \right\}$$
$$+ \frac{1}{s^o(x_s')} \int_{\delta\mathfrak{S}(x_s')} k \frac{\partial u}{\partial n}\, ds, \text{ with mes } \delta\mathfrak{S} = O(h^2), \qquad (5.2.46)$$

and this allows us to modify the estimation regions at the boundary Γ_h, cf. HEINRICH [7]. Error splittings similar to (5.2.46) will be discussed below for FDS $A_h y = F_h$ of the type (MD). Moreover, the primary triangles Δ can also be used for the error estimation of PB-schemes, cf. HEINRICH [7]. If the function $u \in W_2^2(\Omega)$ is extended by Geymonat, cf. Theorem 2 in Appendix EX, then the "rounding" of seminorms to the full norm is not necessary, i.e. we could take the estimates $|u|_{k,\tilde{\Omega}} \leq C_E' |u|_{k,\Omega}$, $\Omega \subset \tilde{\Omega}$. ≡

The global estimate of the error functional $\varkappa$ from (5.2.1) will be described in the following

Lemma 5.11. Let $\bar{\omega}$ be a grid defined by an irregular network of triangles and (or) rectangles, which is equipped with a secondary network of the type (PB). Furthermore, let Condition 2.17. - VO"(PB,ω^1) and $0 < \theta_0 \leq \theta \leq \frac{\pi}{2}$ (cf. (5.2.45)) be satisfied. Then, for the error $\varkappa$ from (5.1.32), with the splitting (5.2.1), the following estimates hold:

$$\|\varkappa\|_0' \leq C_0 \begin{cases} h\ \|u\|_{2,\Omega} & k \in C^1(\bar{\Omega}),\ u \in W_2^2(\Omega),\ \bar{\omega} \text{ irregular}, \qquad (5.2.47) \\ h^{3/2} \|u\|_{3,\Omega} \text{ for } & k \in C^2(\bar{\Omega}),\ u \in W_2^3(\Omega),\ \bar{\omega} \text{ locally irregular}, \\ h^2\ \|u\|_{3,\Omega} & k \in C^2(\bar{\Omega}),\ u \in W_2^3(\Omega),\ \bar{\omega} \text{ regular and } \Gamma = \Gamma_1. \end{cases}$$

The norm $\|\varkappa\|_0'$ is given in (5.1.32) and, the locally irregular grids $\bar{\omega}$ are taken from Definition 2.10., 2.11. or 2.12. The constant C_0 is a universal constant which is independent of u and h, $h \leq h_0$. ≡

Proof: The estimates obviously follow from (5.2.24), (5.2.27), (5.2.29), (5.2.42), (5.2.43) and (5.2.44), with $\varkappa$ from (5.2.1); cf. also Remark 5.9.

5.3. The error $\varkappa$ of the principal part of MD-operators

5.3.1. Splittings and estimates via centres of gravity. The error estimation of the principal part of MD-operators A_h is essentially similar to that of PB-operators, i.e., the discrete L_2-norm of $\varkappa$ will be estimated and bounded in terms of some constants and powers of h. According to the a priori error estimates (5.1.33) and (5.1.34) from Theorem 5.3., the norms $\|\varkappa\|_0''$ and $'\|\varkappa\|_0$ must be evaluated. This requires a separate discussion of the estimation of $\varkappa$ from (5.1.21) with respect to the centres of gravity $x'' \in \bar{\omega}''$, on the one hand, and of $\varkappa$ from (5.1.22) with respect to the flux points $x' \in \omega' + \gamma_{23}'$ (mesh midpoints), on the other hand. It should be mentioned that this distinction is essential for the assumptions to be made on the grid regularity and for the asymptotic behaviour of the error bounds. Thus, the smaller bounds will be obtained for the x'-splitting (5.1.22), but the x"-splitting (5.1.15) -

(5.1.18) requires only the grid regularity condition VO(MD, $\bar{\omega}$). Since a detailed analysis of such estimates was already presented in the paper HEINRICH [7], the author confines himself to giving a survey of the results.

This section is devoted to estimates of $\varkappa$ from (5.1.21) via the norm $\|\varkappa\|_0''$ from (5.1.33), i.e., the x"-splitting and $\varkappa$ from (5.1.21) are taken into account. The scalar components $\varkappa_1$, $\varkappa_2$ of the vector functional $\varkappa = (\varkappa_1, \varkappa_2)^T \in R^2$ contain bilinear functionals of k_{ij} and u. Obviously, $\varkappa_1$, $\varkappa_2$ are well-defined for $k_{ij} \in C(\bar{\Omega})$ and $u \in W_2^2(\Omega)$, since $u \in C(\bar{\Omega})$ and $|\nabla u| \in L_2(\mathfrak{S}^{\pm})$ hold, where $\mathfrak{S}^{\pm}(.,x")$ are parts of the medians of $\Delta(x")$, cf. (5.1.21). By analogy with (5.2.1), the vector functional $\varkappa$ will be split into three parts, i.e.

$$\varkappa(x") = \{\varkappa^{(k)} + \varkappa^{(s)} + \varkappa^{(u)}\}(x") \quad \text{for } x" \in \bar{\omega}", \tag{5.3.1}$$

where $\varkappa^{(k)}$, $\varkappa^{(s)}$ and $\varkappa^{(u)}$ involve approximation errors of the coefficients, the boundary Γ and the first order derivations of u. For the representation and estimation of the errors $\varkappa^{(.)}$ from (5.3.1), a finer splitting will be introduced. This splitting will be expressed by the following vector functionals a, $\tilde{a}$, with respect to the subsets ω_i'' (i=0,1,2) of $\bar{\omega}"$ introduced in (5.1.19):

$$x" \in \omega_0'': a(x",K,u) := \frac{1}{\text{mes}\,\Delta(x")} \Big\{ \Big[\int_{\mathfrak{S}^-(x,x")} (K\nabla u, n)ds \Big] h + \Big[\int_{\mathfrak{S}^-(\xi,x")} (K\nabla u, n)ds \Big] (h^+ - h)$$

$$- \Big[\int_{\mathfrak{S}^-(\xi^+,x")} (K\nabla u, n)ds \Big] h^+ \Big\}, \tag{5.3.2a}$$

$$x" \in \omega_1'': a(x",K,u) := \frac{1}{\text{mes}\,\Delta(x")} \Big\{ \Big[\int_{\mathfrak{S}(x,x")} (K\nabla u, n)ds \Big] h^+ + \Big[\int_{\mathfrak{S}(\xi,x")} (K\nabla u, n)ds \Big] (h^+ - h) \Big\}, \tag{5.3.2b}$$

$$x" \in \omega_2'': a(x",K,u) := \frac{1}{\text{mes}\,\Delta(x")} \Big[\int_{\mathfrak{S}(x,x")} (K\nabla u, n)ds \Big] h; \tag{5.3.2c}$$

$x" \in \omega_i''$ (i=0,1,2):

$a(x",K_h,u)$: just like $a(x",K,u)$, but with the constant matrix $K_h = K_h(x")$ from (3.3.9a) instead of K, (5.3.2d)

$\tilde{a}(x",K_h,u)$: just like $a(x",K_h,u)$, but with the modified integration intervals $\tilde{\mathfrak{S}}^-(.,x")$ or $\tilde{\mathfrak{S}}(.,x")$ instead of $\mathfrak{S}^-(.,x")$ or $\mathfrak{S}(.,x")$, cf. (2.3.16). (5.3.2e)

The symbol $\tilde{\mathfrak{S}}$ denotes the modification of $\mathfrak{S}$ from $\tilde{\Delta}(x")$ with respect to the straight triangle $\Delta(x")$, cf. Section 2.3.3. and Figs. 8b, c. The function $u \in W_2^2(\Omega)$ is extended by Calderon to $u \in W_2^2(\tilde{\Omega})$, with $\tilde{\Omega} \supset \Omega \cup \Omega_h$ for $h \leq h_o$ and fixed $\tilde{\Omega}$. Then, $\tilde{a}(.)$ in (5.3.2e) is always defined. An extension of $K = (k_{ij})_{2x2}$ to $\tilde{\Omega}$ is not necessary, and for further estimates, $k_{ij} \in C^1(\bar{\Omega})$ is assumed to be fulfilled. Finally, the errors $\varkappa^{(.)}$ in (5.3.1) are given by

<u>Definition 5.12.</u> Let $\varkappa^{(.)}$ denote the following partial errors:

$$\varkappa^{(k)}(x''):= \varkappa^{(k)}(x'',K,u):= a(x'',K,u) - a(x'',K_h,u),$$

$$\varkappa^{(s)}(x''):= \varkappa^{(s)}(x'',K_h,u):= a(x'',K_h,u) - \tilde{a}(x'',K_h,u),$$

$$\varkappa^{(u)}(x''):= \varkappa^{(u)}(x'',K_h,u):= \tilde{a}(x'',K_h,u) - g(x'',K_h,u),$$

with a, $\tilde{a}$ from (5.3.2) and (5.3.3)

$$g(x'',K_h,u):=\begin{cases} (K_h \nabla_{\hbar} u)(x'') & \text{for } x'' \in \omega_o'', \\ \frac{1}{\text{mes}\,\Delta(x'')} \; (K_h \nabla_{\hbar} u, \frac{\hbar^{\oplus}-\hbar^{o}}{2})\hbar^{+} + (K_h \nabla_{\hbar} u, \frac{-\hbar^{\oplus}}{2})(\hbar^{+}-\hbar) & \\ & \text{for } x'' \in \omega_1'', \\ \frac{1}{\text{mes}\,\Delta(x'')} \; (K_h \nabla_{\hbar} u, \frac{\hbar^{\oplus}-\hbar^{o}}{2})\hbar & \text{for } x'' \in \omega_2''. \blacksquare \end{cases}$$

Indeed, summing up the partial errors $\varkappa^{(.)}$, identity (5.3.1) proves true. After some estimations, the following inequalities will be obtained:

$$|\varkappa^{(k)}(x'')| \leq M_1 h^{1/2} \max_{\eta,\pm} \left\{ \| |\nabla u| \|_{o,\tilde{\zeta}^{\pm}} + h^{1/2} \| |\nabla u| \|_{o,\delta\zeta^{\pm}} \right\}$$

$$\text{for } x'' \in \bar{\omega}'', \qquad (5.3.4)$$

$$|\varkappa^{(s)}(x'')| \leq M_2 \max_{\eta,\pm} \| |\nabla u| \|_{o,\delta\zeta^{\pm}} \quad \text{or} \quad \varkappa^{(s)}(x'') = 0$$

with $\tilde{\zeta}^{\pm} := \tilde{\zeta}^{\pm}(\eta, x'')$, $\delta\zeta^{\pm} := \delta\zeta^{\pm}(\eta, x'')$, $\delta\zeta^{\pm} = \tilde{\zeta}^{\pm} \setminus \zeta^{\pm}$ or $\delta\zeta^{\pm} = \zeta^{\pm} \setminus \tilde{\zeta}^{\pm}$ in the cases $\tilde{\zeta}^{\pm} \supset \zeta^{\pm}$ or $\zeta^{\pm} \supset \tilde{\zeta}^{\pm}$, respectively, $\delta\zeta^{\pm} = \emptyset$ for $\tilde{\zeta}^{\pm} = \zeta^{\pm}$ (at least, for all Δ: $\Delta \cap \Gamma_h = \emptyset$), and η runs through the vertices $x, \xi, \xi^{+} \in \partial^2\Delta$ for $x'' \in \omega_o''$, or through x, ξ for $x'' \in \omega_1''$, and $\eta = x$ for $x'' \in \omega_2''$. For $\delta\zeta^{\pm} \neq \emptyset$, we have mes $\delta\zeta^{\pm} = O(h^2)$, cf. (3.3.3). In analogy to Section 5.2., the "line functionals" $\| |\nabla u| \|_o$ noted in (5.3.4) must be bounded by Sobolev norms of u on two-dimensional finite regions ("estimation regions"). Appropriate estimation regions are here given by the triangles $\Delta(x'')$ of the primary triangulation or by the triangles with two vertices from $\Delta(x'')$ and x" as the third vertex. Due to VO(MD, $\bar{\omega}$), these triangles are regular, uniformly with respect to h, $h \leq h_o$, which guarantees the uniform regularity of F, F^{-1} from (5.2.13). It should be noted that the affine images of medians in triangles Δ are also medians in the reference triangle $\hat{\Delta}$, which implies that integration intervals

$\varsigma^\pm$, $\tilde{\varsigma}^\pm$, $\delta\varsigma^\pm$ lying on medians are mapped onto subsets of reference medians. If $\delta\varsigma \not\subset \Delta(x'')$ occurs at the boundary Γ, cf. Fig. 8c, an "exterior auxiliary triangle" $\Delta_e(x''_e)$ can be defined such that $\delta\varsigma \subset \Delta_e(x''_e)$ and $\Delta(x'') \cap \Delta_e(x''_e) = h(x')$, $h(x') \subset \Gamma_h$, holds. Moreover, $\delta\varsigma := \delta\varsigma(.,x'')$ lies on a median of Δ_e, and the arc $\widehat{x\varsigma_\Gamma}$ is contained entirely in Δ_e, cf. Fig. 11c. If there are reentrant corners on Γ, then an overlap of some triangles within a small, convergent boundary strip of width O(h) cannot be excluded. But this will not affect the results given in Lemma 5.13. However, all integration intervals $\tilde{\varsigma}^\pm$, $\delta\varsigma^\pm$ are subsets of medians of a triangle.
We now transform the "estimation triangles" onto $\hat{\Delta}$, apply the Trace Theorem (Theorems 2, 3 from Appendix IM), perform the backward transformation and obtain local bounds of $|\varkappa^{(k)}(x'')|$, $|\varkappa^{(s)}(x'')|$, e.g.

$$|\varkappa^{(k)}(x'')| \le M_3\{|u|^2_{1,\Delta} + h^2|u|^2_{2,\Delta}\}^{1/2} \quad \text{for } \Delta = \Delta(x''),$$

$$|\varkappa^{(s)}(x'')| \le M_4 h^{-1/2}\{|u|^2_{1,\Box} + h^2|u|^2_{2,\Box}\}^{1/2} \text{ for } \Box = \Delta(x'') \cup \Delta_e(x''_e), \quad \Delta_e = \emptyset \text{ may occur.} \tag{5.3.5}$$

Furthermore, we consider the error $\varkappa^{(u)}$ from (5.3.3), which results from the approximation of $\nabla\mu$ by $\nabla_h u$. The idea of estimating $\varkappa^{(u)}$ is similar to that proposed in Section 5.2.4., but here the primary triangles $\Delta(x'')$ are involved. We shall apply the Bramble-Hilbert Lemma to the components $\varkappa_i^{(u)}$ $(i=1,2)$ and obtain local bounds of the type

$$|\varkappa^{(u)}(x'')| \le M_5|u|_{2,\Delta} \quad \text{for } \Delta = \Delta(x''). \tag{5.3.6}$$

Summing up local bounds of the type (5.3.5), (5.3.6) and using Theorem 4 from Appendix IM, Theorem 1 from Appendix EX and some elementary estimates, we arrive at

<u>Lemma 5.13.</u> For the partial errors $\varkappa^{(k)}$, $\varkappa^{(s)}$ and $\varkappa^{(u)}$ from (5.3.1), (5.3.3) and on irregular grids $\bar{\omega}$ (networks of triangles, with VO(MD, $\bar{\omega}$)), the following estimates hold:

$$\| |\varkappa^{(l)}| \|^*_0 \le M_6 h \|u\|_{2,\Omega} \quad \text{for } l = k, s, u, \tag{5.3.7}$$

under the smoothness assumptions $u \in W^2_2(\Omega)$, $k_{ij} \in C(\bar{\Omega})$ and, especially, $k_{ij} \in C^1(\bar{\Omega})$ for $l = k$, the constant M_6 is independent of u and h, $h \le h_o$. ■
<u>Proof:</u> See HEINRICH [7].

<u>5.3.2. Splittings and estimates via mesh midpoints.</u> We now consider the functional $\varkappa(x')$ defined in (5.1.22), which was introduced as a consequence of the x'-splitting and which will be needed in the estimate

(5.1.34), with the norm $\|\varkappa\|_0$. The introductory remarks of Section 5.3.1., preceding relation (5.3.1), can be transferred to this section, and we are in position to start with the splitting

$$\varkappa(x') = \{\varkappa^{(k)} + \varkappa^{(s)} + \varkappa^{(u)}\}(x') \text{ for } x' \in \omega' + \gamma'_{23}, \qquad (5.3.8)$$

$\varkappa$ from (5.1.22), and ω', γ'_{23} from (2.3.14). Moreover, the "forward" and "backward" parts $\varkappa^{\pm}$, $\varkappa^{(k)}_{\pm}$ and $\varkappa^{(u)}_{\pm}$ are distinguished.

<u>Definition 5.14.</u> Let $\varkappa^{(k)}$, $\varkappa^{(s)}$ and $\varkappa^{(u)}$ be given in the following form,

<u>$x' := x'_{\xi} \in \omega'$:</u> two triangles $\Delta(x,\xi,\xi^{\pm})$ with the common side $h(x'_{\xi})$, cf. Figs. 6f and 11a, $\varkappa^{(k)} := \varkappa^{(k)}_{-} + \varkappa^{(k)}_{+}$, $\varkappa^{(u)} := \varkappa^{(u)}_{-} + \varkappa^{(u)}_{+}$, $\varkappa^{(s)}(x') = 0$,

$$\varkappa^{(k)}_{\pm}(x'_{\xi}) := \frac{1}{s_0(x')} \int\limits_{S^{\pm}(x,\xi)} (\delta K^{\pm}\nabla u, n)ds, \quad \delta K^{\pm} := K - K^{\pm}_h, \; K^{\pm}_h := K_h(x''_{\pm}),$$

$$\varkappa^{(u)}_{\pm}(x'_{\xi}) := \frac{s^{\pm}(x')}{s_0(x')}\Big\{\frac{1}{s^{\pm}(x')} \int\limits_{S^{\pm}(x,\xi)} (K^{\pm}_h \nabla u, n^{\pm})ds - (K^{\pm}_h \nabla^{\pm}_h u, n^{\pm})\Big\}, \qquad (5.3.9a)$$

with $\nabla^{\pm}_h u := \nabla_h u(x''_{\pm})$, $n^{\pm} := n_{S^{\pm}(x'_{\xi})}$;

<u>$x' := x'_{\xi_\Gamma} \in \gamma'_{23}$:</u> one triangle $\Delta(x,\xi,\xi_\Gamma)$ with the edge $h(x'_{\xi_\Gamma})$ on Γ_h, and $\xi_\Gamma \in \{\xi^{\pm}_\Gamma\} := S'(x) \cap \gamma$ $(x \in \gamma_{23})$, cf. Figs. 6b, c and 11b, c,

$$\varkappa^{(k)}(x'_{\xi^{\pm}_\Gamma}) := \frac{1}{s_0(x'_{\xi^{\pm}_\Gamma})} \int\limits_{S(x,\xi^{\pm}_\Gamma)} (\delta K^{\pm} \nabla u, n)ds, \quad \delta K^{\pm} := K - K^{\pm}_h, \; K^{\pm}_h := K_h(x''_{\pm}),$$

$$\varkappa^{(s)}(x'_{\xi^{\pm}_\Gamma}) := \frac{1}{s_0(x'_{\xi^{\pm}_\Gamma})} \int\limits_{\delta S(x,\xi^{\pm}_\Gamma)} (K^{\pm}_h \nabla u, n)ds, \quad \delta S := \begin{cases} S \setminus \tilde{S} & S \supset \tilde{S}, \\ \tilde{S} \setminus S & \text{for } \tilde{S} \supset S, \end{cases}$$

cf. Figs. 11c, b, (5.3.9b)

$$\varkappa^{(u)}(x'_{\xi^{\pm}_\Gamma}) := \frac{1}{s_0(x'_{\xi^{\pm}_\Gamma})} \int\limits_{\tilde{S}^{\pm}(x,\xi^{\pm}_\Gamma)} (K^{\pm}_h \nabla u, n^{\pm})ds - (K^{\pm}_h \nabla^{\pm}_h u, n^{\pm}),$$

with $\nabla^{\pm}_h u := \nabla_h(x''_{\pm})$, $n^{\pm} := n_{\tilde{S}^{\pm}}$, $\tilde{S}^{\pm}(x,\xi^{\pm}_\Gamma) := \tilde{S}(x, \xi^{\pm}_\Gamma)$,

$$s_0(x'_{\xi^{\pm}_\Gamma}) := |\,\tilde{S}(x,\xi^{\pm}_\Gamma)\,|. \quad \blacksquare$$

If we add all partial errors from (5.3.9), then (5.3.8) is obviously valid. It should be mentioned that the error splitting from Definition 5.14. yields bounds of the errors similar to those in Section 5.3.1. In order to prove higher accuracy for smoother solutions u, e.g. $u \in W^3_2(\Omega)$, and regular or locally irregular grids $\bar{\omega}$, a splitting which is symmetric with respect to the common side $h(x,\xi)$ of two equilateral triangles $\Delta(x,\mathfrak{s},\mathfrak{s}^{+})$ will be introduced.

<u>Definition 5.15.</u> Let $\Delta(x_{\pm}) := \Delta(x,\mathfrak{s},\mathfrak{s}^{\pm})$ denote two equilateral triangles with the common side $\mathfrak{h}(x,\mathfrak{s})$, then the following splitting of $\varkappa$ from (5.1.22), with $\varkappa$ given by

$$\varkappa(x'_{\mathfrak{s}}) := \frac{1}{s_o(x'_{\mathfrak{s}})} \int\limits_{\mathfrak{S}(x,\mathfrak{s})} (K\nabla u,n)ds - \tfrac{1}{2}(K_h^- \nabla_{\mathfrak{h}}^- u + K_h^+ \nabla_{\mathfrak{h}}^+ u, n_{\mathfrak{S}(x,\mathfrak{s})}), \tag{5.3.10a}$$

with $K_h^{\pm} := K_h(x''_{\pm})$, $\nabla_{\mathfrak{h}}^{\pm} u := \nabla_{\mathfrak{h}} u(x''_{\pm})$, $s_o(x') = |\mathfrak{S}(x,\mathfrak{s})|$,

will be introduced:

$$\varkappa = \sum_{i=1}^{4} \varkappa_i, \text{ with } \varkappa_i \text{ defined by} \tag{5.3.10b}$$

$$\varkappa_1(x'_{\mathfrak{s}}) := \varkappa_1(x'_{\mathfrak{s}}, \delta K, u) := \frac{1}{s_o(x'_{\mathfrak{s}})} \int\limits_{\mathfrak{S}(x,\mathfrak{s})} (\delta K \nabla u, n)ds, \quad \delta K := K - K', \quad K' := K(x'_{\mathfrak{s}}),$$

$$\varkappa_2(x'_{\mathfrak{s}}) := \varkappa_2(x'_{\mathfrak{s}}, K', u) := \frac{1}{s_o(x'_{\mathfrak{s}})} \int\limits_{\mathfrak{S}(x,\mathfrak{s})} (K'\nabla u, n)ds - \tfrac{1}{2}(K'[\nabla_{\mathfrak{h}}^- u + \nabla_{\mathfrak{h}}^+ u], n_{\mathfrak{S}(x,\mathfrak{s})})$$

$$= (K(x'_{\mathfrak{s}})\Big\{\frac{1}{s_o(x'_{\mathfrak{s}})} \int\limits_{\mathfrak{S}(x,\mathfrak{s})} \nabla u ds - \tfrac{1}{2}[\nabla_{\mathfrak{h}}^- u + \nabla_{\mathfrak{h}}^+ u]\Big\}, n_{\mathfrak{S}(x,\mathfrak{s})}),$$

$$\varkappa_3(x'_{\mathfrak{s}}) := \varkappa_3(x'_{\mathfrak{s}}, K, u) := \tfrac{1}{2}(K'[\nabla_{\mathfrak{h}}^- u + \nabla_{\mathfrak{h}}^+ u] - [K_h^- \nabla_{\mathfrak{h}}^- u + K^+ \nabla_{\mathfrak{h}}^+ u], n_{\mathfrak{S}(x,\mathfrak{s})}),$$

with $K^{\pm} := K(x''_{\pm})$, $K' := K(x'_{\mathfrak{s}})$,

$$\varkappa_4(x'_{\mathfrak{s}}) := \varkappa_4(x'_{\mathfrak{s}}, K, u) := \tfrac{1}{2}([K^- - K_h^-]\nabla_{\mathfrak{h}}^- u + [K^+ - K_h^+]\nabla_{\mathfrak{h}}^+ u, n_{\mathfrak{S}(x,\mathfrak{s})}). \equiv$$

In Definition 5.15., the error functional $\varkappa(x')$ is not split into two parts $\varkappa^{\pm}$, since the estimation of some elementary errors $\varkappa_i$ is treated on the rhombus consisting of two equilateral triangles, cf. Fig. 11d. Thus, the symmetry and the orthogonality of $\mathfrak{S}(x,\mathfrak{s})$ with respect to $\mathfrak{h}(x,\mathfrak{s})$ will lead to the phenomenon that certain error terms vanish and, therefore, the error will become considerably smaller.

To determine local bounds of $|\varkappa^{(k)}|$, $|\varkappa^{(s)}|$ and $|\varkappa^{(u)}|$, some "line functionals" and "point functionals" must be estimated by Sobolev seminorms of u over finite two-dimensional regions of $\tilde{\Omega}$, $\tilde{\Omega} \supset \Omega \cup \Omega_h$. Here the primary triangles $\Delta(x'') \subset \Omega_h$ and some "exterior triangles" $\Delta_e \not\subset \Omega_h$ can be taken and, for the special splitting (5.3.10b), we shall apply the rhombus $\square(x'_{\mathfrak{s}}) := \Delta(x,\mathfrak{s},\mathfrak{s}^-) \cup \Delta(x,\mathfrak{s},\mathfrak{s}^+)$. Thus, using the technique of error estimation shown in Sections 5.2. and 5.3.1., Lemma 5.13. remains

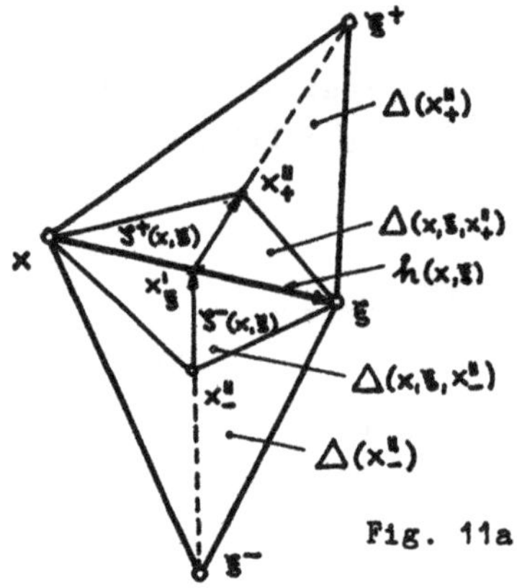

Fig. 11a

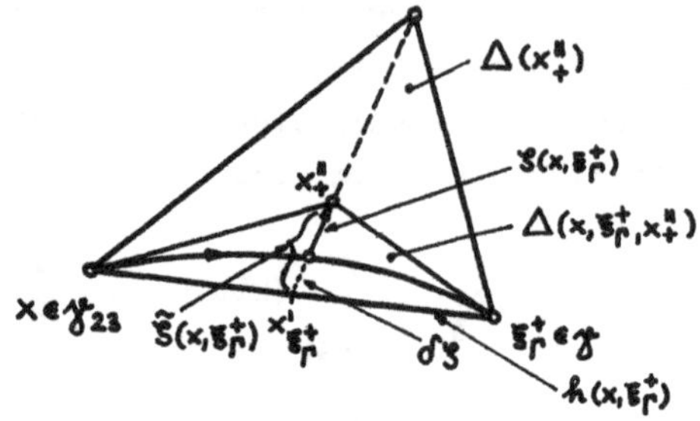

Fig. 11b

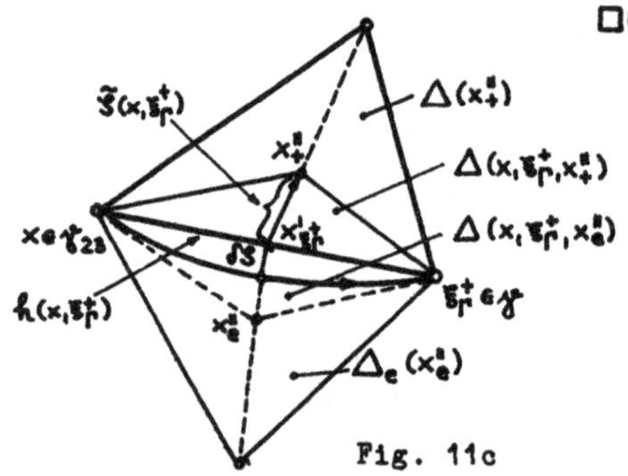

Fig. 11c

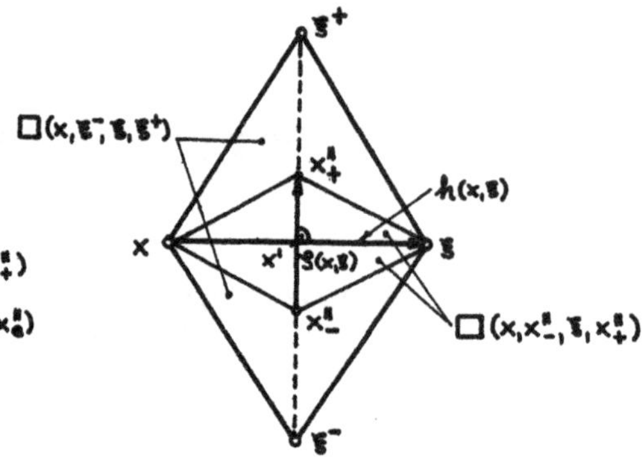

Fig. 11d

valid for the x'-splitting via Definition 5.14., i.e.

$$'\|\varkappa^{(l)}\|_0 \leq M_6' h \|u\|_{2,\Omega} \quad \text{for } l = k, s, u, \tag{5.3.11}$$

with the smoothness assumptions from (5.3.7).
In analogy to (5.2.25), we introduce the following subsets of $\overline{\omega}'$:

$$\overset{\circ}{\omega}' := \{x' \in \omega' : \Delta(x''_{\pm}) \text{ over } h(x') \text{ are equilateral}\},\ \overset{*}{\omega}' := \omega' \setminus \overset{\circ}{\omega}', \tag{5.3.12}$$

with ω' from (2.3.14). Consequently, the triangles associated with $x' \in \overset{*}{\omega}' + \gamma_{23}'$ are, in general, not equilateral. In the following, estimates of $|\varkappa^{(s)}|$, $|\varkappa^{(k)}|$ and $|\varkappa^{(u)}|$ near the boundary Γ_h are considered and collected such that global estimates with respect to a convergent boundary strip ω_h of width O(h) will be obtained, where ω_h includes the triangles adjacent to the mesh segment $h(x')$ for $x' \in \overset{*}{\omega}' + \gamma_{23}'$ with $\operatorname{mes} \omega_h = O(h)$.
First of all, starting from (5.3.9b) and taking into account $|\nabla u| \in C(\overline{\Omega})$ due to $u \in W_2^3(\Omega)$ and Theorem 1 from Appendix IM, we get

$$|\varkappa^{(s)}(x'_{s_{\Gamma}^{\pm}})| \leq \frac{C_1}{h} \max_{\bar{\Omega}} |\nabla u| \int_{\delta S(x, s_{\Gamma}^{\pm})} ds \leq C_2 h \|u\|_{3,\Omega}. \tag{5.3.13}$$

Estimate (5.3.13), $\varkappa^{(s)}(x') = 0$ for $x' \in \omega'$ and $\sum_{x' \in \gamma'_{23}} s_o(x')h(x') \leq C'_2 h$ yield

$$'\|\varkappa^{(s)}\|_o^2 = \sum_{x' \in \gamma'_{23}} \{\varkappa^{(s)}(x')\}^2 \, s_o(x')h(x') \leq C_3 h^3 \|u\|^2_{3,\Omega} . \tag{5.3.14}$$

The local bound of $|\varkappa^{(k)}(x')|$ is of the same type as in (5.3.5), perhaps with an "exterior triangle" Δ_e. Therefore,

$$'\|\varkappa^{(k)}\|^2_{\overset{*}{\omega}' + \gamma'_{23}} := \sum_{x' \in \overset{*}{\omega}' + \gamma'_{23}} \{\varkappa^{(k)}(x')\}^2 s_o(x')h(x')$$
$$\leq C_4 \, h^2 \{ |u|^2_{1,\omega_h} + h^2 |u|^2_{2,\omega_h} \} \leq C_5 \, h^3 \|u\|^2_{2,\Omega} \tag{5.3.15}$$

can be derived, since $|u|^2_{1,\omega_h} \leq C'_5 h \|u\|^2_{2,\Omega}$ holds, cf. also Theorem 4 in Appendix IM and Theorem 1 in Appendix EX. Using local bounds of $|\varkappa^{(u)}(x')|$, which are of the same type as described in (5.3.6) (but also for deltoids $\square(x') := \Delta(x''_-) \cup \Delta(x''_+)$), the estimate

$$'\|\varkappa^{(u)}\|^2_{\overset{*}{\omega}' + \gamma'_{23}} := \sum_{x' \in \overset{*}{\omega}' + \gamma'_{23}} \{\varkappa^{(u)}(x')\}^2 s_o(x')h(x') \leq C_6 h^2 |u|^2_{2,\omega_h}$$
$$\leq C_7 h^3 \|u\|^2_{3,\Omega} \tag{5.3.16}$$

will be obtained, where $|u|^2_{2,\omega_h} \leq C'_7 h \|u\|^2_{3,\Omega}$ was inserted. The estimation of the functional $\varkappa$ from (5.3.10) is now to be considered for the case $x' \in \overset{\circ}{\omega}'$, $\overset{\circ}{\omega}'$ from (5.3.12). The primary triangles $\Delta(x''_\pm)$ with the common side $h(x')$, $x' \in \overset{\circ}{\omega}'$, are equilateral. Therefore, the rhombus $\square(x') := \Delta(x''_-) \cup \Delta(x''_+)$ can be used for further estimations, since $\square$ is the affine image of a reference rhombus $\hat{\square}$. After some calculations and estimations, cf. HEINRICH [7], which are similar to the techniques previously utilized for PB- and MD-schemes, we get the following local bounds at $x' \in \overset{\circ}{\omega}'$, with $\square := \square(x')$, and under the assumption $u \in W_2^3(\Omega)$, $k_{ij} \in C^2(\bar{\Omega})$:

$$|\varkappa_1(x')| \leq C_8 h \{ |u|^2_{1,\square} + |u|^2_{2,\square} + h^2 |u|^2_{3,\square} \}^{1/2}, \; |\varkappa_2(x')| \leq C_9 h |u|_{3,\square},$$
$$|\varkappa_3(x')| \leq C_{10} h \{ |u|_{2,\square} + h \|u\|_{3,\Omega} \}, \; |\varkappa_4(x')| \leq C_{11} h^2 \|u\|_{3,\Omega}, \tag{5.3.17}$$

where the constants C_i $(i=8,9,10,11)$ are independent of u and h, but C_8, C_{10} and C_{11} depend on $\|k_{ij}\|_{C^2(\bar{\Omega})}$. The global evaluation of $\varkappa(x')$ by means of the local bounds (5.3.17) leads to

$$'\|\varkappa\|_{\tilde{\omega}'} := \Big\{ \sum_{x' \in \tilde{\omega}'} \varkappa^2(x') s_o(x') h(x') \Big\}^{1/2} \le C_{12} h^2 \|u\|_{3,\Omega} \quad \text{for } u \in W_2^3(\Omega) \tag{5.3.18}$$

and $k_{ij} \in C^2(\bar{\Omega})$.

Lemma 5.16. For the norm $'\|\varkappa\|_o$ of the error $\varkappa(x')$ defined for the x'-splitting, cf. (5.1.34) and (5.1.22), the following estimates hold for grids $\bar{\omega}$ with triangles Δ

$$'\|\varkappa\|_o \le C_{13} h \|u\|_{2,\Omega}, \ \bar{\omega} \text{ is irregular}, \ k_{ij} \in C^1(\bar{\Omega}), \ u \in W_2^2(\Omega),$$

$$\left.\begin{array}{l} '\|\varkappa\|_o \le C_{14} h^{3/2} \|u\|_{3,\Omega}, \ \bar{\omega} \text{ is locally irregular} \\ '\|\varkappa\|_o \le C_{12} h^2 \|u\|_{3,\Omega}, \ \bar{\omega} \text{ is regular}, \ \Gamma = \Gamma_1 \end{array}\right\} \begin{array}{l} \text{and } k_{ij} \in C^2(\bar{\Omega}), \\ u \in W_2^2(\Omega). \end{array} \tag{5.3.19}$$

Proof: See HEINRICH [7]. The estimates noted in (5.3.19) can be derived directly from inequalities (5.3.11), (5.3.14) - (5.3.18). The norm $'\|\varkappa\|_o$ is taken from (5.1.34), and the relation $'\|\varkappa\|_o^2 = {}'\|\varkappa\|_{\tilde{\omega}'}^2 + {}'\|\varkappa\|_{\tilde{\omega}'+\gamma_{23}'}^2$ is employed.

5.4. The error ψ_N for PB- and MD-schemes

5.4.1. ψ_N-estimates for grid points $x \in \omega$. We now consider $\psi_N^j(x)$, $x \in \omega$, given in (5.1.8), (5.1.9) for method (PB), and in (5.1.16) for method (MD). Clearly, ψ_{cu} and ψ_f from (5.1.9) are functionals of the type

$$\psi_v(x) := \frac{1}{H(x)} \int_{\varkappa(x)} v(\bar{x})\, d\bar{x} - v(x) \quad \text{for } x \in \omega, \ v := cu \text{ or } v := f. \tag{5.4.1}$$

Due to $x \in \varkappa(x)$, the box $\varkappa(x)$ can serve as an estimation region for $\psi_v(x)$, and $\psi_v(x)$ describes the error of a cubature formula with the cubature point x. The box $\varkappa(x)$ consists of a finite number of triangles Δ which are (uniformly with respect to h) regular, if VO"(PB, ω^1) or VO(MD, $\bar{\omega}$) are satisfied. Since the results for the cases (PB) and (MD) are quite similar, we confine our explanations to the case (PB).

Lemma 5.17. For $\psi_v(x)$ from (5.4.1), the following estimates hold:

$$|\psi_v(x)| \le M_1 \{|v|_{1,\varkappa(x)}^2 + h^2 |v|_{2,\varkappa(x)}^2\}^{1/2} \quad \text{for } v \in W_2^2(\varkappa), \ S(x) \text{ irregular}, \tag{5.4.2}$$

$$|\psi_v(x)| \le M_2 h |v|_{2,\varkappa(x)} \quad \text{for } v \in W_2^2(\varkappa), \ S(x) \text{ regular}, \tag{5.4.3}$$

for $x \in \omega$; S(x) denotes the difference star. The constants M_1 and M_2 are independent of v, h ($h \le h_o$) and x. ■

Proof: First of all, we split $\psi_v(x)$ with respect to the triangles $\Delta(x,x'_\xi) := \Delta(x,p_\xi^-,p_\xi^+)$, where $p_\xi^\pm$ denote the circumcentres of the primary elements Δ, $\square$, cf. Fig. 9f, and $\mathcal{K}(x) = \bigcup_{\xi\in S'(x)} \Delta(x,x'_\xi)$ holds. Thus, we get

$$\psi_v(x) = \frac{1}{H(x)} \sum_{\xi\in S'(x)} \text{mes}\ \Delta(x,x'_\xi)\ l_v(x,x'_\xi),$$

$$l_v(x,x'_\xi) := \frac{1}{\text{mes}\,\Delta} \int_\Delta v(\bar{x})\ d\bar{x} - v(x), \quad \Delta := \Delta(x,x'_\xi). \tag{5.4.4}$$

By means of the affine transformation $\hat{\Delta} = F^{-1}(\Delta)$, with the reference triangle $\hat{\Delta}$ and F^{-1} from (5.2.13b), we get

$$l_v(x,x') = \hat{l}_{\hat{v}}(\hat{x},\hat{x}') = \frac{1}{\text{mes}\,\hat{\Delta}} \int_{\hat{\Delta}} \hat{v}(\hat{\bar{x}})\ d\hat{\bar{x}} - \hat{v}(\hat{x}), \tag{5.4.5}$$

cf. Appendix TR. Obviously, the functional $\hat{l}_{\hat{v}}$ is linear and bounded on $W_2^2(\hat{\Delta})$: $|\hat{l}_{\hat{v}}| \le (\text{mes}\ \hat{\Delta})^{-\frac{1}{2}} \|\hat{v}\|_{0,\hat{\Delta}} + \|\hat{v}\|_{C(\hat{\Delta})} \le M_3 \|\hat{v}\|_{2,\hat{\Delta}}$, and, $\hat{l}_{\hat{v}}$ vanishes on the space of all polynomials of degree zero. Thus, applying Theorem 2 from Appendix ES, which is a modification of the Bramble-Hilbert Lemma, the estimate $|\hat{l}_{\hat{v}}(\hat{x},\hat{x}')| \le M_4 \{|\hat{v}|^2_{1,\hat{\Delta}} + h^2 |\hat{v}|^2_{2,\hat{\Delta}}\}^{1/2}$ will be obtained. After backward transformation $\Delta = F(\hat{\Delta})$ according to (5.2.13a) and application of transformation rules of the norms from Appendix TR, we are led to

$$|l_v(x,x')|^2 \le M_5 \|B\|^2 |\det B|^{-1} \{|v|^2_{1,\Delta} + \|B\|^2 |v|^2_{2,\Delta}\}$$

$$\le M_6 \{|v|^2_{1,\Delta} + h^2 |v|^2_{2,\Delta}\} \text{ for } v \in W_2^2(\Delta),\ \Delta \text{ from (5.4.4)}. \tag{5.4.6}$$

Combining (5.4.4) and (5.4.6), estimate (5.4.2) is obvious. If the difference star S(x) is regular, then the box $\mathcal{K}(x)$ is a regular polygon, and x is the centre of gravity of $\mathcal{K}(x)$. Therefore, $\mathcal{K}(x)$ may be transformed by an affine mapping F^{-1} of the type $\hat{x} = F^{-1}(x) = B^{-1}x - B^{-1}b$ onto a regular reference box $\hat{\mathcal{K}}(\hat{x})$, with $\hat{x} = \hat{0}$ (origin). The transformed functional $\hat{\psi}_{\hat{v}}(\hat{x}) = \psi_v(x)$ is bounded on $W_2^2(\hat{\mathcal{K}})$ and vanishes on the space of all polynomials $\hat{p}_1(\hat{\bar{x}}) := \hat{a}_0 + (\hat{a},\hat{\bar{x}})_{R^2}$ $(\hat{a}_0 \in R^1$ and $\hat{a}$, $\hat{\bar{x}} \in R^2)$, since $\hat{x} = \hat{0}$ is the centre of gravity of $\hat{\mathcal{K}}(\hat{x})$. Thus, applying the Bramble-Hilbert Lemma from Appendix ES and the backward transformation $\mathcal{K} = F(\hat{\mathcal{K}})$, we get

$$|\psi_v(x)| = |\hat{\psi}_{\hat{v}}(\hat{x})| \le M_7 |\hat{v}|_{2,\hat{\mathcal{K}}(\hat{x})} \le M_8 h |v|_{2,\mathcal{K}(x)} \text{ for } v \in W_2^2(\mathcal{K}) \tag{5.4.7}$$

and regular S(x), i.e., (5.4.3) is proved.

The estimation of ψ_{cu} or ψ_f can now be reduced to that of ψ_v, if $v \in W_2^2(\Omega)$ holds. If the degree of smoothness of c and u differs, e.g. $c \in C^1(\bar{\Omega})$ and $u \in W_2^2(\Omega)$, ψ_{cu} from (5.1.9) is split as follows:

$$\psi_{cu}(x) = \frac{1}{H(x)} \int_{\varkappa(x)} \{c(\bar{x})-c(x)\}\, u(\bar{x})\, d\bar{x} + c(x) \frac{1}{H(x)} \int_{\varkappa(x)} \{u(\bar{x})-u(x)\} d\bar{x} \tag{5.4.8}$$

By estimation and using Taylor's expansion of c at x, the inequality

$$|\psi_{cu}(x)| \leq M_9 \|u\|_{0,\varkappa(x)} + M_{10}|\psi_u(x)|, \quad \psi_u \text{ from (5.4.1) } (u=v), \tag{5.4.9}$$

will be obtained, if $c \in C^1(\bar{\Omega})$, $u \in W_2^2(\Omega)$ are posited. The constants M_9 and M_{10} depend respectively on $\|\partial^1 c\|_{C(\bar{\Omega})}$ and $\|c\|_{C(\bar{\Omega})}$, but they are independent of u, h ($h \leq h_0$) and x.
We are now in position to give bounds of $\|\psi_N^j\|_0$ with respect to the set of interior grid points ω, i.e. for

$$\|\psi_N^j\|_{0,\omega} := \Big\{ \sum_{x\in\omega} (\psi_N^j(x))^2\, H(x) \Big\}^{1/2}, \quad \psi_N^j \text{ from (5.1.8), (5.1.16).} \tag{5.4.10}$$

Lemma 5.18. Let VO"(PB,ω^1) or VO(MD,$\bar{\omega}$) be satisfied for grids $\bar{\omega}$ and secondary networks of the type (PB) or (MD). Then, the following estimates hold:

$$\|\psi_N^j\|_{0,\omega} \leq \begin{cases} M_{11} h \{\|u\|_{2,\Omega} + \|f\|_{C^1(\bar{\Omega})}\}, & \omega \text{ irregular,} \quad (5.4.11a)\\ M_{12} h^{3/2} \{\|u\|_{2,\Omega} + \|f\|_{C^2(\bar{\Omega})}\}, & \omega \text{ locally irregular,} \quad (5.4.11b)\\ M_{13} h^2 \{\|u\|_{2,\Omega} + \|f\|_{C^2(\bar{\Omega})}\}, & \omega \text{ regular,} \quad (5.4.11c)\end{cases}$$

where the smoothness needed for u and f is indicated by the norms used in (5.4.11). Moreover, $c \in C^1(\bar{\Omega})$ is assumed to be fulfilled in (5.4.11a), otherwise $c \in C^2(\bar{\Omega})$. If the assumptions $f \in C^s(\bar{\Omega})$, $s \in \{1,2\}$, are replaced by $f \in W_2^2(\Omega)$, then $\|f\|_{C^s(\bar{\Omega})}$ must be substituted by $\|f\|_{2,\Omega}$.
All types of networks (with triangles, rectangles) are admitted, but for secondary networks (MD) in the case (5.4.11b), locally irregular grids $\bar{\omega}$ from (2.2.9) will be taken. The constants M_{11}, M_{12} and M_{13} especially depend on $\|c\|_{C^s(\bar{\Omega})}$, $s \in \{1,2\}$, but they are independent of u, f and h, $h \leq h_0$. ∎

Proof: Using (5.4.2), (5.4.3) and (5.4.9), we get local bounds of $|\psi_{cu}|$ and $|\psi_f|$ by setting v:= cu, v:= f or v:= u. It is obvious that $c \in C^2(\bar{\Omega})$ and $u \in W_2^2(\Omega)$ imply $cu \in W_2^2(\Omega)$. If $v \in C^s(\bar{\Omega})$ holds, then a bound of $|\psi_v|$ may be given in terms of the C-norm of the derivatives of v. Here, Taylor's expansion suffices to verify this. For simplicity, some seminorms are "rounded", i.e., the seminorms are replaced by the corresponding full norms. Thus, we get the following estimates for $x \in \omega$:

$$|\psi_{cu}(x)| \leq \begin{cases} M_{14}^{(1)} \{\|u\|_{1,\varkappa(x)} + h|u|_{2,\varkappa(x)}\} \text{ for } c \in C^1(\bar{\Omega}),\ u \in W_2^2(\Omega), & (5.4.12a) \\ M_{14}^{(2)} h\|u\|_{2,\varkappa(x)} \text{ for } c \in C^2(\bar{\Omega}),\ u \in W_2^2(\Omega),\ S(x) \text{ regular}, & (5.4.12b) \end{cases}$$

$$|\psi_f(x)| \leq M_{15} h^l \|f\|_{C^l(\Omega)} \text{ for } f \in C^l(\Omega),\ l = 1,\ l = 2 \text{ and} \qquad (5.4.12c)$$

$S(x)$ regular; for $\psi_f(x)$ and $f \in W_2^2(\Omega)$, see (5.4.2), (5.4.3).

The constant $M_{14}^{(1)}$ depends on $\|c\|_{C^1(\bar{\Omega})}$. In the case of locally irregular grids $\bar{\omega}$, the boxes $\varkappa(x)$ and difference stars $S(x)$ are, in general, regular for $x \in \mathring{\omega}$ and irregular for $x \in \overset{*}{\omega}$. The irregular boxes $\varkappa(x)$ are gathered in a convergent boundary strip ω_h of width $O(h)$. Then, (5.4.12a) and Theorem 4 from Appendix IM allow us to write

$$\sum_{x \in \overset{*}{\omega}} \psi_{cu}^2(x)\, H(x) \leq M_{16} h^2 \{\|u\|_{1,\omega_h}^2 + h^2 |u|_{2,\omega_h}^2\} \leq M_{17} h^3 \|u\|_{2,\Omega}^2. \qquad (5.4.13a)$$

Moreover, from (5.4.12c), case $l = 1$, we get

$$\sum_{x \in \overset{*}{\omega}} \psi_f^2(x)\, H(x) \leq M_{15}^2 h^2 \|f\|_{C^1(\bar{\Omega})}^2 \sum_{x \in \overset{*}{\omega}} H(x) \leq M_{18} h^3 \|f\|_{C^1(\bar{\Omega})}^2. \qquad (5.4.13b)$$

Taking the estimates (5.4.13) for $\overset{*}{\omega}$ and (5.4.12b), (5.4.12c) ($l=2$) for the regular difference stars $S(x)$ at $x \in \mathring{\omega}$, which are associated with a regular box $\varkappa(x)$, and using $\|\psi_N\|_{0,\omega}^2 = \|\psi_N\|_{0,\mathring{\omega}}^2 + \|\psi_N\|_{0,\overset{*}{\omega}}^2$, we see that inequalities (5.4.11) hold. For $f \in W_2^2(\Omega)$, the conclusions are quite similar.

<u>Remark 5.19.</u> The proof of Lemma 5.18. was performed for PB-schemes under the assumption VO"(PB,ω^1). If another technique of estimation is used, where irregular boxes $\varkappa_p(x)$ are imbedded geometrically in small rectangles $\square$ which serve as "estimation region", then VO(PB,$\bar{\omega}$) suffices to prove the estimates noted in (5.4.11), with constants which are, in general, greater than M_{11}, M_{12}, but also independent of h ($h \leq h_o$), u and f. ■

This technique of using "majorant estimation regions" will be described in Section 6.3.2.

<u>5.4.2. ψ_N-estimates for grid points $x \in \gamma_{23}$.</u> We now consider the functional $\psi_N^j(x)$ for $x \in \gamma_{23}$, $j \in \{0,1\}$ and the methods (PB) and (MD); cf. (5.1.13), (5.1.14) and (5.1.18). Here, a splitting of ψ_N^j into $\psi_{\alpha u}^j$ and ψ_g^j has already been defined. Let us recall that $\alpha, g \in PC^s(\Gamma_{23})$, $s \in \{1,2\}$, $\alpha|_{\Gamma_2} = 0$ hold and that the points from $\Gamma_{23} \setminus \Gamma_1$ where α, g have jumps are

located on γ_{23}. Hence, α and g are of the type α, $g \in C^s$ on the arcs $\varsigma^{\pm}(x)$, $\widehat{x\xi_\Gamma^{\pm}}$, cf. Figs. 5b, 6b. Since the approach of estimating ψ_N^j and the results obtained for PB- and MD-schemes are very similar - with only one exception caused by the asymptotic behaviour of mes $\varsigma^{\pm}(x)$ - we restrict ourselves to the case (PB) and shall emphasize below the modifications for the case (MD).

In the following, we take into account the estimation regions $\varsigma^{\pm}(x)$, $\varsigma(x)$, $\widehat{x\xi_\Gamma^{\pm}}$, $\mathcal{H}(x)$, $\tilde{\mathcal{H}}(x)$ for $x \in \gamma_{23}$ (cf. Definition 2.15., Figs. 5b, c) and split ψ_g^j into further elementary errors:

$$\psi_g^j(x) := \psi_f(x) + \varphi(x,g,j) \quad \text{for } j = 0 \text{ and } j = 1, \tag{5.4.14a}$$

$$\psi_f(x) := \frac{1}{h(x)} \Big\{ \int_{\mathcal{H}(x)} f(\bar{x})\, d\bar{x} - \tilde{H}(x)\, f_h(x) \Big\}, \tag{5.4.14b}$$

$$\varphi(x,g,j) := \varphi^-(x,g,j) + \varphi^+(x,g,j), \; \varphi^{\pm}(x,g,j) := \frac{1}{h(x)} \sum_{i=1}^{4} \varphi_i^{\pm}(x,g,j), \tag{5.4.14c}$$

$$\varphi_1^{\pm}(x,g,j) := \int_{\varsigma^{\pm}(x)} g(s)\,ds - s^{\pm}(x) g^{\pm}(x) \mp j \frac{(s^{\pm}(x))^2}{2} \cdot \frac{dg(x_\Gamma^{\pm})}{ds}, \quad s^{\pm}(x) := \text{mes}\, \varsigma^{\pm}(x),$$

$$\varphi_2^{\pm}(x,g,j) := \Big\{ s^{\pm}(x) - \frac{h(x,\xi_\Gamma^{\pm})}{2} \Big\} g^{\pm}(x) \pm j \Big\{ \frac{(s^{\pm}(x))^2}{2} - \frac{h^2(x,\xi_\Gamma^{\pm})}{8} \Big\} \frac{dg(x_\Gamma^{\pm})}{ds},$$

$$\varphi_3^{\pm}(x,g,j) := \pm j \frac{h^2(x,\xi_\Gamma^{\pm})}{8} \Big\{ \frac{dg(x_\Gamma^{\pm})}{ds} - \frac{\delta g(x_\Gamma^{\pm})}{\delta b} \Big\}, \tag{5.4.14d}$$

$$\frac{\delta g(x_\Gamma^{\pm})}{\delta b} := \begin{cases} \dfrac{g^-(\xi_\Gamma^+) - g^+(x)}{\text{mes}\, \widehat{x\xi_\Gamma^+}}, \\[2ex] \dfrac{g^-(x) - g^+(\xi_\Gamma^-)}{\text{mes}\, \widehat{x\xi_\Gamma^-}}, \end{cases}$$

$$\varphi_4^{\pm}(x,g,j) := \pm j \frac{h^2(x,\xi_\Gamma^{\pm})}{8} \Big\{ \frac{\delta g(x_\Gamma^{\pm})}{\delta b} - \frac{\delta g(x'_{\xi_\Gamma^{\pm}})}{\delta h} \Big\}, \quad \frac{\delta g(x'_{\xi_\Gamma^{\pm}})}{\delta h} := \frac{\text{mes}\, \widehat{x\xi_\Gamma^{\pm}}}{h(x,\xi_\Gamma^{\pm})} \frac{\delta g(x_\Gamma^{\pm})}{\delta b}$$

The error splitting (5.4.14) naturally corresponds to the steps of elementary approximations proposed in Section 3.2.1.; cf. especially (3.2.8), (3.2.9), (3.2.12), but also (3.3.16), (3.3.17) and (5.1.13), (5.1.14). Indeed, summing up the partial errors from (5.4.14), we get ψ_g^j. The error functional $\psi_{\alpha u}^j$ from (5.1.14) will be also split into several partial errors, where certain functionals already defined in (5.4.14) will be employed:

$$\psi_{\alpha u}^j(x) := \psi_{ou}(x) + \chi(x,\alpha u,j) \quad \text{for } j = 0 \text{ and } j = 1, \tag{5.4.15a}$$

$$\psi_{cu}(x) := \frac{1}{h(x)} \left\{ \int_{\mathcal{X}(x)} c(\bar{x})u(\bar{x})d\bar{x} - \widetilde{H}(x)c(x)u(x) \right\}, \tag{5.4.15b}$$

$$\chi(x,\alpha u,j) := \chi^-(x,\alpha u,j) + \chi^+(x,\alpha u,j),\ \chi^{\pm}(x,\alpha u,j) := \frac{1}{h(x)} \sum_{i=1}^{8} \chi_i^{\pm}(x,\alpha u,j), \tag{5.4.15c}$$

$$\chi_i^{\pm}(x,\alpha u,j) := \varphi_i^{\pm}(x,\alpha u,j) \text{ for } i = 1,2,\ \chi_i^{\pm}(x,\alpha u,j) := \alpha_h(x'_{s\Gamma}{}^{\pm})\varphi_i^{\pm}(x,u,j) \text{ for } i = 3,4,$$

$$\chi_5^{\pm}(x,\alpha u,j) := u(x)\,\varphi_3^{\pm}(x,\alpha,j),\ \chi_6^{\pm}(x,\alpha u,j) := u(x)\,\varphi_4^{\pm}(x,\alpha,j),$$

$$\chi_7^{\pm}(x,\alpha u,j) := \pm\frac{j}{8}\, h^2(x,s_\Gamma^{\pm})\,\frac{du(x_\Gamma^{\pm})}{ds}\left\{\alpha(x_\Gamma^{\pm}) - \alpha_h(x'_{s\Gamma}{}^{\pm})\right\}, \tag{5.4.15d}$$

$$\chi_8^{\pm}(x,\alpha u,j) := \pm\frac{j}{8}\, h^2(x,s_\Gamma^{\pm})\,\frac{d\alpha(x_\Gamma^{\pm})}{ds}\left\{u(x_\Gamma^{\pm}) - u(x)\right\}.$$

The errors involved in χ_7 and χ_8 originate from (3.2.10). The functionals $\varphi_i^{\pm}$ are taken from (5.4.14), but here for the variables αu, α and u instead of g. That splitting (5.4.15) is correct can be checked by addition.

If the functions g, f, c, α and u are differentiable in the classical sense, then the errors ψ_g^j, $\psi_{\alpha u}^j$ may be estimated by Taylor's expansion. The results are given in HEINRICH [6, pointer (21)]:

$$|\psi_N^j(x)| \le |\psi_g^j(x)| + |\psi_{\alpha u}^j(x)| \le C_o h^{1+j} \text{ for } x \in \gamma_{23},\ j = 0,1, \tag{5.4.16}$$
$$\text{with } \alpha,\ g \in PC^{1+j}(\Gamma_{23}) \text{ and } c,\ f \in C^j(\bar{\Omega}),\ u \in C^{2+j}(\bar{\Omega}),$$

where C_o depends on the $PC^{1+j}(\Gamma_{23})$- and $C^k(\bar{\Omega})$-norms ($k=j$ or $k=2+j$) of the functions noted in (5.4.16). But C_o is independent of h, $h \le h_o$.

We now take into account the case $u \in W_2^{2+j}(\Omega)$ for $j = 0,1$. Then, the trace of u on Γ satisfies $u \in W_2^{1+j}(\Gamma_T)$ (at least) for $j = 0,1$ and each arc $\Gamma_T \in C^2$, cf. Theorem 5 from Appendix IM. Moreover, $\alpha \in C^{1+j}(\Gamma_T)$ implies $\alpha u \in W_2^{1+j}(\Gamma_T)$, for arcs $\Gamma_T \in C^2$. The errors $\varphi(x,g,j)$ and $\chi(x,\alpha u,j)$ essentially depend on the functionals $\varphi_i^{\pm}(x,.,j)$, where "." is the placeholder of g, u, α or αu. Therefore, it is reasonable to estimate, first of all, $\varphi_i^{\pm}(x,v,j)$ for functions $v \in W_2^{1+j}(\Gamma_T)$, but then $\chi_i^{\pm}(x,\alpha u,j)$.

<u>Lemma 5.20. PB-schemes.</u> For $\varphi_i^{\pm}$, $\chi_i^{\pm}$ from (5.4.14), (5.4.15) and under the assumptions $v \in W_2^{1+j}(\widehat{x\,s_\Gamma^{\pm}})$, $\alpha \in C^{1+j}(\widehat{x\,s_\Gamma^{\pm}})$, $u \in W_2^{2+j}(\Omega)$, the following estimates hold for $j = 0,1$ and $x \in \gamma_{23}$ (for brevity, "+" is written instead of "±"):

$$|\varphi_1^{+}(x,v,j)| \le C_1 h^{3/2+j}|v|_{1+j,s^+(x)},\ |\varphi_2^{+}(x,v,j)| \le C_2 h^3 \|v\|_{C^j(s^+(x))}, \tag{5.4.17a}$$
$$|\varphi_3^{+}(x,v,j)| \le jC_3 h^{3/2+j}|v|_{1+j,\widehat{x s_\Gamma^+}},\ |\varphi_4^{+}(x,v,j)| \le jC_4 h^4 \|v\|_{C^j(\widehat{x s_\Gamma^+})},$$

$$|\chi_1^+(x,\alpha u,j)| \leq C_5 h^{3/2+j} \|\alpha\|_{C^{1+j}(\mathfrak{S}^+(x))} \|u\|_{1+j,\mathfrak{S}^+(x)},$$

$$|\chi_2^+(x,\alpha u,j)| \leq C_6 h^3 \|\alpha\|_{C^j(\mathfrak{S}^+(x))} \|u\|_{C^j(\mathfrak{S}^+(x))},$$

$$|\chi_3^+(x,\alpha u,j)| \leq jC_7 h^{3/2+j} \|\alpha\|_{C(\widehat{x\,\mathfrak{s}_\Gamma^+})} |u|_{1+j,\widehat{x\,\mathfrak{s}_\Gamma^+}},$$

$$|\chi_4^+(x,\alpha u,j)| \leq jC_8 h^4 \|\alpha\|_{C(\widehat{x,\mathfrak{s}_\Gamma^+})} \|u\|_{C^j(\widehat{x\,\mathfrak{s}_\Gamma^+})}, \tag{5.4.17b}$$

$$|\chi_5^+(x,\alpha u,j)| \leq jC_9 h^3 \|\alpha\|_{C^{1+j}(\widehat{x\,\mathfrak{s}_\Gamma^+})} |u(x)|,$$

$$|\chi_6^+(x,\alpha u,j)| \leq jC_{10} h^4 \|\alpha\|_{C^j(\widehat{x\,\mathfrak{s}_\Gamma^+})} |u(x)|,$$

$$|\chi_i^+(x,\alpha u,j)| \leq jC_{11} h^3 \|\alpha\|_{C^j(\widehat{x\,\mathfrak{s}_\Gamma^+})} \|u\|_{C^j(\widehat{x\,\mathfrak{s}_\Gamma^+})} \quad \text{for } i = 7,8.$$

The constants $C_1, C_2, \ldots, C_{11}$ are independent of x, v, α, u, j and h, $h \leq h_o$. ∎

<u>Proof:</u> Obviously, $\varphi_1^+(x,v,j)$ is well-defined for $v \in W_2^{1+j}(\widehat{x\,\mathfrak{s}_\Gamma^+})$, since $v \in C^j(\widehat{x\,\mathfrak{s}_\Gamma^+})$ holds for $j = 0,1$; cf. Appendix IM. We map the segment $[s_A, s_E]$ of the arc length s, which is assigned to the arc $\mathfrak{S}^+(x)$, onto the reference segment $[0,1]$:

$s_A \leq s \leq s_E$, $s_A = s(x)$, $s_E = s(x_\Gamma^+)$, where x and x_Γ^+ are the two end points of $\mathfrak{S}^+(x) := \widehat{xx_\Gamma^+}$, with the measure $s^+(x) := \text{mes}\, \mathfrak{S}^+(x)$,

$$0 \leq \hat{s} \leq 1, \quad \hat{s} := \frac{s-s_A}{s_E-s_A}, \quad ds = (s_E-s_A)d\hat{s} = s^+(x)d\hat{s}. \tag{5.4.18}$$

Then, the functional φ_1^+ takes on the form

$$\varphi_1^+(x,v,j) = \hat{\varphi}_1^+(\hat{x},\hat{v},j) = s^+(x)\left\{ \int_0^1 \hat{v}\,d\hat{s} - \hat{v}(\hat{x}) - \frac{j}{2}\frac{d\hat{v}(\hat{x}_\Gamma^+)}{d\hat{s}}\right\}. \tag{5.4.19a}$$

Obviously, $\hat{\varphi}_1^+$ is linear and bounded on $W_2^{1+j}(0,1)$, since

$$|\hat{\varphi}_1^+(\hat{x},\hat{v},j)| \leq s^+(x)\left\{\|\hat{v}\|_{L_2(0,1)} + \|\hat{v}\|_{C^j[0,1]}\right\} \leq s^+(x)\, C_I \|\hat{v}\|_{1+j,(0,1)} \tag{5.4.19b}$$

holds, cf. Appendix IM. The functional $\hat{\varphi}_1^+$ vanishes on the space of all polynomials of degree $\leq j$, i.e. for $\hat{p}_j := \hat{a}_o + j\hat{a}_1\hat{s}$ and $j = 0,1$:

$$\hat{\varphi}_1^+(\hat{x},\hat{p}_j,j) = s^+(x)\left\{\int_0^1 \hat{p}_j ds - \hat{p}_j|_{\hat{s}=0} - \frac{j}{2}\frac{d\hat{p}_j}{d\hat{s}}\Big|_{\hat{s}=1}\right\} = 0. \tag{5.4.19c}$$

Thus, (5.4.19) verifies the assumptions of the Bramble-Hilbert Lemma

(Theorem 1, Appendix BS) for the case p = 2, k = 1 + j. After application of this lemma and backward transformation, cf. (5.4.18), we get the estimates

$$|\varphi_1^+(x,v,j)| = |\hat{\varphi}_1^+(\hat{x},\hat{v},j)| \leq s^+(x) C_I C_{BH} |\hat{v}|_{1+j,(0,1)} \leq C_1 h^{3/2+j} |v|_{1+j,\mathfrak{s}^+(x)}.$$

Thus, the first estimate or (5.4.17) is confirmed and the remaining inequalities can be proved similarly or by simple estimates involving some norms as well as some measures on the boundary. For instance, for φ_2^+ we apply $|s^+(x) - \frac{1}{2}h(x,\mathfrak{s}_\Gamma^+)| \leq K_2 h^3$ from (3.2.12), for φ_4^+ the estimate $\operatorname{mes} \widehat{x\mathfrak{s}_\Gamma^+} - h(x,\mathfrak{s}_\Gamma^+) \leq K_2 h^3$, cf. Appendix GE. The estimates of $|\chi_1^+|$ are based on that of $|\varphi_1^+|$. We also realise that

$|\alpha u|_{1+j,\mathfrak{s}^+(x)} \leq C_{12} \|\alpha\|_{C^{1+j}(\mathfrak{s}^+)} \|u\|_{1+j,\mathfrak{s}^+}$ and

$|\alpha(x_\Gamma^+) - \alpha_h(x_{\mathfrak{s}_\Gamma^+}^{\prime})| \leq C_{13} h \|\alpha\|_{C^1(\widehat{x\mathfrak{s}_\Gamma^+})}$, $|u(x_\Gamma^+) - u(x)| \leq C_{14} h \|u\|_{C^1(\mathfrak{s}^+(x))}$

hold, and the smoothness $u \in W_2^{1+j}(\widehat{x\mathfrak{s}_\Gamma^+})$ due to $u \in W_2^{2+j}(\Omega)$ is always sufficient. The detailed derivation of all estimates (5.4.17) can be found in HEINRICH [7].

Remark 5.21. MD-schemes. The estimates of the functionals $\varphi_i^\pm$, $\chi_i^\pm$ defined for MD-schemes are the same as those noted for PB-schemes in Lemma 5.20., with the only exception that the bounds of $|\varphi_2^\pm|$, $|\chi_2^\pm|$ are, in general, greater, viz.

$$|\varphi_2^+(x,v,j)| \leq C_2' h^k \|v\|_{C^j(\mathfrak{s}^+(x))}, \text{ with } k = \begin{cases} 2 & \text{for} \quad \operatorname{int} h(x,\mathfrak{s}_\Gamma^+) \cap \bar{\Omega} = \emptyset, \\ 3 & \quad h(x,\mathfrak{s}_\Gamma^+) \subset \bar{\Omega}. \end{cases} \quad (5.4.20)$$

Hence, the term h^3 in the estimate of $|\chi_2^+|$ from (5.4.17b) must be replaced by h^k, too. The value of k is also given by (5.4.20). ■

Estimate (5.4.20), with $k \in \{2,3\}$, is caused by the definition of the MD-boxes near the boundary Γ. Due to the second inequality in (3.3.8), we have $|s^\pm(x) - \frac{1}{2}h(x,\mathfrak{s}_\Gamma^\pm)| \leq K_2 h^k$, with k according to (5.4.20).

Since the functionals φ and χ from (5.4.14a) and (5.4.15a), respectively are estimated, we shall proceed with the estimation of ψ_f and ψ_{cu}. For these functionals, the natural estimation region $\mathcal{K}(x)$ is partially curvilinear. Nevertheless, for f, c, $u \in C^1(\bar{\Omega})$, we are able to derive the following estimates by Taylor's expansion:

$$|\psi_f(x)| \leq M_1 h^{1+l}, \; |\psi_{cu}(x)| \leq M_2 h^{1+l} \quad \text{for } x \in \mathfrak{F}_{23}, \; l \in \{0,1\}, \qquad (5.4.21)$$

where M_1 and M_2 depend on $\|f\|_{C^1(\mathcal{K})}$ and $\|cu\|_{C^1(\mathcal{K})}$, respectively. In order to find estimates of $|\psi_f|$ and $|\psi_{ou}|$ for $f \in W_2^2(\Omega)$, $ou \in W_2^2(\Omega)$, we consider the functional

$$\psi_v(x) := \frac{1}{h(x)} \left\{ \int_{\mathcal{X}(x)} v(\bar{x})d\bar{x} - \tilde{H}(x)v(x) \right\} \quad \text{for } v \in W_2^2(\Omega),\ x \in \gamma_{23}. \quad (5.4.22)$$

To estimate the functional $\psi_v(x)$, which is associated with the curvilinear estimation region $\mathcal{X}(x)$, $x \in \gamma_{23}$, the box $\mathcal{X}(x)$ could be split into straight and curved triangles or imbedded into a finite "majorant region", for instance, a rectangle. But in the following, we shall utilize the polygonal box $\tilde{\mathcal{X}}(x)$, $x \in \gamma_{23}$, since $\tilde{\mathcal{X}}(x)$ may be completely decomposed into straight triangles, and the technique developed in Section 5.4.1. can be applied.

Let $\tilde{\Omega}$ be a fixed domain which has the same smoothness properties as Ω and includes $\Omega \cup \Omega_h$ for $h \leq h_o$, i.e. $\tilde{\Omega} \supset \Omega \cup \Omega_h$. Because of $\tilde{\mathcal{X}} \subset \tilde{\Omega}$, $\mathcal{X} = (\mathcal{X} \cap \tilde{\mathcal{X}}) \cup (\mathcal{X} \setminus \tilde{\mathcal{X}})$ and $\tilde{\mathcal{X}} = (\mathcal{X} \cap \tilde{\mathcal{X}}) \cup (\tilde{\mathcal{X}} \setminus \mathcal{X})$, cf. Figs. 5b,c, the functional $\psi_v(x)$ from (5.4.22) can be expressed as follows:

$$\psi_v(x) = \frac{1}{h(x)} \left\{ \int_{\mathcal{X} \setminus \tilde{\mathcal{X}}} - \int_{\tilde{\mathcal{X}} \setminus \mathcal{X}} \right\} v(\bar{x})d\bar{x} + \frac{\tilde{H}(x)}{h(x)}\, \tilde{\psi}_v(x),$$

$$\tilde{\psi}_v(x) := \frac{1}{\tilde{H}(x)} \int_{\tilde{\mathcal{X}}(x)} v(\bar{x})d\bar{x} - v(x), \quad \text{for } v \in W_2^2(\tilde{\Omega}). \quad (5.4.23)$$

The inequalities $|H(x)-\tilde{H}(x)| \leq K_o h^3$ from (3.2.12) or (3.3.8) and $\|v\|_{C(\bar{\tilde{\Omega}})} \leq C_I \|v\|_{2,\tilde{\Omega}}$, which follows from Theorem 1 in Appendix IM, allow us to estimate $|\psi_v(x)|$ by

$$|\psi_v(x)| \leq M_3 h^2 \|v\|_{2,\tilde{\Omega}} + M_4 h |\tilde{\psi}_v(x)|, \text{ with } \psi_v,\ \tilde{\psi}_v \text{ from (5.4.23).} \quad (5.4.24)$$

The functional $\tilde{\psi}_v(x)$ is of the same type as $\psi_v(x)$ from (5.4.1), because $\tilde{\mathcal{X}}(x)$, $x \in \gamma_{23}$, is a polygonal box. Hence, $\tilde{\psi}_v$ taken from (5.4.23) obeys an estimate like (5.4.2), i.e.,

$$|\tilde{\psi}_v(x)| \leq M_5 \{|v|^2_{1,\tilde{\mathcal{X}}(x)} + h^2 |v|^2_{2,\tilde{\mathcal{X}}(x)}\}^{1/2} \quad \text{for } v \in W_2^2(\tilde{\mathcal{X}}),\ x \in \gamma_{23}. \quad (5.4.25)$$

<u>Lemma 5.22.</u> The error functionals $\psi_v(x)$ from (5.4.22), $\psi_f(x)$ from (5.4.14b) and $\psi_{ou}(x)$ from (5.4.15b) satisfy the estimates

$$|\psi_v(x)| \leq M_3 h^2 \|v\|_{2,\tilde{\Omega}} + M_4 M_5 h \{|v|^2_{1,\tilde{\mathcal{X}}(x)} + h^2 |v|^2_{2,\tilde{\mathcal{X}}(x)}\}^{1/2} \quad \text{for } x \in \gamma_{23} \quad (5.4.26)$$

and $v \in W_2^2(\tilde{\Omega})$, i.e. also for $v := f \in W_2^2(\tilde{\Omega})$,

$$|\psi_{ou}(x)| \leq M_6 h^{1+l} \|u\|_{2,\tilde{\Omega}} + M_7 h \{|u|^2_{1,\tilde{\mathcal{X}}(x)} + h^2 |u|^2_{2,\tilde{\mathcal{X}}(x)}\}^{1/2} \quad \text{for } x \in \gamma_{23} \quad (5.4.27)$$

and $o \in C^l(\bar{\tilde{\Omega}})$, $l \in \{0,1\}$, $u \in W_2^2(\tilde{\Omega})$. ■

Proof: Combining (5.4.24) and (5.4.25), we get (5.4.26). Using the splitting

$$\psi_{cu}(x) = \frac{1}{h(x)} \int_{\mathcal{X}(x)} \{c(\bar{x})-c(x)\} u(\bar{x})d\bar{x}+c(x)\,\psi_u(x), \quad \psi_u(x) \text{ from } (5.4.22) \tag{5.4.28}$$

(v:=u), estimating $|\psi_{cu}(x)|$ by means of Taylor's expansion of c in (5.4.28) and applying (5.4.26) for v:= u, we obtain (5.4.27). The extension of c to $\tilde{\Omega}$ or $\tilde{\mathcal{X}}$ was not needed. The constants M_6 and M_7 depend on $\|c\|_{C^1(\bar{\Omega})}$ and $\|c\|_{C(\bar{\Omega})}$ respectively, but they are independent of u, x and h, $h \leq h_o$.

The global evaluation of the error ψ_N^j on γ_{23} will be performed in the norm $\|\psi_N^j\|_{o,\gamma_{23}}$ defined by

$$\|\psi_N^j\|_{o,\gamma_{23}} := \Big\{\sum_{x\in\gamma_{23}} (\psi_N^j(x))^2 h(x)\Big\}^{1/2}, \ \psi_N^j \text{ from } (5.1.13),\ (5.1.18). \tag{5.4.29}$$

Lemma 5.23. Let VO"(PB,ω^1) or VO(MD, $\bar{\omega}$) be satisfied for irregular grids $\bar{\omega}$ and secondary networks of the type (PB) or (MD). Then, the following estimates hold for $h \leq h_o$ and $j = 0, 1$:

$$\text{case (PB):}\quad \|\psi_N^j\|_{o,\gamma_{23}} \leq (M_8+M_9\|u\|_{2+j,\Omega})h^{1+j}, \tag{5.4.30}$$

$$\text{case (MD):}\quad \|\psi_N^j\|_{o,\gamma_{23}} \leq (M_{10}+M_{11}\|u\|_{2+j,\Omega})\begin{cases} h^{1+j} & \text{for } k(x,s\tfrac{+}{\Gamma})\subset\bar{\Omega} \text{ at each } x\in\gamma_{23}, \\ h & \text{for } j = 0. \end{cases} \tag{5.4.31}$$

Here we suppose that the functions u, α, g, c and f satisfy the smoothness assumptions $u \in W_2^{2+j}(\Omega)$ or $u \in W_2^2(\Omega) \cap PW_2^{1+j}(\Gamma_{23})$, $\alpha \in PC^{1+j}(\Gamma_3)$, $g \in PC^{1+j}(\Gamma_{23})$ or $g \in PW_2^{1+j}(\Gamma_{23})$, $c \in C^j(\bar{\Omega})$, $f \in C^j(\bar{\Omega})$ or $f \in W_2^2(\Omega)$.

The constants M_i (i=8,9,10,11) are independent of u and h, $h \leq h_o$, but they depend on the norms of α, g, c and f in the spaces dealt with above, especially: $M_i = M_i(f,g)$ for i = 8,10, and $M_i = M_i(\alpha,c)$ for i = 9,11. ∎

Proof: Taking into account the splitting of ψ_N^j defined in (5.1.13), (5.1.14), (5.1.18), (5.4.14) and (5.4.15), we get

$$\|\psi_N^j\|^2_{o,\gamma_{23}} \leq 2\sum_{x\in\gamma_{23}}\{\psi_g^2(x) + \psi_{\alpha u}^2(x)\}h(x) \leq 4\sum_{x\in\gamma_{23}}\{\psi_f^2(x) + \varphi^2(x,g,j) + \psi_{cu}^2(x) + \chi^2(x,\alpha u,j)\}h(x), \tag{5.4.32}$$

with $\|\psi_N^j\|_{0,\gamma_{23}}$ from (5.4.29). By means of Taylor's expansion and Lemma 5.20., we may derive the following estimates for PB-schemes:

$$\psi_f^2(x) + \varphi^2(x,g,j) \le M_{12}h^{2+2j} \quad \text{for } g\in PC^{1+j}(\Gamma_{23}),\ f\in C^j(\bar\Omega),$$

$$\psi_f^2(x) + \varphi^2(x,g,j) \le M_{13}\{h^4\|f\|^2_{2,\tilde\Omega} + h^2|f|^2_{1,\tilde\Omega(x)}\}$$
$$+ M_{14}\{h^{1+2j}\|g\|^2_{PW_2^{1+j}(\widehat{\xi_\Gamma^- \xi_\Gamma^+})} + h^4\|g\|^2_{PC^j(\widehat{\xi_\Gamma^- \xi_\Gamma^+})}\}$$
$$\text{for } g\in PW_2^{1+j}(\Gamma_{23}),\ f\in W_2^2(\Omega),$$

$$\psi_{cu}^2(x) + \chi^2(x,\alpha u,j) \le M_{15}\{h^{2+2j}\|u\|^2_{2,\tilde\Omega} + h^2|u|^2_{1,\tilde\Omega(x)}\} \tag{5.4.33}$$
$$+ M_{16}\{h^{1+2j}\|u\|^2_{PW_2^{1+j}(\widehat{\xi_\Gamma^- \xi_\Gamma^+})} + h^4\|u\|^2_{PC^j(\widehat{\xi_\Gamma^- \xi_\Gamma^+})}\}$$
$$\text{for } c\in C^j(\bar\Omega),\alpha\in PC^{1+j}(\Gamma_{23}),\ u\in W_2^{2+j}(\Omega).$$

In the corresponding estimates for MD-schemes, inequality (5.4.20) produces modified bounds for $\varphi^2(x,g,j)$ and $\chi^2(x,\alpha u,j)$. For instance, relations (5.4.33) are valid for MD-schemes, if $\hbar(x,\xi_\Gamma^\pm)\subset\bar\Omega$ holds for all $x\in\gamma_{23}$. Otherwise, (5.4.33) may be easily replaced by another set of inequalities, which take into account (5.4.20) for $|\varphi_{\frac{\pm}{2}}|$ and $|\chi_{\frac{\pm}{2}}|$. Summing up inequalities (5.4.33) or their modified versions (case (MD)), taking into account imbedding, trace and extension theorems (Theorems 1, 4 and 5 from Appendix IM, Theorem 1 from Appendix EX) as well as some simple estimates, we arrive at inequalities (5.4.30), (5.4.31). For instance, the inequalities

$$\|u\|_{2,\tilde\Omega}\le C_E\|u\|_{2,\Omega},\ \|u\|_{C^j(\Gamma_{23}^i)} \le C_{I1}\|u\|_{1+j,\Gamma_{23}^i} \le C_{I2}\|u\|_{2+j,\Omega}$$

$(i=1,\dots,m)$ for $j\in\{0,1\}$, $\sum_{x\in\gamma_{23}}|u|^2_{1,\tilde\Omega(x)} \le h\, C^2_{EI}\|u\|^2_{2,\Omega}$ were employed. For more details, cf. HEINRICH [7,8].

<u>5.4.3. The modification and the weakening of some assumptions.</u> First of all, we shall discuss some modifications which are of practical interest. Sometimes it may happen that the local error $\psi_N^j(x)$, $x\in\gamma_{23}$, is "bad" only at a finite number $m = O(1)$ of grid points $x\in\gamma_{23}$. Then, the asymptotic behaviour of the error norm $\|\psi_N^j\|_{0,\gamma_{23}}$, cf. (5.4.29), need not necessarily be "bad" on the whole. This is due to

<u>Remark 5.24.</u> There is a one-dimensional analogue of Theorem 4 from Appendix IM, which asserts that

$$\|u\|^2_{o,\Gamma_e} \leq C\, h \|u\|^2_{1,\Gamma_T} \quad \text{for } u \in W^1_2(\Gamma_T) \text{ and } \Gamma_e \subset \Gamma_T \tag{5.4.34}$$

holds, if $\Gamma_T \in C^2$ (Γ_T fixed) and mes $\Gamma_e = O(h)$ are assumed to be fulfilled. ≡

The proof of (5.4.34) can be carried out by elementary means of differential and integral calculus.

Remark 5.25. If special standard domains are covered with a regular network, e.g. a rectangle with a regular grid of rectangles, then $x \in \gamma_{23}$ is the midpoint of the boundary segment $S(x)$ and the quadrature formula (3.2.8), for $j = 0$ and $r \in C(S)$, i.e. $\int_{S(x)} r\,ds \approx r(x)$ mes $S(x)$, becomes the "midpoint formula". For $r \in C^2(S)$, the accuracy of the finite difference approximation, which is based on the "midpoint formula", is the same as in the case (3.2.8), for $j = 1$. Sometimes, due to corners on Γ_{23} or jumps of r, a finite number $m = O(1)$ of grid points is provided with a quadrature formula ($j=0$) of lower accuracy only. But in such cases, Remark 5.24. allows us to state $\|\psi_N\|_{o,\gamma_{23}} = O(h^{3/2})$. ≡

We now proceed with the derivation of the right-hand side F_h of the FDS $A_h y = F_h$, if the smoothness assumption on the right-hand side f from (2.1.1a) varies. In Theorems 3.1. and 3.4., $f \in C(\bar{\Omega})$ was used. But now the assumptions on f which are given in (2.1.6) will be taken into account. Obviously, some of these assumptions are weaker than $f \in C(\bar{\Omega})$. The right-hand side F_h of the FDS $A_h y = F_h$ is defined by $F_h(x) := f_h(x)$ for $x \in \omega$, and $f_h(x)$ is an approximation of the functional

$$\tilde{f}(x) := \frac{1}{H(x)} \int_{\mathcal{X}(x)} f(\bar{x})d\bar{x} \quad \text{for } x \in \omega,\ f \text{ from (2.1.1a), (2.1.6).} \tag{5.4.35}$$

At boundary grid points $x \in \gamma_{23}$, the right-hand side f is also involved in $F^j_h(x) := g^j_h(x)$, but here as an approximation of the functional

$$\tilde{f}(x) := \frac{1}{h(x)} \int_{\mathcal{X}(x)} f(\bar{x})d\bar{x} \quad \text{for } x \in \gamma_{23},\ f \text{ from (2.1.1a), (2.1.6),} \tag{5.4.36}$$

cf. also (3.2.2), (3.2.16) and (3.2.18).

In the following, we shall define approximations $f_h(x)$ which can be employed in Theorem 3.1., (3.2.18), or in Theorem 3.4., (3.3.19). At the same time, the asymptotic behaviour of the local approximation error $\psi_f(x)$ will be noted.

Assumption 5.26. - $f \in C^s(\bar{\Omega})$. Let $f_h(x)$ be defined as follows,

$$\text{for } s \geq 0{:}\ f_h(x) := f(x) \quad \text{for } x \in \omega + \gamma_{23}, \tag{5.4.37}$$

i.e., $f_h(x) \approx \tilde{f}(x)$ for $x \in \omega$, and $f_h(x) \approx h(x)\tilde{H}^{-1}(x)\tilde{f}(x)$ for $x \in \gamma_{23}$. ≡

The error $\psi_f(x)$ from (5.1.9) ($x \in \omega$) and (5.4.14b) ($x \in \gamma_{23}$) is given by

$$\psi_f(x) = \begin{cases} o(1) & s = 0 \\ O(h) \quad \text{for} & s = 1 \\ O(h^2) & s = 2, \\ & \varkappa(x) \text{ regular} \end{cases} \quad (x \in \omega) \qquad \text{and} \quad \psi_f(x) = \begin{cases} O(h) & s = 0, \\ \quad \text{for} & \\ O(h^2) & s = 1, \end{cases} \quad (x \in \gamma_{23}) \tag{5.4.38}$$

which can be shown by Taylor's expansion.

Assumption 5.27. - $f \in W_2^s(\Omega)$. Let $f_h(x)$ be given by (5.4.39), i.e.

for $s = 2$: $\quad f_h(x) := f(x)$ for $x \in \omega + \gamma_{23}$,

$$s = 0,1: \quad f_h(x) := \tilde{f}(x) \text{ for } x \in \omega, \; f_h(x) := \frac{h(x)}{\tilde{H}(x)} \tilde{f}(x) \text{ for } x \in \gamma_{23}. \equiv \tag{5.4.39}$$

For $s = 2$, due to Theorem 1 from Appendix IM, f can be regarded as a continuous function. But for $s \in \{0,1\}$, $\Omega \subset R^2$, the value $f(x)$ is not convenient for the approximation of the functionals considered here. In such cases, other approximations must be found, for instance, projections of f on a set of known test functions. But this will not be discussed here, and for simplicity, we take exact values, cf. (5.4.39). Thus, we have $\psi_f(x) = 0$ for $x \in \omega + \gamma_{23}$, $s \in \{0,1\}$. For $s = 2$, we get

$$\psi_f(x) = \begin{cases} O(|f|_{1,\varkappa(x)} + h|f|_{2,\varkappa(x)}) \\ O(h|f|_{2,\varkappa(x)}), \; \varkappa(x) \text{ regular} \end{cases} \quad (x \in \omega), \quad \psi_f(x) \text{ for } x \in \gamma_{23} \text{ can be taken from (5.4.26).} \tag{5.4.40}$$

Assumption 5.28. - $f \in PC^s(\bar{\Omega})$. Let $f_h x$ be defined in the following way,

for $s \geq 0$: $\quad f_h(x) := \frac{1}{\tilde{H}(x)} \{f(x-0)\tilde{H}^-(x) + f(x+0)\tilde{H}^+(x)\}$ for $x \in \omega + \gamma_{23}$,

with $\tilde{H}(x) = H(x)$ for $x \in \omega$, $x \pm 0 \in \bar{\Omega}^{\pm}$, $\tilde{H}^{\pm}(x) := \text{mes } \tilde{\varkappa}^{\pm}(x)$,

$\tilde{\varkappa}^{\pm}(x)$: polygonal modifications of $\varkappa^{\pm}(x) := \varkappa(x) \cap \bar{\Omega}^{\pm}$. $\equiv$ (5.4.41)

We assume that f is discontinuous along a curve $\Gamma_0(f)$, and $\Gamma_0(f)$ cuts Ω into two portions $\Omega^{\pm}$, where the domains $\Omega^{\pm}$ are of the same type as Ω, i.e. $\partial\Omega^{\pm} \in C^{0,1} \cap PC^2$. If more than two partial domains $\Omega^{\pm}$ occur, then (5.4.41) and the following considerations can be easily modified. The curve $\Gamma_0(f)$ defines the portions $\varkappa^{\pm}(x) := \varkappa(x) \cap \bar{\Omega}^{\pm}$, and $f \in PC^s(\bar{\Omega})$ implies $f \in C^s(\bar{\Omega}^{\pm})$, i.e. $f \in C^s(\varkappa^{\pm})$. Due to $\int_{\varkappa} \cdot = \int_{\varkappa^-} \cdot + \int_{\varkappa^+} \cdot$, it seems to be reasonable to take

$$f_h(x) := \frac{1}{H(x)} \{f(x^-)H^-(x) + f(x^+)H^+(x)\}, \; x^{\pm} \in \varkappa^{\pm}(x), \tag{5.4.42}$$

where $x^{\pm}$ are fixed points in $\varkappa^{\pm}(x)$, $H^{\pm}(x) := \text{mes } \varkappa^{\pm}(x)$. For practical purposes, $x^{\pm}$ should be replaced by grid points and $H^{\pm}(x)$ by $\tilde{H}^{\pm}(x)$, $\tilde{H}^{\pm}(x)$ from (5.4.41). This and the application of networks with $\Gamma_0(f)$-adaption will be described in the following.

Assumption 5.29. Let γ, γ_0 be sets of grid points on Γ, Γ_0, respectively, such that the points of $\Gamma \cap \Gamma_0(f)$ (if $\Gamma \cap \Gamma_0 \neq \emptyset$) are also contained in γ. Since $\partial\Omega^{\pm} \in C^{0,1} \cap PC^2$ is assumed to be fulfilled, the sets γ_0, γ define polygonal lines Γ_{oh}, Γ_h and polygonally bounded domains $\bar{\Omega}_h^{\pm}$, which approximate $\bar{\Omega}^{\pm}$ in the same sense as $\bar{\Omega}_h$ approximates $\bar{\Omega}$, cf. Section 2.2.1. Now the domains $\bar{\Omega}_h^{\pm}$ will be equipped with the primary and secondary networks described in Sections 2.2. and 2.3. ∎

Thus, Γ_0, Γ_{oh} and γ_0 are common boundary portions of $\bar{\Omega}^{\pm}$, $\bar{\Omega}_h^{\pm}$ and $\bar{\omega}^{\pm} := \bar{\omega} \cap \bar{\Omega}^{\pm}$, respectively, and each grid point $x \in \gamma_0 = \omega \cap \Gamma_0(f)$ and $x \in \gamma \cap \Gamma_0(f)$ (if $\Gamma \cap \Gamma_0(f) \neq \emptyset$) satisfies the relation $x \in \mathcal{H}^{\pm}(x) := \mathcal{H}(x) \cap \bar{\Omega}^{\pm}$. Therefore, instead of $x^{\pm}$ and $f(x^{\pm})$ in (5.4.42), we may take the grid point x and the values $f(x \pm 0)$, respectively, with $x \pm 0 \in \bar{\Omega}^{\pm}$, more precisely: $f(x \pm 0) := \lim_{\xi^{\pm} \to x} f(\xi^{\pm})$, $\xi^{\pm} \in \Omega^{\pm}$. Since $\Gamma_0 \in PC^2$ holds, it is reasonable to take $\tilde{H}(x)$, $\tilde{H}^{\pm}(x)$ as an approximation of $H(x)$, $H^{\pm}(x)$, where $\tilde{\mathcal{H}}(x)$ is the polygonal modification of $\mathcal{H}(x)$ $(x \in \gamma_{23})$, $\tilde{\mathcal{H}}^{\pm}(x) := \mathcal{H}(x) \cap \bar{\Omega}_h^{\pm}$ (for $x \in \gamma_{23}$: $\tilde{\mathcal{H}}(x)$ instead of $\mathcal{H}(x)$). Obviously, $H^{\pm}(x) - \tilde{H}^{\pm}(x) = O(h^3)$ holds. The local approximation error $\psi_f(x)$ for $f_h(x)$ from (5.4.41) is given by

$$\underset{(x \in \omega)}{\psi_f(x)} = \begin{cases} o(1) & s = 0 \\ O(h) \quad \text{for} & s = 1 \\ O(h^2) & s = 2, \\ & \mathcal{H}(x) \text{ regular and } x \notin \gamma_0 \end{cases} \quad \text{and} \quad \underset{(x \in \gamma_{23})}{\psi_f(x)} = \begin{cases} O(h) & s = 0, \\ & \text{for} \\ O(h^2) & s = 1, \end{cases} \tag{5.4.43}$$

if Assumption 5.29. holds, i.e., (5.4.43) coincides with (5.4.38) for $x \in (\omega \setminus \gamma_0) + \gamma_{23}$. But for $x \in \gamma_0$ and $s \geq 1$, we only get $\psi_f(x) = O(h)$, since $\mathcal{H}^{\pm}(x)$ cannot be regular. Assumption (5.4.42) yields similar results.

Finally, the global error $\|\psi_N\|_0$ should be discussed in a shortened form. For $f \in C^s(\bar{\Omega})$, $s \in \{1,2\}$, and $f \in W_2^2(\Omega)$, the estimates of $\|\psi_N\|_{0,\omega}$ and $\|\psi_N^j\|_{0,\gamma_{23}}$ are given in Section 5.4.1., 5.4.2. In the case $f \in C(\bar{\Omega})$ (s=0), the relation $\|\psi_f\|_{0,\omega} = o(1)$ is obvious. For $f \in W_2^s(\Omega)$, $s \in \{0,1\}$, the error ψ_f vanishes everywhere, since "exact approximation" is assumed. Under Assumption 5.29. and for $f \in PC^1(\bar{\Omega})$ instead of $f \in C^1(\bar{\Omega})$, we get $\|\psi_f\|_{0,\omega} = O(h)$, $\|\psi_f\|_{0,\gamma_{23}} = O(h^2)$. Moreover, for locally irregular grids $\bar{\omega}$ from Definition 2.13. and $f \in PC^2(\bar{\Omega})$, the relation $\|\psi_f\|_{0,\omega} = O(h^{3/2})$ holds, cf. (5.4.43) and the technique of deriving inequalities of the type (5.4.13) (with γ_0 instead of $\overset{*}{\omega}$).
Lastly, it remains to discuss the smoothness assumptions $c \in W_2^2(\Omega)$ and $\alpha \in PW_2^{1+j}(\Gamma_3)$ (j=0,1) from (2.1.6). By virtue of Theorem 1 from Appendix IM, these assumptions imply $c \in C(\bar{\Omega})$ and $\alpha \in C^j(\Gamma_3^i)$ for each arc $\Gamma_3^i \in C^2$ $(i=1,2,\ldots,m_3)$. Assuming $u \in W_2^{2+j}(\Omega)$, the relations $cu \in W_2^2(\Omega)$ and

$\alpha u \in PW_2^{1+j}(\Gamma_3)$ ($j=0,1$) are also obvious. Therefore, all functionals used for the derivation of the FDSs $A_h y = F_h$ of Theorems 3.1. and 3.4. are defined, and error estimates can be proved. Thus setting $v := cu$ and taking into account (5.4.1) - (5.4.3), we get local error bounds of $|\psi_v(x)|$ which also lead to (5.4.11). Analogously, all estimations carried out for $\chi(x,\alpha u,j)$ under the assumption $\alpha \in PC^{1+j}(\Gamma_3)$ can also be made for $\alpha \in PW_2^{1+j}(\Gamma_3)$. The estimates of $\varphi_i^{\pm}(x,v,j)$ for $v \in PW_2^{1+j}(\Gamma_3)$ are given in (5.4.17a) and yield bounds of $|\chi_i^{\pm}|$ ($i=1,2,\dots,6$). In order to estimate $|\chi_i^{\pm}|$ for $i = 7,8$, we may utilize $\alpha \in C^1(\Gamma_3^i)$. Then, the global estimates are quite similar to that of Lemma 5.23.

5.5. Convergence for $W_2^2(\Omega)$-solutions

5.5.1. Convergence in the discrete W_2^1-norm. Some results on the convergence $y \to u$ for classical solutions $u \in C^{2+j}(\bar{\Omega})$, $j \in \{0,1\}$, have been given in Section 5.1.4., cf. also HEINRICH [6,7]. Here, the notion "convergence" is understood in the sense that y approaches u as the mesh size parameter h tends to zero, and y, u denote the unique solutions of the FDS $A_h y = F_h$ on $\bar{\omega}$ and the BVP $Au = F$ on $\bar{\Omega}$, respectively. The error $z := y - u$ will be evaluated by the discrete W_2^1-norm, i.e. by $\|y-u\|_1$. Now, error estimates indicating the asymptotic rate of convergence $y \to u$ will be studied for generalized solutions u which belong to the Sobolev space $W_2^{2+j}(\Omega)$, $j \in \{0,1\}$. Taking into account a priori estimates of the type $\|y-u\|_1 \leq C\{\|\varkappa\|_0'^2 + \|\psi_N^j\|_0^2\}^{1/2}$ from Theorem 5.3., we clearly see that convergence results can be stated by estimating the discrete L_2-norms of $\varkappa$ and ψ_N^j, which was the central topic of Sections 5.2.,5.3. and 5.4. Thus, the following Theorem 5.30. holds, which can be proved by summarizing various estimates already derived in the previous sections. At this point we recall that the index j is sometimes omitted, for instance, at y, z and several constants.

Theorem 5.30. PB-schemes. Let $u \in W_2^{2+j}(\Omega)$ ($j=0,1$) be the solution of the BVP $Au = F$ from (3.2.1), and let y be the solution of the PB-schemes $A_h^j y = F_h^j$ from (3.2.18), i.e. from Theorem 3.1., either for $j = 0$ or $j = 1$. Furthermore, let VO(PB, $\bar{\omega}$) be fulfilled and, additionally, condition (5.2.45). Then, for $j \in \{0,1\}$ and the norm $\|\cdot\|_1 := \|\cdot\|_{1,p}$ from (4.3.13), the following estimates verifying the convergence $y \to u$ ($h \to 0$) with respect to the discrete W_2^1-norm $\|\cdot\|_1$ hold:

(i) $\|y-u\|_1 \leq Mh^{1+j/2}$, with $M = O(1)$, for $j = 0,1$ and $h \leq h_o$, (5.5.1)

$u \in W_2^{2+j}(\Omega)$ and k, c, $f \in C^{1+j}(\bar{\Omega})$, $\alpha, g \in PC^{1+j}(\Gamma_{23})$ $(\alpha|_{\Gamma_2}=0)$, $g \in C(\Gamma_1)$, for locally irregular networks from Definitions 2.10., 2.11. and 2.12., which consist of triangles and (or) rectangles. For other smoothness assumptions on f and g, see the subsequent Remark 5.3.1. The constant

M depends on j and on the norms of u, k, c, f, α, g in the spaces previously encountered as well as on further parameters of the network and the BVP (3.2.1). But M is independent of h, $h \le h_o$, for sufficiently small h_o.

(ii) $\qquad \|y-u\|_1 \le Mh^2, \qquad$ (5.5.2)

for regular networks of triangles and (or) rectangles and under the smoothness assumptions from (i) for j = 1, but for the special case $\Gamma = \Gamma_1 (\Gamma_{23} = \emptyset)$, i.e. especially for $u \in W_2^3(\Omega)$ and k, c, $f \in C^2(\bar{\Omega})$.

(iii) $\qquad \|y-u\|_1 \le Mh, \qquad$ (5.5.3)

for irregular networks of triangles and (or) rectangles, with the smoothness assumptions from (i) for j = 0. ■

Remark 5.31.

(i) All estimates of Theorem 5.30. are maintained, if $f \in C^{1+j}(\bar{\Omega})$ is substituted by $f \in W_2^{1+j}(\Omega)$, $j \in \{0,1\}$, or by $f \in L_2(\Omega)$, and if the approximation $f_h(x)$ from (5.4.39) is employed. Moreover, no change in the estimates occurs, if $c \in C^2(\bar{\Omega})$ is replaced by $c \in W_2^2(\Omega)$.

(ii) Estimates (5.5.1) and (5.5.3) also hold for $f \in PC^{1+j}(\bar{\Omega})$ instead of $f \in C^{1+j}(\bar{\Omega})$, if $f_h(x)$ from (5.4.41), together with Assumption 5.29., or $f_h(x)$ from (5.4.42) are applied; cf. also $\bar{\omega}$ from Definition 2.13.

(iii) If the assumption $g \in PC^{1+j}(\Gamma_{23})$ is substituted by $g \in PW_2^{1+j}(\Gamma_{23})$, i.e. $g \in W_2^{1+j}(\Gamma_{23}^i)$, with $\Gamma_{23}^i \in C^2$, $i = 1,2,\ldots,m_{23}$, cf. (2.1.6), then all estimates of Theorem 5.30. continue to be valid. The same holds, if $\alpha \in PC^{1+j}(\Gamma_3)$ is replaced by $\alpha \in W_2^{1+j}(\Gamma_3)$, $j \in \{0,1\}$. ■

The proof of these assertions is sketched in the proof of Theorem 5.30. or can be partially taken from Section 5.4.3. It should be noted that the smoothness assumptions on the coefficients, right-hand sides and on the boundary $\partial\Omega$, which are assumed to be fulfilled, do not suffice to guarantee, in general, the existence of a solution $u \in W_2^{2+j}(\Omega)$. But they form, together with the assumption $u \in W_2^{2+j}(\Omega)$, a set of smoothness conditions sufficient for the validity of the estimates (5.5.1), (5.5.2) and (5.5.3) from Theorem 5.30. Assumption $f \in W_2^j(\Omega)$ (j=0,1) implies $u \in W_2^{2+j}(\Omega)$ only in some special cases (shift theorem), since the smoothness of u on the whole domain Ω does not depend exclusively on f, but essentially on the geometry of $\partial\Omega$ and on the boundary conditions on $\partial\Omega$, too.

Proof (Theorem 5.30.): Taking into account inequality (5.1.32) and inserting the bounds of the norms $\|\varkappa\|_o'$ and $\|\psi_N^j\|_o^2 = \|\psi_N^j\|_{o,\omega}^2 + \|\psi_N^j\|_{o,\gamma_{23}}^2$, we immediately get all estimates of Theorem 5.30. The bounds of these norms are given in Lemma 5.11., cf. (5.2.47) for $\|\varkappa\|_o'$, in Lemma 5.18., cf. (5.4.11) for $\|\Psi_N^j\|_{0,\omega}$, and in Lemma 5.23., cf. (5.4.30), (5.4.31) for $\|\Psi_N^j\|_{0,\gamma_{23}}$. If various smoothness assumptions on f and g are taken into account, the associated convergence assertions are stated in Remark 5.31. and proved by Lemma 5.18., Lemma 5.23. and the estimates shown in Section 5.4.3.

In the following, MD-schemes $A_h y = F_h$ approximating the BVP $Au = F$ from (2.1.1) will be investigated. Here, more general coefficients k_{il} $(i,l=1,2)$ and grid regularity assumptions than for PB-schemes are admitted. But the estimate $\|y-u\|_1 = O(h^{3/2})$ requires an additional assumption on the curvature of Γ_{23}, which can be regarded as a "convexity condition of Ω in the neighbourhood of Γ_{23}". This condition originates from the asymptotic behaviour of $\partial\mathcal{X}(x)$ for $x \in \gamma_{23}$ and $h \leq h_o$.

<u>Theorem 5.32. MD-schemes.</u> Let $u \in W_2^{2+j}(\Omega)$ $(j=0,1)$ be the solution of the BVP $Au = F$ from (2.1.1), and let y be the solution of the MD-schemes $A_h^j y = F_h^j$ from (3.3.19), i.e. from Theorem 3.4., either for $j = 0$ or $j = 1$. Furthermore, let VO(MD,$\bar{\omega}$) be fulfilled. Then, for $j \in \{0,1\}$ and the norm $\|\cdot\|_1 := \|\cdot\|_{1,m}$ from (4.3.14), the following estimates verifying the convergence $y \to u$ $(h \to 0)$ with respect to the discrete W_2^1-norm $\|.\|_1$ hold:

(i) $\|y-u\|_1 \leq M h^{1+j/2}$, with $M = O(1)$, for $j = 0,1$ and $h \leq h_o$, (5.5.4)

$u \in W_2^{2+j}(\Omega)$ and $k_{il}, c, f \in C^{1+j}(\bar{\Omega})$, α, $g \in PC^{1+j}(\Gamma_{23})$ $(\alpha|_{\Gamma_2} = 0)$, $g \in C(\Gamma_1)$, for locally irregular networks of triangles from Definition 2.10. Moreover, the Condition 2.17. - VO"(PB,$\bar{\omega}$) is assumed to be fulfilled for that part of the network which consists of non-equilateral triangles. Furthermore, for the boundary Γ_{23}, the relation $\hbar(x') \subset \bar{\Omega}$ for all $x' \in \gamma_{23}'$ must be satisfied. For other smoothness assumptions on f and g, see Remark 5.31. The constant M is of the same type as in Theorem 5.30.

(ii) $\|y-u\|_1 \leq M h^2$, (5.5.5)

for regular networks of equilateral triangles and under the smoothness assumptions from (i) for $j = 1$, but for the special case $\Gamma = \Gamma_1$ $(\Gamma_{23} = \emptyset)$, i.e. especially for $u \in W_2^3(\Omega)$ and $k_{il}, c, f \in C^2(\bar{\Omega})$.

(iii) $\|y-u\|_1 \leq M h$, (5.5.6)

for networks of arbitrary triangles satisfying VO(MD,$\bar{\omega}$), and under the smoothness assumptions from (i) for $j = 0$. ■

Remark 5.31. assigned to Theorem 5.30. can be repeated verbatim for Theorem 5.32. and estimates (5.5.4) - (5.5.6).

Proof: The proof is quite similar to that of Theorem 5.30. In order to show (5.5.6), we start from inequality (5.1.33), which holds for networks provided with VO(MD, $\bar{\omega}$), and we insert the bounds of $\|\varkappa\|_o''$ and $\|\psi_N^j\|_o^2 = \|\psi_N^j\|_{o,\omega}^2 + \|\psi_N^j\|_{o,\gamma_{23}}^2$. These bounds are to be taken from Lemma 5.13., cf. (5.3.7) and (5.3.1), and from Lemma 5.18., (5.4.11a), Lemma 5.23., (5.4.31). To prove (5.5.4) and (5.5.5), we take into account the x'-splitting for MD-schemes and inequality (5.1.34). The bounds of $\|\varkappa\|_o$ were derived in Lemma 5.16., cf. (5.3.19). Lemma 5.18., (5.4.11b, c), and Lemma 5.23., (5.4.31), yield the bounds of $\|\psi_N^j\|_o$.

5.5.2. Convergence in the discrete C-norm. The following theorem is based on the results of Section 5.5.1. and contains an assertion about the rate of convergence $y \to u$ $(h \to 0)$ with respect to the discrete C-norm.

Theorem 5.33. Under the assumptions of Theorem 5.30. (PB-schemes) or of Theorem 5.32. (MD-schemes), together with Remark 5.31., the convergence $y \to u$ $(h \to 0)$ with respect to the discrete C-norm $\|\cdot\|_{C(\bar{\omega})}$ from (4.3.9) can be guaranteed. Here, u and y denote respectively the solutions of the BVP Au = F and the FDSs $A_h^j y = F_h^j$ $(j=0,1)$. Moreover, the following estimates hold:

$$\|y-u\|_{C(\bar{\omega})} \le M_1 |\ln h|^{1/2} \begin{cases} h^{1+j/2} & j = 0,1,\ \bar{\omega} \text{ locally irregular, case (i)}, \\ h^2 & \text{for } j = 1,\ \bar{\omega} \text{ regular},\ \Gamma = \Gamma_1, \text{ case (ii)}, \\ h & j = 0,\ \bar{\omega} \text{ irregular, case (iii)}, \end{cases} \tag{5.5.7}$$

where the three rows in (5.5.7) correspond directly to the three cases (i), (ii), (iii) of Theorem 5.30. (PB-schemes) or of Theorem 5.32. (MD-schemes). ■

Proof: Since the grid regularity conditions VO(MD, $\bar{\omega}$) or VO(PB, $\bar{\omega}$) are satisfied, Theorem 4.36. or Corollary 4.37. yield the estimates $\|y-u\|_{C(\bar{\omega})} \le C_r |\ln h|^{1/2} \|y-u\|_{1,r}$, for r = m or r = p. After combining this inequality with the estimates from Theorem 5.32. (r=m) or Theorem 5.30. (r=p), the proof is complete.

6. FINITE DIFFERENCE SCHEMES FOR NONSYMMETRIC PROBLEMS

6.1. Construction of finite difference approximations

6.1.1. Boundary value problem and balance equations. We now consider second order differential equations containing terms $b_i \frac{\partial u}{\partial x_i}$ (i=1,2), i.e.,

$$L^b u := -\sum_{i,j=1}^{2} \frac{\partial}{\partial x_i}\left(k_{ij}\frac{\partial u}{\partial x_j}\right) + \sum_{i=1}^{2} b_i \frac{\partial u}{\partial x_i} + cu = f \text{ in } \Omega, \tag{6.1.1a}$$

$$lu := \begin{cases} u = g & \Gamma_1, \\ & \text{on} \\ \sum_{i,j=1}^{2} k_{ij}\frac{\partial u}{\partial x_j} n_i + \alpha u = g & \Gamma_{23} \; (\alpha|_{\Gamma_2}=0). \end{cases} \tag{6.1.1b}$$

Taking $\nabla u := \left(\frac{\partial u}{\partial x_1}, \frac{\partial u}{\partial x_2}\right)^T$ and the vector $b = (b_1,b_2)^T$ of the coefficients $b_i(x)$ (i=1,2) defined for $x \in \bar{\Omega}$, the term with the derivatives of first order may be expressed by

$$(b, \nabla u) := \sum_{i=1}^{2} b_i \frac{\partial u}{\partial x_i}, \quad (b, \nabla u) = |b|(\mathring{b}, \nabla u) = |b| \frac{\partial u}{\partial b}, \tag{6.1.2}$$

with $|b| := (b_1^2+b_2^2)^{1/2}$, $\mathring{b} := |b|^{-1} b$, i.e. $|\mathring{b}| = 1$,

$(.,.)$: scalar product in R^2, and $\frac{\partial u}{\partial b}$ denotes the derivative with respect to the direction b.

Equations of the type (6.1.1a) are often called convection-diffusion equations, since the terms $L_0 u$, cf. (2.1.3), and $(b, \nabla u)$ correspond to the diffusion process and the convection effect, respectively, in some flow models. Since the BVP $Au = F$ from (2.1.1), (2.1.2) differs from (6.1.1) in the term $(b, \nabla u)$ only, the following notations are appropriate.

Assumption 6.1. Assume that

(i) the symbols $Au = F$, $A = (L,l)^T$ and $F = (f,g)^T$ are reserved for the BVP (2.1.1), (2.1.2) (with formally selfadjoint operator), and the symbols

$$A^b u = F, \; A^b = (L^b,l)^T, \; F = (f,g)^T \tag{6.1.3a}$$

will be taken for the BVP (6.1.1) (with non-selfadjoint operator), and

(ii) R denotes the difference of A^b and A, viz.

$$Ru := A^b u - Au, \; R = (L^b - L, 0)^T, \text{ with } L^b u - Lu = (b, \nabla u), \tag{6.1.3b}$$

and 0 represents the zero operator for functions defined on Γ. Concerning the properties of the coefficients $b_i(x)$ from (6.1.1a), we assume

$$b_i \in C^s(\bar{\Omega}) \text{ for } s \in \{0,1,2\} \text{ and } |b_i(x)| \leq b_o \text{ for } x \in \bar{\Omega},\ i=1,2. \qquad (6.1.4)$$

In the following, the symbols A_h, A_h^b, L_h, L_h^b and R_h will denote finite difference approximatious of the corresponding operators without the subscript h. ∎

For the analysis of problems such as (6.1.1), see e.g. AZIZ [1], GILBARG/TRUDINGER [1], GOERING et al. [1], LADYZHENSKAYA/URAL'TSEVA [1], LADYZHENSKAYA [1], MEYER [1], NEČAS [1], TEMAM [2]. Studies on numerical methods for such problems can be found in ASTRAKHANTSEV [1], AXELSSON [1], IKEDA [1], MITCHELL/GRIFFITHS [1,2], NÄVERT [1], OGANESYAN et al. [1], OGANESYAN/RUKHOVETS [1], SAMARSKIĬ/ANDREEV [1], SAMARSKIJ [1], THOMASSET [1], VAINIKKO [1] and in many other papers.
For the BVP $A^b u = F$, maximum principles hold even in the case of "dominant convection", i.e. for great values of b_o in comparison with k_o, where b_o, k_o are taken from (6.1.4), (2.1.7a), respectively, cf. GILBARG/TRUDINGER [1], MEYER [1]. Mainly for the sake of stability of the FDS $A_h^b y = F_h$ approximating the BVP $A^b u = F$, one tries to get the monotonicity property $M(A_h^b, \bar{\omega})$, cf. (4.2.1), for A_h^b and realistic values of the mesh size parameter h. For getting such properties, the approaches of "upwind schemes" with finite differences (or elements) and "artificial viscosity schemes" are used, see e.g. IKEDA [1], NÄVERT [1], SAMARSKIĬ/ANDREEV [1], THOMASSETT [1].
For $|b| \neq 0$, the operator L^b from (6.1.1a) is not selfadjoint, in general. Thus, one cannot expect that the difference operator A_h^b is symmetric on D_o. Nevertheless, for sufficiently small b_o in comparison with k_o, the positive definiteness of the operator A_h^b on D_o, D_o from (2.2.8), can be proved; cf. LADYZHENSKAYA [1] for the "continuous case".
In the following, irregular grids $\bar{\omega}$ and secondary networks (PB) and (MD) will be used for the derivation of FDSs $A_h^b y = F_h$, and PB-schemes are considered only for special coefficients $k := k_{11} \equiv k_{22}$ and $k_{12} \equiv 0$.
Integrating $L^b u = f$ over $\mathfrak{K}(x)$, cf. (6.1.1a), where $u \in C^2(\bar{\Omega})$ or Assumption 2.3. - V(u) is supposed to be fulfilled, and taking into account (3.1.1) - (3.1.6), we now get the relations (6.1.5) instead of (3.1.7), i.e.,

$$\int_{\partial\mathfrak{K}(x)\setminus\mathfrak{S}(x)} (K\nabla u,n)ds + \int_{\mathfrak{K}(x)} (b,\nabla u)d\bar{x} + \int_{\mathfrak{K}(x)} cu\, d\bar{x} + \int_{\mathfrak{S}(x)} \alpha u\, ds = \int_{\mathfrak{K}(x)} f\, d\bar{x} + \int_{\mathfrak{S}(x)} g\, ds \qquad (6.1.5)$$

for $x \in \omega + \gamma_{23}$, with $\mathfrak{S}(x) = \emptyset$ for $x \in \omega$ and $\mathfrak{S}(x) = \partial\mathfrak{K}(x) \cap \Gamma$ for $x \in \gamma_{23}$, and with $u(x) = g(x)$ for $x \in \gamma_1$.

This is the "system of balance equations" assigned to (6.1.1) on the grid $\omega + \gamma_{23}$ and under the restriction $u = g$ on γ_1. Since the balance equations from (6.1.5) and (3.1.7) differ in the term $\int_{\mathfrak{K}(x)}(b, \nabla u)d\bar{x}$ only, all approximations proposed in Chapter 3 can be utilized and further finite difference approximations must be developed for $\int_{\mathcal{K}(x)}(b,\nabla u)d\bar{x}$ by means of cubature formulas and difference quotients $\nabla_h u$.

6.1.2. Upwind difference quotients. Taking the grid point $x\in\omega+\gamma_{23}$ itself as the cubature point for the numerical integrations over $\mathcal{K}(x)$, then, by means of the one-point cubature formula from (3.2.11), we obtain the approximation

$$\int_{\mathcal{K}(x)}(b,\nabla u)d\bar{x}\approx(b(x),\nabla u(x))\,H(x) = |b(x)|\frac{\partial u}{\partial b}(x)H(x) \text{ for } x\in\omega+\gamma_{23}, \qquad b_i\in C(\bar{\Omega}),\ u\in C^1(\bar{\Omega}), \tag{6.1.6}$$

with $|b|$ and $\frac{\partial u}{\partial b}$ from (6.1.2). Furthermore, the derivatives at x can be approximated as follows:

$$\nabla u(x)\approx\nabla_h u(x''),\ \frac{\partial u}{\partial x_1}(x)\approx(e_1,\nabla_h u(x'')),\ \frac{\partial u}{\partial b}(x)\approx(\mathring{b}(x),\nabla_h u(x'')), \tag{6.1.7}$$

with $\nabla_h u$ from (3.3.14), $e_1 = (1,0)^T$, $e_2 = (0,1)^T$ and $\mathring{b}$ from (6.1.2),

where the special choice of x" from the set S"(x) will be defined subsequently, by means of so-called "upwind triangles". The choice of such upwind triangles $\Delta(x''):=\Delta(x,\mathfrak{s},\mathfrak{s}^+)$ depends on the vector b(x) and will now be discussed under several assumptions.

Assumption 6.2. - upwind triangle $\Delta^b(x,x'')$. There is a triangle $\Delta^b(x,x''):=\Delta(x,\mathfrak{s},\mathfrak{s}^+)$ called "upwind triangle with respect to b(x)"(for $|b|\neq 0$) which is assigned to the grid point x and to the vector b(x) such that the projections of the vector -b(x) onto $h(x,\mathfrak{s})$ and $h^+(x,\mathfrak{s})$ ($h^+(x,\mathfrak{s}):=h(x,\mathfrak{s}^+)$) are nonnegative and the straight line g_b spanned by b(x) meets the triangle side opposite to x, i.e.

$$(-b,h)\geq 0,\ (-b,h^+)\geq 0 \text{ and } g_b\cap[\mathfrak{s},\mathfrak{s}^+]\neq\emptyset, \text{ cf. Fig. 12a.} \ \blacksquare \tag{6.1.8}$$

If h or h^+ have the same direction as -b(x), then it may happen that two triangles Δ^b satisfy (6.1.8). In such cases, one of these triangles will be selected.

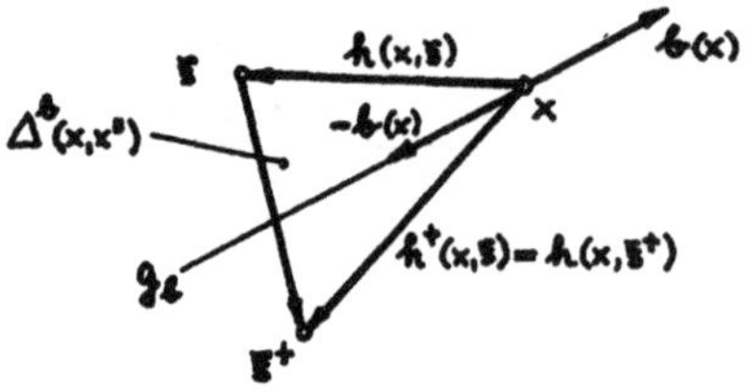

Fig. 12a

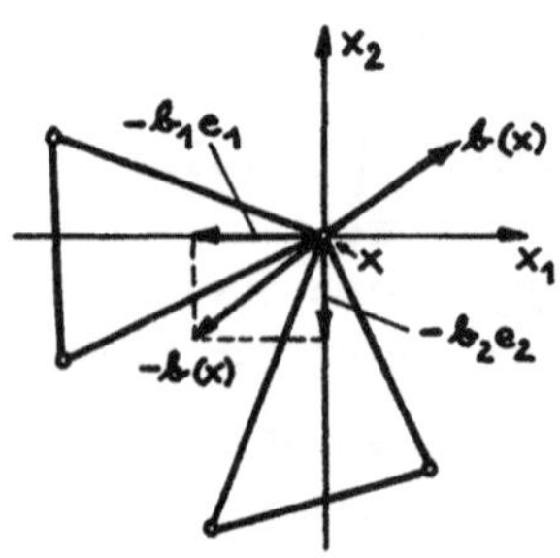

Fig. 12b

Remark 6.3. If $\bar{\Omega}$ is approximated by $\bar{\Omega}_h = \bigcup_{x''} \Delta(x'')$, and the angles θ of all triangles Δ satisfy the inequality $\theta \le \pi/2$, then Assumption 6.2. is fulfilled at least for interior grid points $x \in \omega$. Moreover, let Γ_{23} be smooth in some neighbourhood of $x \in \gamma_{23}$, and, let $(b,n) > 0$ be satisfied on this part of Γ_{23}, where n denotes the outward normal to Γ_{23}. Then, for sufficiently small h_o, Assumption 6.2. can be satisfied for $x \in \gamma_{23}$, too. ∎

Starting from (6.1.6), (6.1.7) and assuming the existence of upwind triangles $\Delta^b(x,x'')$ for $x \in \omega + \gamma_{23}$, we are in a position to define the so-called <u>upwind approximation with respect to b</u>:

$$\int_{\mathcal{K}(x)} (b, \nabla u)(\bar{x})d\bar{x} \approx (b(x), \nabla_h^b u(x)) \cdot \begin{cases} H(x) & x \in \omega, \\ & \text{for} \\ \tilde{H}(x) & x \in \gamma_{23}, \end{cases} \tag{6.1.9}$$

with $\nabla_h^b u(x) := \nabla_h u(x'')$ taken on the upwind triangle $\Delta^b(x,x'')$; for $\Delta^b(x,x'')$ and $\nabla_h u(x'')$ see Assumption 6.2. and (3.3.14), respectively.

Upwind approximations can also be defined with respect to the coordinate axes.

<u>Assumption 6.4. - upwind triangle $\Delta^{b_i}(x,x_i'')$.</u> There is a triangle $\Delta^{b_i}(x,x_i'') := \Delta(x,\xi,\xi^+)$ called "upwind triangle with respect to $b_i(x)e_i$" (for $b_i(x) \neq 0$, e_i from (6.1.7)) which is assigned to the grid point x and to the vector $b_i(x)e_i$ such that the projections of the vector $-b_i(x)e_i$ onto $h(x,\xi)$ and $h^+(x,\xi)$ $(h^+(x,\xi) := h(x,\xi^+))$ are nonnegative and the x_i-axis g_i meets the triangle side opposite to x, i.e.

$$b_i(-e_i,h) \geq 0, \; b_i(-e_i,h^+) \geq 0 \text{ and } g_i \cap [\xi,\xi^+] \neq \emptyset, \text{cf. Fig. 12b.} \blacksquare \tag{6.1.10}$$

Obviously, a remark similar to Remark 6.3. might be added here. Assuming the existence of upwind triangles $\Delta^{b_i}(x,x'')$ for $i = 1,2$ and $x \in \omega + \gamma_{23}$, we are able to define the so-called <u>upwind approximation with respect to the coordinate axes</u>:

$$\int_{\mathcal{K}(x)} (b, \nabla u)(\bar{x})d\bar{x} \approx \sum_{i=1}^{2} (b_i(x)e_i, \nabla_h^i u(x)) \cdot \begin{cases} H(x) & x \in \omega, \\ & \text{for} \\ \tilde{H}(x) & x \in \gamma_{23}, \end{cases} \tag{6.1.11}$$

with $\nabla_h^i u(x) := \nabla_h u(x_i'')$ taken on the upwind triangle $\Delta^{b_i}(x,x_i'')$; for $\Delta^{b_i}(x,x_i'')$ and $\nabla_h u(x_i'')$ see Assumption 6.4. and (3.3.14), respectively.

We clearly see that (6.1.9) requires one upwind triangle $\Delta^b(x,x'')$ for one direction b(x), but two upwind triangles $\Delta^{b_i}(x,x_i'')$ (i=1,2) are needed for the two directions $b_i(x)e_i$ (i=1,2).

If the network consists of rectangles, i.e., it is formed by mesh lines which are parallel to the x_1- or x_2-axis, then approximation (6.1.11) will be replaced by

$$\int_{\partial\mathcal{X}(x)} (b, \nabla u)(\bar{x})d\bar{x} \approx \sum_{i=1}^{2} b_i(x) \frac{\delta u}{\delta x_i}(x) \cdot \begin{cases} H(x) & \\ & \text{for} \\ \tilde{H}(x) & \end{cases} \begin{matrix} x \in \omega, \\ \\ x \in \gamma_{23}, \end{matrix}$$

with

$$\frac{\delta u}{\delta x_i}(x) := \begin{cases} \dfrac{u(\xi)-u(x)}{h(x,\xi)} & \text{for } b_i(x) > 0, \text{ cf. figure: } \\ \dfrac{u(x)-u(\xi)}{h(x,\xi)} & \text{for } b_i(x) < 0, \text{ cf. figure: } \end{cases} \quad (6.1.12)$$

The difference quotients $\frac{\delta u}{\delta x_i}(x)$ used in (6.1.12) are the so-called "backward" and "forward" difference quotients for the grid point x, see e.g. MITCHELL/GRIFFITHS [1] or SAMARSKIǏ/ANDREEV [1]. Moreover, if the network consists of rectangles and triangles, then a convenient combination of formulas (6.1.11) and (6.1.12) can be made. Thus, upwind approximations will also be obtained for these composite networks. It is worth noting that the difference quotient $(e_i, \nabla_h^i u(x))$ from (6.1.11) yields exactly the difference quotients $\frac{\delta u}{\delta x_i}$ applied in (6.1.12), if one triangle side of $\Delta^{b_i}(x,x_i^*)$ lies on the x_i-axis.

<u>6.1.3. Further difference quotients and the FDSs $A_h^b y = F_h$.</u> The difference quotients in (6.1.9), (6.1.11) and (6.1.12) correspond either to backward or forward finite difference quotients. Let us now consider some other difference quotients and cubature formulas which were not mentioned in Section 6.1.2. Thus, we can also introduce difference quotients which are more or less "central" of "weighted":

a) central difference quotient for a mesh line parallel to the x_i-axis:

$$\frac{\delta u}{\delta x_i}(x) := \frac{u(\xi^-)-u(\xi^+)}{\text{mes } [\xi^-, \xi^+]}, \text{ cf. figure: } \quad (6.1.13a)$$

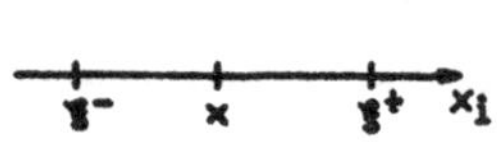

with ξ^-, $\xi^+ \in S'(x)$. For $h(x,\xi^+) = h(x,\xi^-)$, $\frac{\delta u}{\delta x_i}$ is central in the true sense, otherwise, i.e. for $h(x,\xi^+) \neq h(x,\xi^-)$, $\frac{\delta u}{\delta x_i}$ is more or less central.

b) $\frac{\delta u}{\delta x_i}(x) := (e_i, \nabla_h u)$, and $\nabla_h u$ is defined on the triangle $\Delta(\xi_1,\xi_2,\xi_3)$ (6.1.13b)

given by certain vertices $\xi_i \in S(x)$ $(i=1,2,3)$, and $\Delta(\xi_1,\xi_2,\xi_3)$ is regular, uniformly with respect to h, $h \leq h_o$.

c) central weighted difference quotient for triangles associated with x:

$$\frac{\delta u}{\delta x_1}(x) := (e_1, \bar{\nabla}_h u), \text{ with } \bar{\nabla}_h u(x) := \frac{1}{H(x)} \sum_{x'' \in S''(x)} \operatorname{mes}(\mathcal{X}(x) \cap \Delta(x'')) \nabla_h u(x''), \tag{6.1.13c}$$

$\nabla_h u(x'')$ from (3.3.14).

For $x \in \gamma_{23}$, the box $\mathcal{X}(x)$ will be replaced by $\tilde{\mathcal{X}}(x)$, cf. (2.3.3c). The difference operator $\bar{\nabla}_h u$ from (6.1.13c) is the weighted mean of the difference quotients $\nabla_h u(x'')$ on $\Delta(x'')$ for all $x'' \in S''(x)$, and $\operatorname{mes}(\mathcal{X}(x) \cap \Delta(x''))$ denotes the weight assigned to x". These weights are easily calculable for method (MD), since $\operatorname{mes}(\mathcal{X}(x) \cap \Delta(x'')) = \operatorname{mes} \Delta(x'')/3$ holds. For method (PB), the condition V1(PB, $\bar{\omega}$) from (2.3.12) should be fulfilled for getting reasonable weights. Furthermore, we may also introduce other weights which depend on the orientation of $b = (b_1, b_2)^T$ such that $\bar{\nabla}_h u$, more or less, acquires an upwind character. In (6.1.13b), the difference quotient over $\Delta(\xi_1, \xi_2, \xi_3)$ is defined in analogy to $\nabla_h u(x'')$ over $\Delta(x'')$, cf. (3.3.14). Since the difference quotients in (6.1.13a, c) are more or less central, their practical use is restricted to BVPs $A^b u = F$, where b_0/k_0 is sufficiently small.

Up to now we have applied the cubature formula (6.1.6), where one cubature point, the grid point x, was involved. But for the numerical integration of $\int_{\mathcal{X}(x)} (b, \nabla u) d\bar{x}$ we may also employ several cubature points which can be taken from int $\mathcal{X}(x)$ or (and) $\partial\mathcal{X}(x)$. For instance, taking into account MD-boxes $\mathcal{X}(x)$, we see that $S''(x) \subset \partial\mathcal{X}(x)$ holds. Therefore, it is reasonable to choose the centres of gravity $x'' \in S''(x)$ to be the cubature points, i.e.,

$$\int_{\mathcal{X}(x)} (b, \nabla u) d\bar{x} = \sum_{x'' \in S''(x)} \int_{\mathcal{X}(x) \cap \Delta(x'')} (b, \nabla u) d\bar{x} \approx \sum_{x'' \in S''(x)} (b(x''), \nabla u(x'')) \operatorname{mes}(\mathcal{X}(x) \cap \Delta(x'')). \tag{6.1.14}$$

The values $b_i(x'')$ (i=1,2) needed in (6.1.14) can be calculated directly at x" or approximately by linear interpolation with the values of b_i at the grid points x, ξ and ξ^+, i.e. $b_h(x'') := \frac{1}{3}\{b(x) + b(\xi) + b(\xi^+)\}$. The derivative $\nabla u(x'')$ will be approximated by $\nabla_h u(x'')$ from (3.3.14). Clearly, the difference approximation resulting from (6.1.14) has a more central character.

We now turn to the construction of FDSs $A_h^b y = F_h$ approximating the BVP $A^b u = F$ from (6.1.3a), with the convection term $(b, \nabla u)$. Several FDSs $A_h y = F_h$ assigned to the BVPs $Au = F$ from (2.1.1) or (3.2.1), i.e., without the term $(b, \nabla u)$, were constructed in Chapter 3. Therefore, discrete analogues $A_h^b y = F_h$ of $A^b u = F$ will be obtained by adding difference expressions $R_h y$ assigned to Ru from (6.1.3b), i.e.

$$A_h^b y = F_h \text{ on } \bar{\omega} = \omega + \gamma, \text{ with } A_h^b y = A_h y + R_h y. \tag{6.1.15}$$

Applying the general approach to deriving finite difference analogues from the balance equations (6.1.5), which is made by analogy to that of Section 3.2.1., cf. (3.2.13) - (3.2.16), we clearly see that $R_h u$ must be an approximation of the following terms:

$$(R_h u)(x) \approx \int_{\mathcal{X}(x)} (b, \nabla u)(\bar{x})d\bar{x} \cdot \begin{cases} \frac{1}{H(x)} & x \in \omega, \\ & \text{for} \\ \frac{1}{h(x)} & x \in \gamma_{23}, \end{cases} \qquad (R_h u)(x) = 0 \text{ for } x \in \gamma_1. \tag{6.1.16}$$

Here, the cubature formulas (6.1.6) and (6.1.14) as well as the difference quotients from (6.1.7), (6.1.9) and (6.1.11) - (6.1.13) yield a great number of approaches to approximating the integral term in (6.1.16). In the following, some difference operators R_h will be presented.

Definition 6.5. The following approximations are taken into account:

Case 1 (upwinding with respect to the direction b):The terms in (6.1.16) are approximated by means of (6.1.9), and R_h is defined by

$$(R_h u)(x) := \begin{cases} (b(x), \nabla_h^b u(x)) & x \in \omega, \\ \frac{\tilde{H}(x)}{h(x)} (b(x), \nabla_h^b u(x)) \text{ for } & x \in \gamma_{23}, \quad \nabla_h^b u(x) \text{ from (6.1.9)}, \\ 0 & x \in \gamma_1, \end{cases} \tag{6.1.17}$$

where Assumption 6.2. is supposed to be fulfilled for $x \in \omega + \gamma_{23}$.

Case 2 (upwinding with respect to the coordinate axes x_i, $i = 1,2$): Applying (6.1.11) to (6.1.16), R_h is given by

$$(R_h u(x) := \begin{cases} \sum_{i=1}^{2} (b_i(x)e_i, \nabla_h^i u(x)) & x \in \omega, \\ \frac{\tilde{H}(x)}{h(x)} \sum_{i=1}^{2} (b_i(x)e_i, \nabla_h^i u(x)) \text{ for } & x \in \gamma_{23}, \quad \nabla_h^i u(x) \text{ from (6.1.11)}, \\ 0 & x \in \gamma_1, \end{cases} \tag{6.1.18}$$

where Assumption 6.4. is made for $x \in \omega + \gamma_{23}$ and $i = 1,2$.

Case 3 (central weighted difference quotient): Using (6.1.13c) for the approximation of the terms in (6.1.16), R_h is defined by

$$(R_h u)(x) := \begin{cases} (b(x), \overline{\nabla}_h u(x)) & x \in \omega, \\ \frac{\tilde{H}(x)}{h(x)} (b(x), \overline{\nabla}_h u(x)) \text{ for } & x \in \gamma_{23}, \quad \overline{\nabla}_h u \text{ from (6.1.13c)}, \\ 0 & x \in \gamma_1, \end{cases} \tag{6.1.19}$$

where no special assumptions on upwind triangles are needed (but V1(PB, $\bar{\omega}$) for PB-schemes should be fulfilled).

Case 4 (upwinding with respect to the coordinate axes, for rectangles): Inserting (6.1.12) in (6.1.16), R_h is defined by

$$(R_h u)(x) := \begin{cases} \sum_{i=1}^{2} b_i(x) \frac{\delta u}{\delta x_i}(x) & x \in \omega, \\ \frac{\tilde{H}(x)}{h(x)} \sum_{i=1}^{2} b_i(x) \frac{\delta u}{\delta x_i}(x) & \text{for } x \in \gamma_{23},\ \frac{\delta u}{\delta x_i}(x) \text{ from (6.1.12)}, \\ 0 & x \in \gamma_1, \end{cases} \quad (6.1.20)$$

if there are mesh segments $h(x,s)$ according to (6.1.12).

Case 5 (central difference quotient for rectangles): The terms in (6.1.16) are approximated by the cubature formula from (6.1.12), but with $\frac{\delta u}{\delta x_i}(x)$ from (6.1.13a) at $x \in \omega$:

$$(R_h u)(x) := \begin{cases} \sum_{i=1}^{2} b_i(x) \frac{\delta u}{\delta x_i}(x) & x \in \omega\ ,\ \frac{\delta u}{\delta x_i} \text{ from (6.1.13a)}, \\ \frac{\tilde{H}(x)}{h(x)} \sum_{i=1}^{2} b_i(x) \frac{\delta u}{\delta x_i}(x) & \text{for } x \in \gamma_{23},\ \frac{\delta u}{\delta x_i} \text{ from (6.1.12)}, \\ 0 & x \in \gamma_1, \end{cases} \quad (6.1.21)$$

if segments $h(x,s^-)$, $h(x,s^+)$ according to (6.1.13a) exist, or $h(x,s)$ in case (6.1.12). ∎

Since networks of triangles are of central interest in this monograph, we shall prefer to discuss difference operators on triangles, i.e. R_h from (6.1.17) - (6.1.19). The difference operators R_h on networks of rectangles, i.e. R_h from (6.1.20) and (6.1.21), are well known and will be treated only in passing. If the network consists of triangles and rectangles, then the difference formulas of Definition 6.5. can be conveniently combined. This is due to the fact that the difference operator R_h can be taken individually at each grid point $x \in \omega + \gamma_{23}$.

On the basis of Theorems 3.1. and 3.4. from Chapter 3, we are in a position to formulate

Theorem 6.6. Let the assumptions of Theorems 3.1. and 3.4. concerning the BVP $Au = F$ and the networks be fulfilled. Moreover, $V(u)$ is assumed for the BVP $A^b u = F$ from (6.1.3), too, and (6.1.4) is added to $V(A,F)$. Furthermore, let Assumptions 6.2. and 6.4. ($i=1,2$) be satisfied for some or all $x \in \omega + \gamma_{23}$. Then, some FDSs $A_h^b y := A_h y + R_h y = F_h$ on $\bar{\omega}$ can be constructed via method (PB) or method (MD), where A_h, F_h are taken from Theorem 3.1. (PB) or Theorem 3.4. (MD), and R_h is given by Definition 6.5. or originates from (6.1.16) by using other cubature and difference formulas proposed in Sections 6.1.2., 6.1.3. These FDS are discrete analogues of the BVP $A^b u = F$ from (6.1.1), (6.1.3). ∎

The proof of Theorem 6.6. is based on that of Theorems 3.1. and 3.4. as well as on the derivation of $R_h u$ approximating the normed balance terms from (6.1.16). The resulting "point functionals" involved in $R_h u$ are well-defined. If there are no upwind triangles satisfying Assumptions 6.2. or 6.4., the difference approximations $R_h u$ in (6.1.17) and (6.1.18) will be defined by means of $\nabla_h u$ on some other triangles $\Delta(x,x'')$ which have one vertex at x. Analogous remarks hold for networks of rectangles and composite networks (rectangles and triangles), e.g., if there are no upwind mesh segments at x, other mesh segments $h(x,\xi)$ (or triangles) may be employed, etc. But a finite difference operator R_h can always be defined.

For the approximation error of $R_h \approx R$, see Section 6.3. It should be borne in mind that the coefficients $b_i (i=1,2)$ occuring only in L^b cause "weak" additional terms in the difference operator l_{h23} which is assigned to the grid boundary γ_{23}.

6.2. Properties of the difference operators A_h

6.2.1. Monotonicity and a priori estimates. In the following Lemma 6.7., some assumptions will be formulated which guarantee the monotonicity condition $M(A^b,\bar{\omega})$ from (4.2.1), where the notation A_h was replaced by A_h^b. Due to $A_h^b := A_h + R_h$ the properties of A_h^b can be derived from those of A_h and R_h.

Lemma 6.7. Consider $A^b u = F$ from (6.1.1), (6.1.3) and networks under the following assumptions: $k_{11} \equiv k_{22} =: k$, $k_{12} \equiv 0$, and Condition 2.16.- $VO(PB,\bar{\omega})$ is fulfilled. If MD-schemes for $k \neq$ const are taken into account, then $VO(PB,\bar{\omega})$ is replaced by $VO''(PB,\bar{\omega})$, cf. Condition 2.17. These assumptions imply $M(A_h,\bar{\omega})$ for the part A_h of A_h^b, and, moreover, the following assertions can be stated for $A_h^b := A_h + R_h$:

a) If Assumptions 6.2. or 6.4. (i=1,2) hold for all $x \in \omega + \gamma_{23}$, and the upwind operator R_h is taken from (6.1.17) or (6.1.18), respectively, then A_h^b satisfies $M(A_h^b,\bar{\omega})$ for $h \leq h_o$ and any b_o from (6.1.4), and the value of b_o does not impose restrictions on h_o. The same assertion holds, if rectangles and the difference operator R_h from (6.1.20) is taken into account, under the assumption that, for each $x \in \omega + \gamma_{23}$, the upwind mesh segments $h(x,\xi)$ from (6.1.12) exist. An analogous assertion can be stated for networks which consist of triangles and rectangles.

b) If Assumptions 6.2., 6.4. or the existence of upwind mesh segments cannot be guaranteed, but Condition 2.17. - $VO''(PB,\bar{\omega})$ and $h \leq h_o$ are fulfilled, then $M(A_h^b,\bar{\omega})$ holds for all difference operators A_h^b from Theorem 6.6., if h_o is chosen sufficiently small.

c) If A_h^b satisfies the monotonicity condition $M(A_h^b,\bar{\omega})$, then the assertions 2., 3., 4. and 5. from Theorem 4.3. are also valid for the difference operator A_h^b. Moreover, there exists a unique solution y of the FDS $A_h^b y = F_h$. ∎

<u>Proof:</u> The properties of $a(x,\xi)$, especially $M(A_h,\bar{\omega})$, were proved in Theorem 4.3. (PB-schemes) and in Theorem 4.7. (MD-schemes). For brevity, we focus our attention on the difference operators R_h from (6.1.17) - (6.1.19). The remaining operators R_h can be treated similarly.
Consider the canonical representation of the difference operators R_h and $A_h^b := A_h + R_h$:

$$(R_h y)(x) := \sum_{\xi \in S(x)} r(x,\xi)\, y(\xi), \quad (A_h^b y)(x) := \sum_{\xi \in S(x)} a^b(x,\xi)\, y(\xi) \qquad (6.2.1)$$

for $x \in \bar{\omega}$, with $a^b(x,\xi) := a(x,\xi) + r(x,\xi)$, $a(x,\xi)$ from (4.1.1).
Furthermore, the following identities are taken into account:

$$(b(x), \nabla_{\hbar}^b y(x) = \frac{1}{2 \text{ mes } \Delta} \{ y(\xi)(b(x),\hbar^{\bullet}) + y(\xi^+)(-b(x),\hbar^{\circ})$$

$$+ y(x)(-b(x),\hbar^{\bullet}-\hbar^{\circ}) \}, \qquad (6.2.2)$$

with $\nabla_{\hbar}^b$ from (6.1.17), and

$$(b_1(x)e_1, \nabla_{\hbar}^1 y(x)) = \frac{b_1(x)}{2 \text{ mes } \Delta} \{ y(\xi)(e_2,\hbar^+) + y(\xi^+)(e_2,-\hbar) + y(x)(e_2,\hbar-\hbar^+) \}$$

$$(6.2.3)$$

as well as a similar relation for $(b_2(x)e_2, \nabla_{\hbar}^2 y(x))$, with $\nabla_{\hbar}^1 y(x)$ from (6.1.18). These obvious identities help to clarify the sign properties of $r(x,\xi)$. For upwind triangles Δ^b used in (6.1.17), we get

$$(b(x),\hbar^{\bullet}) \leq 0, \; (-b(x),\hbar^{\circ}) \leq 0 \text{ and } (-b(x),\hbar^{\bullet}-\hbar^{\circ}) > 0 \text{ for } b(x) \neq 0. \qquad (6.2.4)$$

If upwind triangles Δ^{b_1} are employed in (6.1.18), the inequalities

$$b_1(x)(e_2,\hbar^+) \leq 0, \; b_1(x)(e_2,-\hbar) \leq 0 \text{ and } b_1(x)(e_2,\hbar-\hbar^+) > 0, \text{ for } b_1(x) \neq 0,$$

$$(6.2.5)$$

as well as corresponding inequalities for upwind triangles Δ^{b_2}, with $b_2(x) \neq 0$, are fulfilled. Inserting (6.2.4) and (6.2.5) in (6.2.2) and (6.2.3), respectively, and taking into account R_h from (6.1.17) or (6.1.18), the following relations for $r(x,\xi)$ from (6.2.1) are obvious:

$$r(x,x) > 0, \; r(x,\xi) \leq 0 \text{ for } \xi \in S'(x), \sum_{\xi \in S(x)} r(x,\xi) = 0 \text{ for } x \in \omega + \gamma_{23},$$

$$\left.\begin{array}{ll} r(x,\xi) = O(h^{-1}) & x \in \omega \\ & \text{for} \\ r(x,\xi) = O(1) & x \in \gamma_{23} \end{array}\right\} \text{and } \xi \in S(x), \text{ if } r(x,\xi) \neq 0 \text{ holds}, \qquad (6.2.6)$$

$$r(x,x) = 0 \quad \text{for } x \in \gamma_1 \quad (S(x) = \{x\}).$$

If there are no upwind triangles Δ^b or Δ^{b_i} $(i=1,2)$ for $x \in \omega+\gamma_{23}$, then a difference operator R_h can also be defined by means of (6.1.17), (6.1.18) and other triangles $\Delta(x,x'')$ adjacent to x. For these operators R_h and for all other R_h from Theorem 6.6., the relations in (6.2.6), with the exception of $r(x,x) > 0$ and $r(x,\xi) \leq 0$ for $\xi \in S'(x)$, can also be proved. Thus, combining the properties of $a(x,\xi)$ and $r(x,\xi)$ via (6.2.1), we immediately get assertion a) of this lemma. That assertion b) is also valid for restricted values of $h_o(b_o)$ can be derived from the asymptotic behaviour of the coefficients $a(x,\xi)$ and $r(x,\xi)$. Due to VO"(PB, $\bar{\omega}$) the relations $S'(x) = S_r'(x)$ for $x \in \omega+\gamma_{23}$, $|a(x,\xi)| \geq \varepsilon h^{-2} > 0$, for $x \in \omega$ and $\xi \in S(x)$, as well as $|a(x,\xi)| \geq \varepsilon h^{-1} > 0$, for $x \in \gamma_{23}$ and $\xi \in S(x)$, are satisfied, with $\varepsilon > 0$. The asymptotic behaviour of $r(x,\xi)$ is noted in (6.2.6). If h_o is chosen sufficiently small, then the sign of $a^b(x,\xi)$ from (6.2.1) is equal to that of $a(x,\xi)$, for any $h \leq h_o$. The remaining conditions of $M(A_h^b, \bar{\omega})$ are also fulfilled, since $\sum_{\xi \in S(x)} r(x,\xi) = 0$ for all $x \in \omega+\gamma_{23}$ and $M(A_h, \bar{\omega})$ hold. Finally, the proof of assertion c) is quite similar to that of Theorem 4.3.

In Section 4.5.3, a priori estimates of the type $\|y\|_{C(\bar{\omega})} \leq C\|A_h y\|_{C(\bar{\omega})}$ were proved under certain assumptions on the coefficients k_{ij} and the angles θ of the triangles. Unfortunately, the special majorant function w from (4.5.34) is not convenient, in general, for difference operators A_h^b, cf. MICHLIN/SMOLIZKI [1], HEINRICH [3]. But Theorem 4.39. and Corollary 4.42., which involve locally irregular grids and more general majorant functions, can be easily extended to operators A_h^b.

Theorem 6.8. Let the assumptions of Lemma 6.7. be fulfilled, but now with Condition 2.17. - VO"(PB, $\bar{\omega}$) in any case. Let the grid $\bar{\omega}$ be locally irregular. Moreover, the existence of a majorant function v assigned to $A^b = (L^b, l)^T$ from (6.1.3) is assumed, i.e., v obeys $v \in C^2(\bar{\Omega})$, $L^b v \geq \frac{3}{2}$ in Ω, $lv \geq \frac{3}{2}$ on Γ_2. Then, the difference operators A_h^b from Theorem 6.6., with the partition $A_h^b = (A_{h1}^b, \ldots, A_{h6}^b)^T$ according to (4.5.19), satisfies the a priori estimate

$$\|y\|_{C(\bar{\omega})} \leq \sum_{i=1}^{6} N_{hi}^b \|A_{hi}^b y\|_{C(\omega_i)} \leq C^b \|A_h^b y\|_{C(\bar{\omega})} \text{ for } y \in D \qquad (6.2.7a)$$

and $h \leq h_o$, h_o sufficiently small. The constants N_{hi}^b are positive and provided with the asymptotic behaviour

$$N_{hi}^b = O(1) \text{ for } i = 1,2,3,4,\ N_{h5}^b = O(h^2),\ N_{h6}^b = O(h). \blacksquare \qquad (6.2.7b)$$

Proof: Due to Lemma 6.7., the assumptions on the coefficients k_{ii} and on the network guarantee $M(A_h^b, \bar{\omega})$. Since $(R_h v - Rv)(x) = O(h)$ holds for $R_h = A_h^b - A_h$, $x \in \omega + \gamma_{23}$ and the majorant function v, the further steps of the proof can be made for A_h^b in the same manner as described for A_h in

Theorem 4.39. (PB) and Corollary 4.42. (MD).

It should be noted that Theorem 6.8. is essentially based on the following properties or assumptions: $M(A_h^b, \bar{\omega})$, the existence of a majorant function v, the local irregularity of $\bar{\omega}$ and the "coupling condition", cf. Theorem 4.39. Great values of b_o, i.e. "dominant convection terms" in L^b, are not excluded, but the constant C^b depends on b_o.

<u>6.2.2. Positive definiteness and a priori estimates.</u> In general, for $|b| \neq 0$ the difference operators A_h^b from (6.1.15) are not symmetric and positive definite on D_o, D_o from (2.2.8). But the positive definiteness of A_h^b on D_o can be proved in such special cases where b_o is sufficiently small, or $(R_h y, y) = 0$ holds for all $y \in D_o$. Here, the former case will be considered only.[1)]

<u>Theorem 6.9.</u> Let $A_h^b := A_h + R_h$ be the difference operator from (6.1.15), where A_h^b is a discrete analogue of $A^b := A + R$ from (6.1.1), (6.1.3). Furthermore, let b_o from (6.1.4) be sufficiently small. Then, A_h^b is positive definite on D_o, uniformly with respect to $h \leq h_o$, i.e.

$$\|y\|_o^2 \leq C_o (A_h^b y, y) \text{ for } y \in D_o, \tag{6.2.8}$$

and the following inequalities hold:

$$\|y\|_1^2 \leq C'(A_h^b y, y), \quad \|y\|_1 \leq C' \|A_h^b y\|_{-1} \text{ for } y \in D_o. \tag{6.2.9}$$

The constants C_o and C' are independent of $y \in D_o$ and h, $h \leq h_o$. ∎

<u>Proof:</u> Taking the scalar product $(.,.)_r$ from (4.3.1), where the index $r \in \{p,m\}$ will be omitted, and using the inequality $(A_h y, y) \geq C^{-1} \|y\|_1^2$ from (4.4.29), we obviously get the relations

$$(A_h^b y, y) = (A_h y, y) + (R_h y, y) \geq C^{-1} \|y\|_1^2 - |(R_h y, y)| \text{ for } y \in D_o. \tag{6.2.10}$$

Applying the ε-inequality, we may estimate

$$|(R_h y, y)| \leq \|R_h y\|_o \|y\|_o \leq \frac{1}{4\varepsilon} \|R_h y\|_o^2 + \varepsilon \|y\|_o^2 \text{ for } y \in D_o, \ \varepsilon > 0, \tag{6.2.11}$$

with $\|.\|_o$ from (4.3.5), (4.3.16). Bounds of $\|R_h y\|_o$ may be given in terms of $|y|_1$, with $|.|_1$ from (4.3.11) or (4.3.12). For instance, for R_h from (6.1.17) the following estimates are evident:

$$\begin{aligned} \|R_h y\|_o^2 &:= \sum_{x \in \omega} \{(b, \nabla_h^b y)^2 H\}(x) + \sum_{x \in \gamma_{2s}} \{(b, \nabla_h^b y)^2 \tilde{H}^2/h\}(x) \\ &\leq C_K b_o^2 \sum_{x'' \in \omega_b''} |\nabla_h^b y|^2 \operatorname{mes} \Delta\}(x'') \leq C_K b_o^2 |y|_1^2 \text{ for } y \in D_o, \end{aligned} \tag{6.2.12}$$

1) The condition $c \geq \frac{1}{2} \operatorname{div} b$ is not considered here, cf. e.g. NÄVERT [1].

where ω_b'' is a subset of $\bar{\omega}''$ indicating the triangles employed for the definition of $\nabla_h^b y$. The constant C_K is independent of b_o and h, $h \leq h_o$. For other difference operators R_h, we get bounds of the same type as in (6.2.12). Combining (6.2.10), (6.2.11) and (6.2.12), we obtain

$$(A_h^b y,y) \geq C_1\|y\|_o^2 + C_2\|y\|_1^2 \text{ for } y \in D_o,\; C_1 := C^{-1} - \varepsilon,\; C_2 := C^{-1} - \frac{C_K b_o^2}{4\varepsilon}. \tag{6.2.13}$$

Choose ε such that $C_1 > 0$ holds. If b_o is sufficiently small for getting $C_2 > 0$, then the inequalities $(A_h^b y,y) \geq C'\|y\|_1^2 \geq C_o\|y\|_o^2$ follow immediately from (6.2.13), with $C' := \min\{C_1,C_2\} > 0$ and, e.g., $C_o := C'$. Hence, (6.2.8) and the first inequality in (6.2.9) are shown. Finally, applying $(A_h^b y,y) \leq \|A_h^b y\|_{-1}\|y\|_1$ we see that the proof is complete.

6.3. The error ψ_K of the convection term

6.3.1. Error splitting and estimates. The error $z := y - u$ and the local approximation error $\psi := A_h y - A_h u$ of the FDS $A_h y = F_h$ were introduced in Section 5.1.1. This notation will be kept for FDSs $A_h^b y = F_h$ from (6.1.15). Thus, the contributions to the error

$$\psi(x) := (A_h^b y - A_h^b u)(x) = (A_h^b z)(x), \text{ with } z := y - u \in D_o, \tag{6.3.1}$$

are denoted by ψ_H, ψ_N and ψ_K. These partial errors are connected by

$\psi(x) = \{\psi_H + \psi_N^j + \psi_K\}(x)$, with ψ_H, ψ_N^j from (5.1.5) - (5.1.18), for PB- and MD-schemes,

$$\psi_K(x) := \begin{cases} \frac{1}{m(x)} \int\limits_{\varkappa(x)} (b,\nabla u)(\bar{x})d\bar{x} - (R_h u)(x) & \text{for } x \in \omega + \gamma_{23}, \\ 0 & \text{for } x \in \gamma_1, \end{cases} \tag{6.3.2}$$

with $m(x) := H(x)$ for $x \in \omega$ and $m(x) := h(x)$ for $x \in \gamma_{23}$, R_h from (6.1.15). The partial error $\psi_H + \psi_N^j$ originates from the approximations $A_h \approx A$, $F_h \approx F$. But ψ_K is associated with the "convection term" (Konvektionsterm) $(b,\nabla u)$, i.e., ψ_K is caused by $R_h \approx R$. Splitting (6.3.2) enables us to utilize all estimates proved for ψ_H, ψ_N^j, cf. Chapter 5. Hence, we shall be concerned with ψ_K only.

If $A^b u = F$ has a classical solution, with $u \in C^l(\bar{\Omega})$ and $l \geq 2$, then an estimation of the error z with respect to the C-norm can be made by means of inequality (6.2.7a), i.e.,

$$\|z\|_{C(\bar{\omega})} \leq \sum_{i=1}^{6} N_{hi}^b \|\psi\|_{C(\omega_i)}, \text{ with } z, \psi \text{ from (6.3.1), (6.3.2).} \tag{6.3.3}$$

For $b = 0$ and $R_h = 0$, i.e. $\psi_K = 0$, the estimation of ψ has already been made. Therefore, we shall concentrate on the study of the error ψ_K for R_h taken from (6.1.17) - (6.1.19). Obviously, ψ_K satisfies the relations

$$\psi_K(x) := \begin{cases} O(h) & x \in \omega, \\ O(h^2) \text{ for } x \in \gamma_{\delta 23}, \end{cases} \quad b_1 \in C^1(\bar{\Omega}),\ u \in C^2(\bar{\Omega}), \tag{6.3.4}$$

$$\psi_K(x) := O(h^2) \text{ for } x \in \mathring{\omega},\ \mathring{\omega} \text{ from (2.2.9) or (2.2.11), } R_h \text{ from (6.1.19),}$$
$$b_1 \in C^2(\bar{\Omega}),\ u \in C^3(\bar{\Omega}).$$

Here, Taylor's expansion can be used to verify (6.3.4). We note that the error $\psi_K(x)$ associated with R_h from (6.1.20) and (6.1.21) satisfies also (6.3.4), but $\psi_K(x) = O(h^2)$ appears only for $x \in \mathring{\omega}$ and R_h from (6.1.21), and if regular networks of rectangles are employed. In such cases, grids $\mathring{\omega}$ from Definition 2.11. or 2.12. can be applied.

We now derive a priori estimates of the error z with respect to the discrete W_2^1-norm. If b_o from (6.1.4) is sufficiently small such that A_h^b is positive definite, then, due to $\|z\|_1^2 \le C'(A_h^b z, z)$ for $z \in D_o$ and $A_h^b z = \psi = \psi_H + \psi_N^j + \psi_K$ (see (6.2.9), (6.3.1) and (6.3.2)), we immediately get

$$\|z\|_1^2 \le C'(A_h^b z, z) \le C' \left\{ |(\psi_H, z)| + |(\psi_N^j + \psi_K, z)| \right\}. \tag{6.3.5}$$

The weak evaluation of ψ_H by the discrete L_2-norm of the flux approximation error $\varkappa$, which was introduced in Section 5.1.3., cf. Lemma 5.1. and Theorem 5.3., will be used again. Applying the technique of deriving relation (5.1.35), i.e., Cauchy's inequality, and taking into account $\|\psi_N^j + \psi_K\| \le \|\psi_N^j\| + \|\psi_K\|$, we obviously obtain

$$\|z\|_1 \le C \left\{ \|\varkappa\|_o^2 + 2\|\psi_N^j\|_o^2 + 2\|\psi_K\|_o^2 \right\}^{1/2}, \tag{6.3.6}$$

with z and ψ from (6.3.1) and (6.3.2), respectively. Here, $\|\varkappa\|_o$ must be interpreted by $\|\varkappa\|_o'$, $\|\varkappa\|_o''$ or $'\|\varkappa\|_o$ taken from (5.1.32), (5.1.33) or (5.1.34), respectively, and the remaining norms $\|\cdot\|_o := \|\cdot\|_{o,r}$ $\|\cdot\|_1 := \|\cdot\|_{1,r}$, $r \in \{p,m\}$, are given in (4.3.5), (4.3.13) ($r=p$) and (4.3.14) ($r=m$). The constant C from (6.3.6) is connected with C' from (6.3.5) by $C := C'$, if $\|\varkappa\|_o'$, $\|\varkappa\|_o''$ are used, or by $C := C'C_o$, C_o from (5.1.29), if $'\|\varkappa\|_o$ is inserted in (6.3.6).

The estimate (6.3.6) allows us to utilize the global estimates for $\varkappa$ and ψ_N^j, which were derived in Sections 5.2., 5.3. and 5.4., and we have to add only the corresponding estimates of the term ψ_K. For classical solutions u of the BVP $A^b u = F$, with $u \in C^l(\bar{\Omega})$ and $l \ge 2$, the asymptotic behaviour of ψ_K was noted in (6.3.4). Therefore, it remains to study ψ_K under the assumption $u \in W_2^l(\Omega)$, $l \geq 2$, which is the subject of the following sections.

<u>6.3.2. The approach to estimating ψ_K for $u \in W_2^l(\Omega)$ ($l \geq 2$).</u> We shall study the error functional $\psi_K(x)$ from (6.3.2) under the assumption that the coefficients b_i and the solution u of the BVP $A^b u = F$ satisfy the smoothness assumptions

$$b_i \in C^{1+j}(\bar{\Omega}) \ (i=1,2), \ u \in W_2^{2+j}(\Omega) \text{ for } j = 0,1. \tag{6.3.7}$$

In the following, the error ψ_K of the difference operators R_h from (6.1.17) - (6.1.19) will be studied. But the method of estimating ψ_K is also applicable to R_h from (6.1.20), (6.1.21) and to other difference operators R_h, which can be derived from all cubature formulas and difference quotients proposed in Section 6.1.

First, it is reasonable to split ψ_K into the approximation errors of the coefficients b_i and of the derivatives $\frac{\partial u}{\partial x_i}$, which will be denoted by $\psi_K^{(b)}$ and $\psi_K^{(u)}$, respectively. For brevity, the "geometric error" is included in $\psi_K^{(u)}$ and will be separated later in (6.3.21). More precisely, we shall introduce

$$\psi_K(x) = \psi_K^{(b)}(x) + \psi_K^{(u)}(x) \text{ for } x \in \omega + \gamma_{23}, \tag{6.3.8}$$

with

$$\psi_K^{(b)}(x) := \int_{\varkappa(x)} (\delta b, \nabla u)(\bar{x}) d\bar{x} \begin{cases} \frac{1}{H(x)} & x \in \omega, \\ & \text{for} \\ \frac{1}{h(x)} & x \in \gamma_{23}, \end{cases} \tag{6.3.9a}$$

$$\delta b(\bar{x}) := b(\bar{x}) - b(x) \text{ for } x \in \omega + \gamma_{23},\ \bar{x} \in \varkappa(x),\ b = (b_1, b_2)^T,$$

$$\psi_K^{(u)}(x) := \begin{cases} (b(x), l(x,u)) & x \in \omega, \\ & \text{for} \\ \frac{\tilde{H}(x)}{h(x)} (b(x), \tilde{l}(x,u)) & x \in \gamma_{23}. \end{cases} \tag{6.3.9b}$$

The symbol u in $l(x,u)$, $\tilde{l}(x,u)$ will often be omitted. The vector functionals $l = (l_1, l_2)^T$, $\tilde{l} = (\tilde{l}_1, \tilde{l}_2)^T$ will now be specified for various R_h, viz.

<u>Case 1</u>, R_h from (6.1.17):

$$\begin{aligned} l(x) := l(x,u) &:= \frac{1}{H(x)} \int_{\varkappa(x)} \nabla u \, d\bar{x} - \nabla_h^b u(x) & x \in \omega, \\ & & \text{for} \\ \tilde{l}(x) := \tilde{l}(x,u) &:= \frac{1}{\tilde{H}(x)} \int_{\varkappa(x)} \nabla u \, d\bar{x} - \nabla_h^b u(x) & x \in \gamma_{23}, \end{aligned} \tag{6.3.10}$$

Case 2, from (6.1.18), for $i = 1,2$:

$$l_i(x) := l_i(x,u) := \frac{1}{H(x)} \int_{\mathcal{H}(x)} (\nabla u, e_i) d\bar{x} - (\nabla_h^i u(x), e_i) \qquad x \in \omega,$$
$$\text{for} \qquad (6.3.11)$$
$$\tilde{l}_i(x) := \tilde{l}_i(x,u) := \frac{1}{\tilde{H}(x)} \int_{\mathcal{H}(x)} (\nabla u, e_i) d\bar{x} - (\nabla_h^i u(x), e_i) \qquad x \in \gamma_{23},$$

Case 3, R_h from (6.1.19), for $i = 1,2$:

$l_i(x)$, $\tilde{l}_i(x)$ as in (6.3.11), but $\nabla_h^i u(x)$ is replaced by $\bar{\nabla}_h u(x)$, (6.3.12a)

with the further splitting of $l_i(x)$ at $x \in \omega$:

$$l_i(x) = \frac{1}{H(x)} \sum_{x'' \in S''(x)} r_i(x,x''), \; r_i(x,x'') := \int_{\mathcal{H}(x) \cap \Delta(x,x'')} \{(\nabla u, e_i) - (\nabla_h u(x''), e_i)\} d\bar{x}. \qquad (6.3.12b)$$

If $u \in W_2^3(\Omega)$ is fulfilled and R_h from (6.1.19) is taken into account, we additionally consider another splitting of ψ_K, which is given in (6.3.12c). Due to $u \in W_2^3(\Omega)$ and Theorem 1 from Appendix IM, the relation $\frac{\partial u}{\partial x_i} \in C(\bar{\Omega})$ can be employed for defining the following splitting of ψ_K into the cubature error $\psi^{(\varphi)}$ and the error $\psi_K^{(u)}$ of the difference quotient $\bar{\nabla}_h u$:

$$\psi_K(x) = \psi_K^{(\varphi)}(x) + \psi_K^{(u)}(x), \; \psi_K^{(\varphi)}(x) := \frac{1}{H(x)} \int_{\mathcal{H}(x)} \varphi(\bar{x}) d\bar{x} - \varphi(x), \qquad (6.3.12c)$$

with $\varphi := (b, \nabla u)$, $\psi_K^{(u)}(x) := (b(x), \nabla u(x) - \bar{\nabla}_h u(x))$, for $x \in \omega$.

We now turn to the estimation of $\psi_K^{(b)}$ and $\psi_K^{(u)}$ from (6.3.8). The error $\psi_K^{(b)}$ which describes the cubature error with respect to b can be estimated by

$$|\psi_K^{(b)}(x)| \le \int_{\mathcal{H}(x)} |\delta b| |\nabla u| d\bar{x} \begin{cases} \frac{1}{H(x)} \le M_0 |u|_{1,\mathcal{H}(x)} & x \in \omega, \\ & \text{for} \\ \frac{1}{h(x)} \le M_0' h |u|_{1,\mathcal{H}(x)} & x \in \gamma_{23}, \end{cases} \qquad (6.3.13)$$

with $b_i \in C^1(\bar{\Omega})$ $(i=1,2)$ and $u \in W_2^1(\Omega)$.

The estimates on the right-hand side of (6.3.13) may be derived by Taylor's expansion in the term δb, and M_0 as well as M_0' depend on $\|\partial^1 b_i\|_{C(\mathcal{H})}$.

The functional $\psi_K^{(u)}$ contains the error functionals l_i, $\tilde{l}_i$ $(i=1,2)$ of the finite difference approximations of $\frac{\partial u}{\partial x_i}$, cf. (6.3.10) - (6.3.12b). Since $u \in W_2^2(\Omega)$ is assumed to be fulfilled, the technique of estimating linear functionals on Sobolev spaces, which is based on the Bramble-Hilbert Lemma from Appendix BS, needs to be applied. This technique was extensively studied for some functionals in Chapter 5. But here the functionals l_i, $\tilde{l}_i$ $(i = 1,2)$ contain "domain functionals" $\int_{\mathfrak{K}(x)}(\nabla u, e_i)d\bar{x}$ and "point functionals" of the type $(\nabla_h^i u, e_i)$, where $\nabla_h^i u$ involves the values $u(x)$, $u(\xi)$ and $u(\xi^+)$ at the vertices of the triangle $\Delta(x,\xi,\xi^+)$, which may be, for instance, an upwind triangle. Since the union of $\mathfrak{K}(x)$ and $\Delta(x,\xi,\xi^+)$, i.e. $\mathfrak{K}(x) \cup \Delta(x,\xi,\xi^+)$ $(x \in \omega + \gamma_{23})$, cannot be transformed by an affine mapping onto a fixed reference domain (for all $x \in \omega + \gamma_{23}$), we shall define a new "majorant region" which contains $\mathfrak{K}(x) \cup \Delta(x,\xi,\xi^+)$ and is the affine image of a fixed reference domain. For this purpose, a rectangle $\square(x)$ with sides parallel to the coordinate axes is appropriate. We assume that this rectangle $\square(x)$ contains $\mathfrak{K}(x)$ and, at the same time, the (upwind) triangle $\Delta(x,\xi,\xi^+)$ in

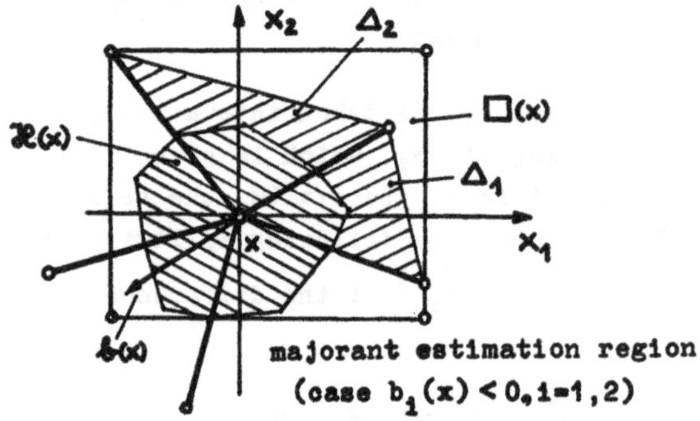

Fig. 13a

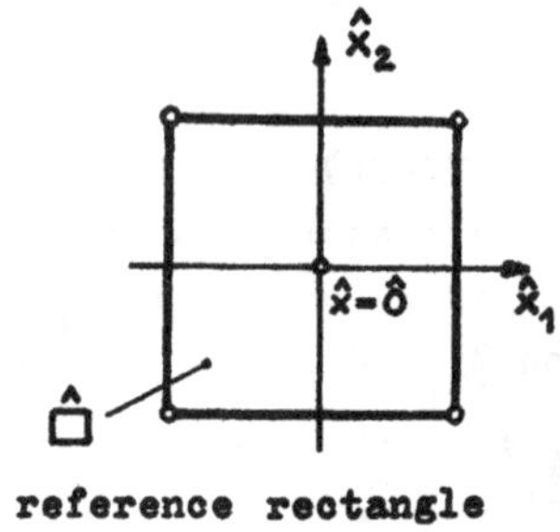

Fig. 13b

Case 1, cf. (6.3.10), or two (upwind) triangles $\Delta_i(x,\xi,\xi^+)$ for $i = 1,2$ in Case 2, cf. (6.3.11), Fig. 13a. But in Case 3, cf. (6.3.12a, b), the functionals l_i, $\tilde{l}_i$ $(i=1,2)$ are split with respect to their portions on the triangles $\Delta(x,x'')$, $x'' \in S''(x)$, and therefore, the estimation can be carried out via $r_i(x,x'')$ on $\Delta(x,x'')$. For $\hat{\square}$, cf. Fig. 13b.

<u>6.3.3. Local and global bounds of ψ_K.</u> We shall prove the following

<u>Lemma 6.10.</u> Let $\bar{\omega}$ be given and provided with the grid regularity conditions VO(PB, $\bar{\omega}$) or VO(MD, $\bar{\omega}$). Consider the difference operators R_h defined in (6.1.17) - (6.1.19) and, additionally, admit that the triangles employed in (6.1.17), (6.1.18) need not necessarily be upwind triangles.

a) Then, for $\psi_K(x)$ defined in (6.3.2) and associated with R_h from (6.1.17) - (6.1.19), the following local estimates hold:

$$|\psi_K(x)| \le M_1 \{|u|^2_{1,\mathfrak{K}(x)} + |u|^2_{2,\square(x)}\}^{1/2} \cdot \begin{cases} 1 & x \in \omega, & (6.3.14a) \\ & \text{for} \\ h & x \in \gamma_{23}, & (6.3.14b) \end{cases}$$

for $b_i \in C^1(\tilde{\Omega})$ $(i=1,2)$, $u \in W_2^2(\tilde{\Omega})$, where Ω is given subsequently, and with some constant M_1 depending on $\|b_i\|_{C^1(\tilde{\Omega})}$ $(i=1,2)$, but not on u, x and h, $h \leq h_o$. The symbol $\square(x)$ denotes some rectangle containing $\mathcal{K}(x)$ and the triangles employed for the construction of R_h. For R_h from (6.1.19), $\square(x)$ must be replaced by $E(x) := \bigcup_{x'' \in S'(x)} \Delta(x,x'')$, since all triangles associated with x are involved.

b) Taking R_h from (6.1.19) for grid points $x \in \omega$ which are regular, i.e. $E(x)$ consists of equilateral triangles only, and assuming $b_i \in C^2(\bar{\Omega})$ as well as $u \in W_2^3(\Omega)$, the estimate

$$|\psi_K(x)| \leq M_2 h \left\{|u|^2_{1,\mathcal{K}(x)} + |u|^2_{2,\mathcal{K}(x)} + |u|^2_{3,E(x)}\right\}^{1/2}, \text{ for regular } x \in \omega, \tag{6.3.15}$$

is fulfilled. The constant M_2 depends on $\|b_i\|_{C^2(\bar{\Omega})}$, but not on u, x and h, $h \leq h_o$. ■

<u>Proof:</u> Firstly, we estimate l_i, $\tilde{l}_i$ $(i=1,2)$ from (6.3.11). Let $\square(x)$ be the smallest rectangle containing $\mathcal{K}(x) \cup \Delta_i(x,\xi,\xi^+)$ for $i = 1,2$, cf. Figs. 12b, 13a. Due to VO(PB, $\bar{\omega}$) or VO(MD, $\bar{\omega}$), the rectangle $\square(x)$ is regular, uniformly with respect to h, $h \leq h_o$. Moreover, $\square(x)$ is the affine image of a fixed reference rectangle $\hat{\square}$ in the $\hat{x}_1$-$\hat{x}_2$-plane, e.g. a square $\hat{\square}$ according to Fig. 13b. The affine mappings F, F^{-1}, with $\square = F(\hat{\square})$ and $\hat{\square} = F^{-1}(\square)$, can be taken from (5.2.13), i.e., we have $x = F\hat{x} = B\hat{x} + b$, and F depends on $\square$. The image $F^{-1}(\mathcal{K}(x) \cup \Delta_i(x,\xi,\xi^+))$ is a closed subset of $\hat{\square}(\hat{x})$, i.e., we have

$$\begin{aligned} &\mathcal{K}(x) \cup \Delta_i(x,\xi,\xi^+) \subset \square(x) \text{ for } i = 1, 2 \text{ and } x \in \omega + \gamma_{23}, \\ &F^{-1}(\mathcal{K}(x) \cup \Delta_i(x,\xi,\xi^+)) \subset \hat{\square}(\hat{x}), \quad \hat{\square} = F^{-1}(\square). \end{aligned} \tag{6.3.16}$$

For grid points x, $\bar{x}$ from $\omega + \gamma_{23}$, which are at a sufficiently small distance from each other, an overlap of the associated rectangles $\square(x)$ and $\square(\bar{x})$ may occur. Moreover, for some x, $\square(x) \not\subset \bar{\Omega}$ is possible. Therefore, we define the closed domain $\bar{\Omega}_h^*$ by

$$\bar{\Omega}_h^* := \bigcup_{x \in \omega + \gamma_{23}} \square(x), \quad \square(x) \text{ from (6.3.16)}, \tag{6.3.17}$$

and extend $u \in W_2^2(\Omega)$ to $u \in W_2^2(\tilde{\Omega})$, where $\tilde{\Omega}$ is a fixed domain, with $\tilde{\Omega} \supset \Omega \cup \Omega_h \cup \Omega_h^*$ for $h \leq h_o$. That some rectangles have a nonempty intersection will not disturb the following local estimation of error functionals. The overlap will only affect the constants in the global estimates.

We now consider $l_i(x,u)$, $x \in \omega$, from (6.3.11). Applying $\hat{\square} = F^{-1}(\square)$ and using the transformation rules for differential and difference quotients from Appendix TR, we get

$$l_1(x,u) = \hat{l}_1(\hat{x},\hat{u}) = \frac{1}{\hat{H}(\hat{x})} \int_{\hat{\varkappa}(\hat{x})} (B^{T-1}\hat{\nabla}\hat{u},e_1)d\hat{\bar{x}} - (B^{T-1}\hat{\nabla}^1_h \hat{u},e_1). \quad (6.3.18)$$

Obviously $\hat{l}_1$ is well defined, linear and bounded on $W_2^2(\hat{\Box})$. Thus, we have

$$|\hat{l}_1(\hat{x},\hat{u})| \le M_3\|B^{-1}\|\{|\hat{u}|_{1,\hat{\varkappa}} + \|\hat{u}\|_{C(\hat{\Delta}_1)}\} \le M_4 h^{-1}\|\hat{u}\|_{2,\hat{\Box}} . \quad (6.3.19a)$$

Moreover, $\hat{l}_1(\hat{x},\hat{u})$ vanishes for all polynomials $\hat{p}_1$ of degree ≤ 1, i.e.

$$\hat{l}_1(\hat{x},\hat{p}_1) = 0 \text{ for } \hat{p}_1(\hat{\bar{x}}) := \hat{a}_o + (\hat{a},\hat{\bar{x}}), \text{ with } \hat{a}_o \in R^1,\ \hat{a},\hat{\bar{x}} \in R^2. \quad (6.3.19b)$$

The relation $\hat{l}_1(\hat{x},\hat{p}_1) = 0$ is due to $\hat{\nabla}^1_h\hat{p}_1 = \hat{\nabla}\hat{p}_1 = \hat{a}$.
Hence, (6.3.19) verifies the assumptions of the Bramble-Hilbert Lemma for p = k = 2, cf. Appendix BS. Applying this lemma and the inverse transformation, we get

$$|l_1(x,u)| = |\hat{l}_1(\hat{x},\hat{u})| \le M_4 h^{-1} C_{BH}|\hat{u}|_{2,\hat{\Box}(\hat{x})} \le M_5|u|_{2,\Box(x)},\ x \in \omega. \quad (6.3.20)$$

Using the inequality $|H(x) - \tilde{H}(x)| \le K_o h^3$ for $x \in \gamma_{23}$, cf. (3.2.12), the estimation of the functional $\tilde{l}_1(x,u)$ from (6.3.11) can be traced back to that of $l_1(x,u)$. This is given by the relations

$$\left.\begin{aligned} |\tilde{l}_1(x,u)| &\le |l_1(x,u)| + M_6 \frac{h}{H(x)} \left|\int_{\varkappa(x)} (\nabla u,e_1)d\bar{x}\right| \\ l_1(x,u) &:= \frac{1}{H(x)} \int_{\varkappa(x)} (\nabla u,e_1)d\bar{x} - (\nabla^1_h u,e_1) \end{aligned}\right\} \text{for } x \in \gamma_{23}. \quad (6.3.21)$$

The functional $l_1(x,u)$ from (6.3.21) obeys also the inequalities in (6.3.20). Hence,

$$|\tilde{l}_1(x,u)| \le M_5|u|_{2,\Box(x)} + M_7|u|_{1,\varkappa(x)} \quad \text{for } x \in \gamma_{23} \quad (6.3.22)$$

is obvious. Taking into account (6.3.8), (6.3.9) and (6.3.11) as well as inequalities (6.3.13), (6.3.20) and (6.3.22), we get the local estimate (6.3.14) for Case 2 from (6.3.11).
We now consider $\psi_K(x)$ for R_h from (6.1.17) and (6.1.19), i.e., Case 1 and Case 3, where $\psi_K(x)$ is given by (6.3.8), (6.3.9) and (6.3.10) (Case 1) or (6.3.12) (Case 3). The functionals l_1, $\tilde{l}_1$ from (6.3.10) can be estimated in the same way as l_1, $\tilde{l}_1$ from (6.3.11). For l_1, $\tilde{l}_1$ from (6.3.12), we estimate $r_1(x,x")$ and use the primary triangles $\Delta(x,x")$ as estimation regions. This approach was principally introduced in Section 5.4.1., and, it can also be applied to the functionals r_1 from (6.3.12b). Thus, for Cases 1 and 2, we also get estimate (6.3.14), but with other constants, and, the rectangle $\Box(x)$ includes $\varkappa(x)$ and one

triangle Δ only (Case 1) or $\square(x)$ must be replaced by $E(x)$ (Case 3).

Finally, taking into account $b_i \in C^2(\bar{\Omega})$, $u \in W_2^3(\Omega)$, splitting (6.3.12c) and regular grid points $x \in \hat{\omega}$ of locally irregular or regular grids $\bar{\omega}$, we shall prove estimate (6.3.15). For the functional $\psi_K^{(\varphi)}(x)$ from (6.3.12c), the box $\mathcal{H}(x)$ is a natural estimation region, which consists of six equilateral triangles. Hence, $\mathcal{H}(x)$ is the affine image of a fixed regular reference hexagon $\hat{\mathcal{H}}(\hat{x})$, with $\hat{x} = \hat{0}$. Since $\varphi := (b, \nabla u) \in W_2^2(\mathcal{H})$ holds, estimate (5.4.3) becomes true for $v := \varphi$ and yields

$$|\psi_K^{(\varphi)}(x)| \leq M_8 h\, |\varphi|_{2,\mathcal{H}(x)} \quad \text{for } \varphi := (b, \nabla u) \in W_2^2(\mathcal{H}),\ x \in \hat{\omega}. \qquad (6.3.23)$$

Furthermore, the functional $\psi_K^{(u)}(x)$ from (6.3.12c) can be treated on its natural estimation region $E(x)$ given in Lemma 6.10., and $E(x)$ is also a regular hexagon. The affine transformation of $E(x)$ onto a reference hexagon $\hat{E}(\hat{x})$, with $\hat{x} = \hat{0}$, yields $\hat{\psi}_K^{(\hat{u})}(\hat{x}) = (\hat{b}(\hat{x}), B^{T-1}[\hat{\nabla}\hat{u}(\hat{x}) - \hat{\nabla}_{\hbar}\hat{u}(\hat{x})])$, cf. Appendix TR. Evidently, $\hat{\psi}_K^{(\hat{u})}$ is linear and bounded for $\hat{u} \in W_2^3(\hat{E})$, and $\hat{\psi}_K^{(\hat{u})}$ vanishes on all polynomials $\hat{p}_2$ of degree ≤ 2. Therefore, the Bramble-Hilbert Lemma from Appendix ES can be applied for $p = 2$, $k = 3$, and, after backward transformation, the inequality

$$|\psi_K^{(u)}(x)| \leq M_9 h\, |u|_{3,E(x)} \quad \text{for } x \in \hat{\omega},\ E(x) = \bigcup_{x'' \in S'(x)} \Delta(x, x''), \qquad (6.3.24)$$

will be obtained. Estimates (6.3.23) and (6.3.24) imply (6.3.15). Lastly, it should be mentioned that the difference operators R_h from (6.1.20) and (6.1.21) yield bounds of $\psi_K(x)$ with the same asymptotic behaviour as in (6.3.14) and (6.3.15), but (6.3.15) only for R_h from (6.1.21) and actual central difference quotients from (6.1.13a) ($h(x, \mathfrak{s}^-) = h(x, \mathfrak{s}^+)$). The functional ψ_K can be estimated as in Case 2, cf. (6.3.11), but $(\nabla_{\hbar}^i u, e_i)$ must be replaced by $\frac{\delta u}{\delta x_i}$, and various estimation regions containing $\mathcal{H}(x)$ as well as $\hbar(x, \mathfrak{s})$, or $\hbar(x, \mathfrak{s}^-) \cup \hbar(x, \mathfrak{s}^+)$, can be used (for instance, suitable deltoids, rhomboids or rectangles).

We now collect the local estimates of $\psi_K(x)$ and derive bounds of the discrete L_2-norm $\|\psi_K\|_o$ employed in estimate (6.3.6).

Lemma 6.11. Under the assumptions of Lemma 6.10., the following global estimates of ψ_K using the discrete L_2-norm $\|\cdot\|_o := \|\cdot\|_{o,r}$ from (4.3.5), $r \in \{p, m\}$, are obtained, viz.

$$\|\psi_K\|_o \leq M_{10} h \|u\|_{2,\Omega} \quad \text{for } b_i \in C^1(\bar{\Omega})\ (i=1,2),\ u \in W_2^2(\Omega), \qquad (6.3.25)$$

and R_h from (6.1.17), (6.1.18) and (6.1.19),

$$\|\psi_K\|_0 \le M_{11}\|u\|_{3,\Omega} \begin{cases} h^{3/2} & \bar{\omega} \text{ locally irregular}, \quad (6.3.26a) \\ & \text{for} \\ h^2 & \omega \text{ regular}, \quad (6.3.26b) \end{cases}$$

$b_1 \in C^2(\bar{\Omega})$, $u \in W_2^3(\Omega)$ and R_h from (6.1.19).

The constants M_{10} and M_{11} depend on $\|b_i\|_{C^1(\bar{\Omega})}$ and $\|b_i\|_{C^2(\bar{\Omega})}$, respectively, but they are independent of u and h, $h \le h_0$. Estimate (6.3.25) is also valid for R_h when using (6.1.20) or (6.1.21) on networks of rectangles, and (6.3.26) can be confirmed for R_h when utilizing (6.1.21) at regular grid points $x \in \omega$. ∎

<u>Proof:</u> Using (6.3.14) for estimating $\|\psi_K\|_0$, we get

$$\begin{aligned} \|\psi_K\|_0^2 &:= \sum_{x\in\omega} \psi_K^2(x)H(x) + \sum_{x\in\gamma_{23}} \psi_K^2(x)h(x) \\ &\le M_{12}h^2\Big\{|u|_{1,\Omega}^2 + \sum_{x\in\omega} |u|_{2,\square(x)}^2 + h\sum_{x\in\gamma_{23}} |u|_{2,\square(x)}^2\Big\}, \end{aligned} \tag{6.3.27}$$

where $\square(x) \cap \square(\bar{x}) \ne \emptyset$ may occur for $x \in \omega + \gamma_{23}$ and some $\bar{x} \in \omega + \gamma_{23}$ which are located in an O(h)-neighbourhood of x ($x \ne \bar{x}$). But each $\square(x)$ has a nonempty intersection at most with n(x,h) rectangles $\square(x^{(i)})$ (i=1,2,...,n), where $\sup_{h\le h_0}\{\max_{\omega+\gamma_{23}} n(x,h)\} \le n_0$ holds. Here, n_0 is a fixed natural number independent of x and h, $h \le h_0$, which can be determined for given sequences of grids $\{\bar{\omega}\}_{h\le h_0}$. Thus, n_0 is approximately of the order of the maximal number of neighbouring points of a grid point $x \in \omega + \gamma_{23}$. If $\tilde{\Omega}$ is chosen sufficiently large, and $u \in W_2^2(\Omega)$ is extended to $u \in W_2^2(\tilde{\Omega})$, we can write

$$\sum_{x\in\omega} |u|_{2,\square(x)}^2 \le n_0|u|_{2,\tilde{\Omega}}^2 \le M_{13}\|u\|_{2,\Omega}^2, \quad \sum_{x\in\gamma_{23}} |u|_{2,\square(x)}^2 \le M_{13}'\|u\|_{2,\Omega}^2. \tag{6.3.28}$$

Combining (6.3.27) and (6.3.28), we get (6.3.25) for R_h from (6.1.17) and (6.1.18). However, for R_h from (6.1.19), the proof of (6.3.25) can be repeated, but now with E(x) instead of $\square(x)$.
If ω is regular and R_h from (6.1.19) is taken into account, then R_h is a central difference quotient for $x \in \omega$. Summing up the local bounds of $|\psi_K(x)|$ given in (6.3.15) for $x \in \omega$ and in (6.3.14b) for $x \in \gamma_{23}$, and applying Theorem 4 (Corollary 4) from Appendix IM, inequality (6.3.26b) for regular ω can be confirmed. For locally irregular grids $\bar{\omega}$, we have, in general, irregular grid points $x \in \overset{*}{\omega} := \omega \setminus \overset{\circ}{\omega}$, cf. (2.2.6), (2.2.9). Here, $\psi_K(x)$ satisfies (6.3.14a) and, thus, the summation of the local bounds and the application of Theorem 4 from Appendix IM yield only (6.3.26a). For networks containing rectangles, the proof can be carried out similarly, cf. also the concluding remarks of the proof of Lemma 6.10. For the use of Theorem 4 (Corollary 4) from Appendix IM, see

Chapter 5, e.g. the proof of (5.4.13a) or Lemma 5.23.

6.4. Convergence for $C^1(\bar{\Omega})$- and $W_2^1(\Omega)$-solutions ($l \geq 2$)

In this section, two theorems will be formulated which are based on the a priori estimates of Theorems 6.8. and 6.9. as well as on the estimates of ψ_K obtained in Section 6.3.

Theorem 6.12. Let u be the classical solution of the BVP $A^b u = F$ from (6.1.1), (6.1.3), with assumption (6.1.4) and for $k_{11} \equiv k_{22} =: k$, $k_{12} \equiv 0$, $u \in C^l(\bar{\Omega})$, $l \geq 2$. Furthermore, let y be the solution of the corresponding FDS $A_h^b y = F_h$ from (6.1.15), i.e. $A_h^b := A_h + R_h = F_h$, where the parts A_h and R_h are taken from Theorems 3.1. (PB), 3.4. (MD) and Theorem 6.6., respectively. Thus, for networks of triangles the difference operators R_h from (6.1.17), (6.1.18) or (6.1.19) are employed. But for rectangles, R_h is based on (6.1.20) or (6.1.21). Suppose that the grid $\bar{\omega}$ is locally irregular, e.g. $\bar{\omega}$ from Definitions 2.10., 2.11. (PB) or 2.12., and that the remaining assumptions of Theorem 6.8. are fulfilled. Then, for $h \leq h_o$ and sufficiently small h_o, the relations (i) and (ii) hold, i.e.,

$$\text{(i)}\ \|y-u\|_{C(\bar{\omega})} \longrightarrow 0 \text{ as } h \longrightarrow 0, \text{ for } u \in C^2(\bar{\Omega}),\ k \in C^1(\bar{\Omega}),$$

$$\text{(ii)}\ \|y-u\|_{C(\bar{\omega})} \leq Mh, \text{ with } 0 < M = O(1),\ u \in C^3(\bar{\Omega}),\ k \in C^2(\bar{\Omega}), \tag{6.4.1}$$

with $b_i \in C^1(\bar{\Omega})$ and further smoothness assumptions on c, α, f, g, e.g. $c, f \in C^1(\bar{\Omega})$ and $\alpha, g \in PC^1(\Gamma_{23})$, at most. ■

Proof: The local approximation error $\psi(x) = (A_h^b z)(x)$ can be evaluated by means of Taylor's expansion. Thus, bounds of $|\psi(x)|$ will be obtained, cf. (4.5.19) and (6.3.4). Then, Theorem 6.8. yields the error estimate (6.3.3), with constants N_{hi} from (6.2.7b). Although we have $\psi(x) = O(1)$ at $x \in \overset{*}{\omega}$, we get (6.4.1) due to the asymptotic behaviour of the constants N_{hi}.

Remark 6.13. In Theorem 6.12., no restrictions on b_o, cf. (6.1.4), are demanded. But the constants M and h_o depend, in general, on b_o, e.g. $M = O(b_o)$. If the assumptions on the existence of upwind triangles can be fulfilled, i.e. Assumption 6.2. or Assumption 6.4. (i=1,2) for all $x \in \omega + \gamma_{23}$, then the restrictions imposed on h_o, which are due to great values of b_o, can be omitted. Thus, upwind approximations can always be defined at least for BVPs $A^b u = F$, where $\Gamma = \Gamma_1$ holds, and the angles θ of all triangles satisfy the inequality $\theta \leq \pi/2$. If rectangles occur, the upwind difference operator R_h from (6.1.20) can be taken.

Remark 6.14. If in Theorem 6.12. additional assumptions are satisfied, e.g. $u \in C^4(\bar{\Omega})$, $k \in C^3(\bar{\Omega})$ and $\Gamma = \Gamma_1$, and moreover, if only central difference quotients approximating R are used, and the grid $\bar{\omega}$ is regular, then $\|y-u\|_{C(\bar{\omega})} \leq Mh^2$ holds. For instance, we can take R_h from (6.1.19) for networks of triangles and, for rectangles, R_h from (6.1.21) is employed.

Remark 6.15. In Theorem 6.12., the smoothness assumptions on c, f, b_i (i=1,2) can be reduced to c, f, $b_i \in C(\bar{\Omega})$, and estimates of the type (6.4.1) via inequality (6.3.3) can be proved, too. This is due to the fact that $\psi(x)$ from (6.3.1) can also represented in the "classical way", i.e. by $\psi(x) = (L^b u - L_h^b u + f_h - f)(x)$ for $x \in \omega$, with $L^b u = f$. For α, g and $\psi(x)$ for $x \in \gamma_{23}$, similar considerations can be made. For this, cf. HEINRICH [6]. ■

Theorem 6.9., together with estimate (6.3.6), and Lemma 6.11. lead to estimates of the error $z := y - u$ with respect to the discrete W_2^1-norm. This will be summarized in

Theorem 6.16. Let $u \in W_2^{2+j}(\Omega)$ (j=0,1) be the solution of the BVP $A^b u = F$ from (6.1.1), (6.1.3). Furthermore, let y be the solution of the corresponding FDS $A_h^b y = F_h$ from (6.1.15), i.e. $A_h^b := A_h + R_h = F_h$, where the parts $A_h := A_h^j$ (j=0,1) are defined in Theorems 3.1. or 3.4. For network of triangles, R_h can be taken from (6.1.17), (6.1.18) or (6.1.19). If rectangles occur, the difference operators R_h from (6.1.20) or (6.1.21) can be applied. Suppose also that the specific additional assumptions concerning the (order of) convergence of PB- and MD-schemes $A_h y = F_h$, which are demanded in Theorems 5.30. and 5.32., respectively, are fulfilled. Finally, let b_o from (6.1.4) be sufficiently small. Then, for $j \in \{0,1\}$ and $\|\cdot\|_1 := \|\cdot\|_{1,r}$, $\|\cdot\|_{1,r}$ from (4.3.13) (r=p) or (4.3.14) (r=m), the subsequent estimates hold, which also imply the convergence $y \to u$ in the discrete W_2^1-norm:

(i) $\|y-u\|_1 \leq Mh^{1+j/2}$, with $M = O(1)$, for $j \in \{0,1\}$ and $h \leq h_o$, (6.4.2)

$u \in W_2^{2+j}(\Omega)$, k_{11}, c, f, $b_i \in C^{1+j}(\bar{\Omega})$ and α, $g \in PC^{1+j}(\Gamma_{23})$, $g \in C(\Gamma_1)$,

for locally irregular networks from Definitions 2.10., 2.11. (PB) or 2.12. For j = 1 and regular grid points $x \in \mathring{\omega}$, the operator R_h is taken from (6.1.19) or (6.1.21);

(ii) $\|y-u\|_1 \leq Mh^2$, (6.4.3)

with the assumptions from (i), but for j = 1, $\Gamma = \Gamma_1$ ($\Gamma_{23} = \emptyset$) and regular grids $\bar{\omega}$;

(iii) $\|y-u\|_1 \leq Mh$, (6.4.4)

with the smoothness assumptions from (i) for j = 0 and for irregular networks. ■

<u>Proof.</u> The assumptions of Theorem 6.16., especially on b_o, enable us to derive a priori estimate (6.3.6). Therefore, it remains to estimate the discrete L_2-norms of $\varkappa$, ψ_N^j and ψ_K. But bounds of $\|\varkappa\|_o$ and $\|\psi_N^j\|_o$ were given in Theorem 5.30. (PB-schemes) and in Theorem 5.32. (MD-schemes), and for $\|\psi_K\|_o$, they were specified in Lemma 6.11. Collecting and inserting these bounds in (6.3.6), we immediately get (6.4.2) - (6.4.4).

<u>Remark 6.17.</u> The assumptions of Theorem 6.16. can be partially weakened or modified. For instance, the remarks made in Section 5.4.3. can be transferred to these more general BVPs $A^b u = F$ and FDSs $A_h^b y = F_h$. Especially, Remark 5.31. continues to be valid. ■

The vonvergence with respect to the C-norm, here considered for generalized solutions $u \in W_2^{2+j}(\Omega)$ and more general networks than in Theorem 6.12., is stated in

<u>Theorem 6.18.</u> Let the assumptions of Theorem 6.16 be fulfilled. Then, for $z := y - u$, with the solution $u \in W_2^{2+j}(\Omega)$ of the BVP $A^b u = F$ from (6.1.1), (6.1.3), and the solution y of the corresponding FDSs $A_h^b y = F_h$ from Theorem 6.16., the following estimates hold:

$$\|y-u\|_{C(\bar\omega)} \le M_1 |\ln h|^{1/2} \cdot \begin{cases} h^{1+j/2} & j = 0,1,\ \bar\omega \text{ locally irregular}, \\ h^2 & \text{for } j = 1,\ \Gamma = \Gamma_1,\ \bar\omega \text{ regular}, \\ h & j = 1,\ \bar\omega \text{ irregular}, \end{cases} \tag{6.4.5}$$

where the three rows in (6.4.5) correspond directly to the three cases (i), (ii), (iii) of Theorem 6.16. The norm $\|\cdot\|_{C(\bar\omega)}$ is taken from (4.3.9). ■

<u>Proof:</u> Due to Theorem 4.36. or Corollary 4.37., we get the inequalities $\|y-u\|_{C(\bar\omega)} \le C_r |\ln h|^{1/2} \|y-u\|_{1,r}$, $r \in \{p,m\}$, which yield, together with (6.4.2) - (6.4.4), estimates (6.4.5).

7. CONCLUDING REMARKS

In the following, some other boundary value problems and topics which are related to the FDM via balance approach, inclusively computational aspects, will be considered.

<u>1° Boundary conditions containing a skew derivative.</u> In HEINRICH [7,10], the FDM via balance approach on irregular networks of triangles and (or) rectangles was also applied to BVPs

$$A^{\mu}u = F, \text{ with } A^{\mu} = (L, l^{\mu})^T, \; F = (f,g)^T,$$

$$(Lu)(x) := -\sum_{i=1}^{2} \frac{\partial}{\partial x_i}\left(k(x)\frac{\partial u}{\partial x_i}\right) + c(x)u = f(x), \; x \in \Omega \subset R^2,$$

$$(l^{\mu}u)(x) := \begin{cases} u = g(x), \; x \in \Gamma_1 \\ \tilde{k}(x)\dfrac{\partial u}{\partial \mu} + \alpha(x)u = g(x), \; x \in \Gamma_{23} \end{cases}, \quad \Gamma = \Gamma_1 \cup \Gamma_{23}, \qquad (7.1)$$

with $\tilde{k}(x) := (\mu,n)^{-1}k(x)$, and k, c, f, α, g, n (outward normal), Γ_1, Γ_{23} ($\alpha = 0$ on Γ_2) as in (3.2.1), μ: skew (oblique) outward direction on Γ_{23}, with $|\mu| = 1$, $(\mu,n)_{R^2} \geq \mu_o > 0$ along Γ_{23}, $(\mu,n)_{R^2} \in PC^s(\Gamma_{23})$, $s \in \{1,2\}$. The notation $\tilde{k}$ is chosen for the sake of convenience, and various other positive coefficients could be transformed into this form. For instance, problems of this type arise in electronics, when the modelling of the Hall-effect in conductors or semiconductors is treated, see e.g. SAMARSKIĬ/ANDREEV [1, p. 15]. For such problems, maximum principles continue to hold even in such cases, where the constant μ_o from (7.1) is small. For the analysis of BVPs $A^{\mu}u = F$, see e.g. AZIZ [1], GILBARG/TRUDINGER [1], MEYER [1], ODEN/REDDY [1] and SAMARSKIĬ/ANDREEV [1]. The discretization of BVPs $A^{\mu}u = F$ on locally irregular grids and by the classical FDM has been dealt with in the literature, cf. the references in HEINRICH [2]. In the author's view, FEM-approximations do not seem to have been considered yet. In HEINRICH [7,10], FDSs $A_h^{\mu}y = F_h$ approximating the BVP $A^{\mu}u = F$ on irregular networks were derived, and the properties of A_h^{μ} as well as the rate of convergence $y \to u$ as $h \to 0$, with respect to the discrete C- and W_2^1-norm, were discussed. For instance, upwind approximations on the boundary, which guarantee the monotonicity condition $M(A_h^{\mu}, \bar{\omega})$ for the difference operators A_h^{μ}, cf. (4.2.1), were given. The tools developed in this monograph and in some previous papers of the author, such as the technique of deriving a priori estimates with the discrete C- and W_2^1-norms as well as the technique of error estimation for solutions $u \in W_2^s(\Omega)$, $s \in \{2,3\}$, were also applied to FDSs $A_h^{\mu}y = F_h$ approximating $A^{\mu}u = F$, cf. HEINRICH [7,10] for the results, which are somehow analogous to that of Chapter 6.

2° Weaker smoothness assumptions on the coefficients. In Section 5.4.3., some modifications of the smoothness assumptions on the coefficients and the right-hand side of the BVP $Au = F$ have been considered explicitly. But FDSs $A_h y = F_h$ or $A_h^b y = F_h$, cf. Chapter 3 and Chapter 6, could also be derived in many cases, where the coefficients k_{ij} $(i,j=1,2)$, b_i $(i=1,2)$ and c are only piecewise smooth, e.g. $k_{ij}, b_i, c \in PC^1(\bar{\Omega})$. We shall only discuss the assumption $(k := k_{11} = k_{22}, k_{12} = 0)$ $k \in PC^1(\bar{\Omega})$, which is often fulfilled for physical objects consisting of various isotropic materials. Where the different materials meet, so-called interfaces $\Gamma_0(k)$ arise, i.e. the coefficient k has a jump along Γ_0. This can cause some trouble due to singularities of the solution u at some points of the interface Γ_0, e.g. at some corners of Γ_0 or, if $\Gamma \cap \Gamma_0 \neq \emptyset$ holds. For the analysis of such singularities, see e.g. DOBROWOLSKI [1], OGANESYAN/RUKHOVETS [2]. For simplicity, let us assume that $k \in PC^1(\bar{\Omega})$, $k \in C^1(\bar{\Omega}^{\pm})$, $\Gamma_0(k) := \bar{\Omega}^- \cap \bar{\Omega}^+$, $\bar{\Omega}^- \cup \bar{\Omega}^+ = \bar{\Omega}$, $\partial\Omega^{\pm} \in C^{0,1} \cap PC^2$, and

$$u|_{\Gamma_0 - 0} = u|_{\Gamma_0 + 0}, \quad k\frac{\partial u}{\partial n}\Big|_{\Gamma_0 - 0} = k\frac{\partial u}{\partial n}\Big|_{\Gamma_0 + 0} \tag{7.2}$$

hold, i.e. we have only one interface $\Gamma_0(k)$ which subdivides Ω into two subdomains Ω^- and Ω^+. In (7.2), u denotes the solution of $Au = b$, n the outward normal to $\partial\Omega^- \cap \Gamma_0$, and $a|_{\Gamma_0 \pm 0}$ is the symbol for limits
$a(x \pm 0) = \lim_{x^{\pm} \to x} a(x^{\pm})$, $x \in \Gamma_0$, $x^{\pm} \in \Omega^{\pm}$.

Furthermore, for the construction of the network, we consider Γ_0 as a part of the boundary Γ and approximate $\bar{\Omega}^{\pm}$ by networks of triangles and (or) rectangles as proposed in Assumption 5.29., cf. also Fig. 4g. The singularities caused by $\Gamma_0(k)$ admit, in general, $u \in W_2^1(\Omega)$ only. But in some cases, solutions u which are regular in the subdomains $\Omega^{\pm}$ may occur, e.g. $u \in W_2^2(\Omega^{\pm})$. Consider a grid point $x \in \gamma_0 \subset \Gamma_0$ and the box $\mathcal{K}(x)$, with the parts $\mathcal{K}^{\pm}(x) := \mathcal{K}(x) \cap \bar{\Omega}^{\pm}$. Then, Gauss' formula (3.1.3) does not hold, in general, if $k \in PC^1(\bar{\Omega})$ is assumed. But taking into account the continuity of the flux, cf. (7.2), Gauss' formula continues to hold, i.e.

$$\int_{\mathcal{K}(x)} (L_0 u)d\bar{x} = -\int_{\partial\mathcal{K}^-(x) \setminus \Gamma_0} (K\nabla u, n)ds - \int_{\partial\mathcal{K}^+(x) \setminus \Gamma_0} (K\nabla u, n)ds, \quad x \in \gamma_0 \subset \Gamma_0. \tag{7.3}$$

The line integrals in (7.3) are of the same type as in (3.1.7) for $x \in \gamma_{23}$, and therefore, they can be approximated in the same way. Thus, for $u \in W_2^2(\Omega^{\pm})$ and under Assumption 5.29., we are in a position to derive FDSs $A_h y = F_h$ which have the same properties as for smooth coefficients. But it should be noted that the grid points $x \in \gamma_0 \subset \Gamma_0$ are to be considered as irregular grid points, which limits the rate of convergence to $\|y-u\|_1 = O(h^{3/2})$, at most, if $u \in W_2^3(\Omega^{\pm})$ holds and locally irregular networks are employed.

3° Weaker smoothness assumptions on the solution u. In this monograph, convergence results have been proved for solutions $u \in W_2^l(\Omega)$, $l > 2$, l integer, or $u \in C^l(\bar{\Omega})$. If $u \in C^{l,\alpha}(\bar{\Omega})$ is admitted, with the exponent α, $0 < \alpha \leq 1$, then the rate of convergence can be expressed in terms of l and α, where Taylor's expansion and the Hölder-continuity of some error terms are employed to show this. If $u \in W_2^s(\Omega)$ is assumed to be fulfilled, with $s > 2$ and s real, the Bramble-Hilbert Lemma from Appendix BS cannot be applied. But the reader interested in this topic can be referred to the paper of DUPONT/SCOTT [1], where corresponding assertions for real s can be found.

If u is the generalized solution of the BVP Au = F, with $u \in W_2^1(\Omega)$ only, and Ω is a rectangle covered with a network of rectangles, some convergence results for FDSs $A_h y = F_h$ can be found in LAZAROV/MAKAROV/WEINELT [1]. Here, the discrete L_2-norm and smoothing procedures $\tilde{u}$ for $u \in W_2^m(\Omega)$ $(m > 0.5)$ are taken into account for proving some rates of convergence $y \to \tilde{u}$ as $h \to 0$. For such problems with weak solutions and related topics of the FDM, the monograph by LAZAROV/MAKAROV/SAMARSKIĬ [1] should be consulted, where basic tools, approaches and results on various differential equations are contained.

4° The FDM via balance approach applied to other problems. The FDM via balance approach for solving variational inequalities with elliptic differential operators of second order, which also describe some classes of free boundary value problems, was investigated by WEINELT [2] for rectangles Ω, networks of rectangles and generalized solutions $u \in W_2^2(\Omega)$. The solution of more general elliptic variational inequalities on domains with curved boundaries is presented in HARTWIG/WEINELT [2], where PB-boxes, locally irregular networks of triangles and (or) rectangles and classical solutions are considered.

For elliptic differential equations of fourth order, the FDM via balance approach was described e.g. in REISSMANN [1,2,3], REISSMANN/HAUG [1] and applied to solving plate problems. Here, the differential equation of fourth order was transformed into a system of two second order elliptic differential equations, and these equations were treated by the FDM via balance approach (PB-boxes) studied in this monograph.

If the elliptic differential operator is nonlinear, e.g. with the principal part $\operatorname{div}(k(u,\nabla u)\operatorname{grad} u)$, then Gauss' formula can also be applied. Thus, assuming that k and u are sufficiently smooth, we get

$$\sum_{i=1}^{N} \int_{\mathcal{X}(x)} \frac{\partial}{\partial x_i} \left(k(u,\nabla u) \frac{\partial u}{\partial x_i}\right) d\bar{x} = \int_{\partial\mathcal{X}(x)} k(u,\nabla u) \frac{\partial u}{\partial n}\, ds, \quad \text{for } \mathcal{X}(x) \subset \bar{\Omega} \subset R^N. \tag{7.4}$$

Starting from the right-hand side of identity (7.4), some nonlinear FDSs $A_h y_h = F_h$ approximating nonlinear BVPs Au = F can be derived.

For instance, see HEIMEIER [1], POHL [1], REICHERT [1], WINSLOW [1] and some other papers in the references, which treat the numerical solution of magnetic field problems by the balance approach. Nonlinearities in the secondary part of the elliptic operator or in the boundary conditions can also be treated, e.g. by lumping approximations.

The FDM via balance approach can also be applied to the construction of FDSs approximating parabolic differential equations. The derivation of such finite difference approximation and the study of convergence has been carried out in some papers, see e.g. FRYASINOV/MASLYANKINA [1], FRYASINOV [2,3,4], SAMARSKIJ [1].

<u>5° Other primary (secondary) networks and the dimension $N \geq 1$.</u> In this monograph, networks of triangles and (or) rectangles have been considered. If convex straight quadrangles are employed, secondary networks can also be defined, e.g. by method (MD). For secondary networks on straight quadrangles and corresponding finite difference quotients, see e.g. GIRAULT [1,2] and SAMARSKIǏ/TISHKIN et al. [1].

The coupling of various types of primary networks, such as networks of triangles (rectangles) with polar-coordinate elements, can be easily made and requires no transformations of coordinates and differential expressions. This is due to the fact that the approximation of the primary part of the elliptic operator is traced back to the approximation of the line integral of the conormal derivative. Moreover, the flexibility in using various types of primary networks allows us to locally refine a given coarse network. After application of such local refinements, all qualitative properties (with respect to monotonicity, symmetry, definiteness) of the FDSs $A_h y = F_h$ can be maintained.

The secondary networks and the boxes $\mathcal{K}(x)$ studied in this monograph are based on the perpendicular bisectors (PB) and the medians (MD) in some primary finite elements.
An alternative procedure of defining secondary networks was introduced by FRYASINOV [3,4]. But here the construction of a box $\mathcal{K}(x)$ depends, in general, on some boxes of the neighbouring grid points, and $\mathcal{K}(x)$ is not centred on regular networks. Obviously, there are further possibilities of defining boxes $\mathcal{K}(x)$, but some of them are not so convenient. For instance, if $\partial\mathcal{K}(x)$ is defined by the straight segments joining directly the centres of gravity of the triangles associated with x, then the portion $\partial\mathcal{K}(x) \cap \Delta(x, \xi, \xi^+)$ depends on the neighbouring triangles $\Delta(x, \xi^+, .)$ and $\Delta(x, ., \xi)$. This will disturb the symmetry. Thus, the portion $\partial\mathcal{K}(x) \cap$ $\cap\, e(x,x")$ should be definable by $e(x,x")$ only.
The FDM via balance approach studied in this monograph on bounded plane domains $\Omega \subset R^2$ can be applied to domains $\Omega \subset R^N$, with $N = 1$ or $N > 2$, in the same way as for $N = 2$. Thus, for $N = 3$, tetrahedrons and (or) parallelepipeds can be taken instead of triangles and (or) rectangles for $\Omega \subset R^2$.

There are three-dimensional generalizations of the secondary networks (PB) and (MD), which are described e.g. in REISSMANN [2]. In engineering problems, such methods are often used on networks consisting of parallelepipeds, see e.g. REICHERT [1], WEILAND [1], WOLFF/MÜLLER [1].

6° Comparison with finite element and other finite difference schemes. For coefficients $k := k_{11} = k_{22}$ and $k_{12} = 0$, the FDSs $A_h y = F_h$ from Chapter 3, but also $A_h^b y = F_h$ from Chapter 6 and $A_h^\mu y = F_h$ from this section (cf. HEINRICH [10]), as well as the resulting systems of linear algebraic equations $\mathfrak{B}_h y = G_h$ on $\omega + \gamma_{23}$, cf. (4.1.6), were analyzed and compared with other schemes in HEINRICH [7]. Thus, a comparison of the schemes obtained by PB- and MD-boxes, and with some classical FDSs on regular networks as well as simple finite element schemes on irregular networks, was carried out. Some results of the comparison of PB- and MD-schemes are presented in Section 4.2.2. The differences of the FDSs originate from the different boxes $\mathfrak{X}(x)$, the different quadrature formulas applied to functions on $\partial\mathfrak{X}(x)$ and from the nonconstant coefficients (right-hand sides), in general. For the derivation of simple finite element schemes, the following cubature and quadrature formulas were considered:

$$\int_\Delta r\, dx \approx \frac{\operatorname{mes}\Delta}{3}\left\{ r(x)+r(\xi)+r(\xi^+)\right\} \quad \text{for } \Delta := \Delta(x,\xi,\xi^+),$$

$$\int_{\hbar(x,\xi)} r\, ds \approx \frac{h(x,\xi)}{2}\left\{ r(x)+r(\xi)\right\} \qquad \text{for } \hbar(x,\xi) \subset \Gamma_{23h}, \tag{7.5}$$

where the triangle Δ and the mesh segment $\hbar$ are taken from Section 2.2. In one case, viz. for MD-systems $\mathfrak{B}_h y = G_h$ and $j = 0$, coincidence with some finite element scheme was observed, but for MD-systems ($j=1$) and PB-systems ($j=0,1$), there is no coincidence. For the systems $\mathfrak{B}_h^b y = G_h$ and $\mathfrak{B}_h^\mu y = G_h$ resulting from $A_h^b y = F_h$ and $A_h^\mu y = F_h$, respectively, none coincidence was encountered or the corresponding finite element schemes are not known from the literature as, in the author's view, e.g. for the BVP $A^\mu u = b$. That MD-schemes are closely related to finite element schemes is due to the fact that $\operatorname{mes}(\mathfrak{X}(x) \cap \Delta) = \frac{1}{3}\operatorname{mes}\Delta$ holds, that the centre of gravity $x'' \in \Delta$ is taken as a quadrature point and that the approximation (3.3.9a) is applied, cf. also (7.5).

7° Computational aspects of the FDM. For the numerical solution of the BVP $Au = F$ by means of the FDM via balance approach, three stages are taken into account

- the generation of the primary networks and grids $\bar{\omega}$,
- the generation of the system of linear algebraic equations $\mathfrak{B}_h y = G_h$, cf. (4.1.6), which corresponds to $A_h y = F_h$,

- the solution of the algebraic system $\mathcal{B}_h y = G_h$.

Obviously, these steps coincide with the general procedure for solving elliptic problems by the FEM. For elliptic problems of the plane and for the practical comparison of various FDSs $A_h y = F_h$, also with simple finite element schemes, a computer program was developed and employed as a research tool (by J. FÖRSTER 1981, extended by A. TOST 1982, using FORTRAN on ES 1040, TH Karl-Marx-Stadt). In this program, Reid's algorithm (REID [1]) for constructing automatically locally irregular networks of triangles was applied, cf. HEINRICH/FÖRSTER [1] for some modifications of this algorithm and the adaption to the FDM via balance approach. Then, using the explicit formulas of the FDSs $A_h y = F_h$ from Theorems 3.1. and 3.4., i.e. for secondary networks of the type (PB) and (MD)(A. TOST 1982), but for the "isotropic" BVP $Au = F$ from (3.2.1) only, the systems of algebraic equations $\mathcal{B}_h y = G_h$ from (4.1.6) were automatically established and solved. For simplicity, the iterative method SOR was employed. The optimal SOR-parameter ω_{opt} $(1 < \omega_{opt} < 2)$ was approximately determined during the iteration process by the method proposed in SCHWARZ et al. [1, pp. 214,215], cf. also HAGEMAN/YOUNG [1]. Clearly, due to the symmetry and positive definiteness of the matrix $\mathcal{B}_h$, various faster iterative methods might be applied.

A program for automatically solving elliptic BVPs $Au = F$ and variational inequalities by means of the FDM via balance approach (PB) on locally irregular networks was developed by K.-H. HARTWIG (1984, TH Karl-Marx-Stadt). Here, SSOR and ATM (alternating triangle method, cf. SAMARSKIĬ/NIKOLAEV [1]) with Chebyshev semi-iterative acceleration were employed for solving the systems of algebraic equations. The corresponding theoretical justification as well as applications to real-life problems can be found in HARTWIG/WEINELT [1,2].

Obviously, the computational problems of the FDM and the FEM used on the same primary networks are quite similar, especially in such cases, where the matrices $\mathcal{B}_h$ of the FDM and FEM have the same nonzero elements. It is advantageous that the FDSs $A_h y = F_h$ proposed in Chapter 3 of our monograph, i.e. the corresponding systems of algebraic equations $\mathcal{B}_h y = G_h$ from (4.1.6), too, do not require the usual calculation procedure (assembly) of the FEM, since the coefficients of $\mathcal{B}_h$ and the right-hand side G_h are given explicitly. The reader who is interested in further computational aspects of the FDM on irregular networks treated in this monograph may also consult the corresponding literature on the FEM. Classical as well as contemporary computational methods for solving elliptic problems by FDM or FEM can be found e.g. in AXELSSON/BARKER [1, Chapter 6 and 7], BANK [1], HAGEMAN/YOUNG [1], RICE/BOISVERT [1], SAMARSKIĬ/NIKOLAEV [1] and THOMASSET [1]. In these papers, complete elliptic solvers, modules for the generation of networks and various finite difference or finite element schemes, and direct as well as iterative methods for solving the system of algebraic equations are described. For instance, multi grid methods or semi-iterative methods with acceleration and preconditioning by incomplete factorization can be employed, see e.g. AXELSSON/BARKER [1].

A P P E N D I C E S

1. Appendix DI. Relations of Differential and Integral calculus, norms

1.1. Norms and seminorms on the Sobolev space $W_p^k(\Omega)$.

For $u \in W_p^k(\Omega)$, the following norms and seminorms are introduced ($k \geq 0$, k integer, $1 \leq p < \infty$):

$$\|u\|_{W_p^k(\Omega)} := \|u\|_{k,p,\Omega} := \Big\{\sum_{|\alpha| \leq k} \int_\Omega |\partial^\alpha u|^p \Big\}^{1/p},$$

$$|u|_{l,p,\Omega} := \Big\{\sum_{|\alpha| = l} \int_\Omega |\partial^\alpha u|^p \Big\}^{1/p} \text{ for } l: 0 \leq l \leq k, \tag{1}$$

with special symbols in the case p = 2:

$$\|u\|_{k,\Omega} := \|u\|_{k,2,\Omega}, \quad |u|_{l,\Omega} := |u|_{l,2,\Omega},$$

and with the multi-index

$$\alpha := (\alpha_1, \alpha_2, \ldots, \alpha_N)^T \in R^N, \ \alpha_i \geq 0, \ \alpha_i \text{ integer}, \ |\alpha| := \sum_{i=1}^N \alpha_i.$$

For p = 2 and l=0,1,2, the seminorms are often written in the form

$$\|u\|_{0,\Omega} := |u|_{0,\Omega} := \Big\{\int_\Omega |u|^2 dx\Big\}^{1/2}, \quad |u|_{1,\Omega} := \Big\{\int_\Omega |\nabla u|^2 dx\Big\}^{1/2},$$
$$|u|_{2,\Omega} := \Big\{\int_\Omega \sum_{i,j=1}^N \Big(\frac{\partial^2 u}{\partial x_i \partial x_j}\Big)^2 dx\Big\}^{1/2}. \tag{2}$$

For Sobolev spaces containing functions defined on the boundary $\partial\Omega$ or subsets $\partial_t\Omega \subset \partial\Omega$, see e.g. GAJEWSKI et al. [1], KUFNER et al. [1], WLOKA [1], but also COSTABEL/STEPHAN [1], ODEN/REDDY [1].

1.2. Some inequalities.

For sufficiently smooth functions $u = u(x)$, $x \in R^N$, and for each vector $\mu \in R^N$, the directional derivative $\frac{\partial u}{\partial \mu}$ can be expressed by

$$\frac{\partial u}{\partial \mu} = (\mu, \nabla u), \text{where } |\mu| := \Big\{\sum_{i=1}^N \mu_i^2\Big\}^{1/2} = 1, \tag{3}$$

and the following inequalities can be verified:

$$\Big|\frac{\partial u}{\partial \mu}\Big| \leq |\nabla u|, \quad \||\nabla u|\|_{1,\Omega} \leq \{|u|_{1,\Omega}^2 + |u|_{2,\Omega}^2\}^{1/2},$$
$$|\nabla|\nabla u|| \leq \Big\{\sum_{|\alpha|=2} |\partial^\alpha u|^2\Big\}^{1/2}. \tag{4}$$

1.3. Some integral identities. Let Ω be a bounded domain, with $\Omega \subset R^N$ and $\partial\Omega \in C^{0,1}$. Furthermore, let $n = (n_1, n_2, \ldots, n_N)^T \in R^N$ denote the outward normal to $\partial\Omega$, and let ds be the surface element of $\partial\Omega$. Then the following integral identities hold:

a) Integration by parts

$$\int_\Omega \frac{\partial u}{\partial x_i} v\,dx = \int_{\partial\Omega} u v n_i\,ds - \int_\Omega u \frac{\partial v}{\partial x_i}\,dx \text{ for } i = 1,2,\ldots;N,\ u,v \in W_2^1(\Omega), \quad (5)$$

b) Green's formula

$$\int_\Omega \sum_{i,j=1}^N k_{ij} \frac{\partial u}{\partial x_i} \frac{\partial v}{\partial x_j}\,dx = \int_{\partial\Omega} \sum_{i,j=1}^N k_{ij} \frac{\partial u}{\partial x_j} n_i v\,ds - \int_\Omega \sum_{i,j=1}^N \frac{\partial}{\partial x_i}\left(k_{ij} \frac{\partial u}{\partial x_j}\right) v\,dx, \quad (6)$$

for $k_{ij} \in C^1(\bar{\Omega})$ $(i,j=1,2,\ldots,N)$, $u \in W_2^2(\Omega)$, $v \in W_2^1(\Omega)$.

c) Gauss' formula

$$\int_\Omega \sum_{i,j=1}^N \frac{\partial}{\partial x_i}\left(k_{ij} \frac{\partial u}{\partial x_j}\right) dx = \int_{\partial\Omega} \sum_{i,j=1}^N k_{ij} \frac{\partial u}{\partial x_j} n_i\,ds \text{ for } k_{ij} \in C^1(\bar{\Omega}),\ u \in W_2^2(\Omega), \quad (7)$$

especially, for $k \in C^1(\bar{\Omega})$, $u \in W_2^2(\Omega)$:

$$\int_\Omega \sum_{i=1}^N \frac{\partial}{\partial x_i}\left(k \frac{\partial u}{\partial x_i}\right) dx = \int_{\partial\Omega} k \frac{\partial u}{\partial n}\,ds, \quad \int_\Omega \Delta u\,dx = \int_{\partial\Omega} \frac{\partial u}{\partial n}\,ds \quad \text{for } k = 1. \quad (8)$$

For $N = 2$, the following special identities hold:

$$\int_\Omega \frac{\partial u}{\partial x_1}\,dx = \int_{\partial\Omega} u\,dx_2, \quad \int_\Omega \frac{\partial u}{\partial x_2}\,dx = -\int_{\partial\Omega} u\,dx_1 \quad \text{for } u \in W_2^1(\Omega). \quad (9)$$

2. Appendix BS: Estimation of functionals on Sobolev spaces

2.1. The Bramble-Hilbert Lemma

Theorem 1 (Bramble/Hilbert)

(i) Let Ω be a bounded domain, with $\Omega \subset R^N$ and $\partial\Omega \in C^{0,1}$. For some integer $k \geq 1$ and some number $p \in [0,\infty)$, let l be a continuous linear functional on the space $W_p^k(\Omega)$.

(ii) Let the functional l fulfil

$$l(q) = 0 \text{ for each } q \in P_{k-1} := \left\{ a : a(x) := \sum_{\gamma} a_\gamma x^\gamma,\ |\gamma| \leq k-1 \right\}. \tag{1}$$

Then there exists a constant $C_{BH}(\Omega)$, which is independent of u, such that the inequality

$$|l(u)| \leq C_{BH} \|l\|^*_{k,p,\Omega} |u|_{k,p,\Omega} \quad \text{for each } u \in W_p^k(\Omega) \tag{2}$$

holds, where $\|\cdot\|^*_{k,p,\Omega}$ denotes the norm in the dual space of $W_p^k(\Omega)$. ■
For Theorem 1, see BRAMBLE/HILBERT [1], CIARLET [2, p. 192] or ODEN/REDDY [1, p. 276]. The symbol P_{k-1} denotes the space of all polynomials in $x_1, x_2, \ldots, x_N$ of degree $|\gamma|$, with $0 \leq |\gamma| \leq k-1$, and γ is a multi-index defined in Appendix DI.

Remark 1. For many purposes it is sufficient to take an upper bound C_l of the norm $\|l\|^*_{k,p,\Omega}$ in the sense of

$$|l(u)| \leq \|l\|^*_{k,p,\Omega} \|u\|_{k,p,\Omega} \leq C_l \|u\|_{k,p,\Omega}, \tag{3}$$

where C_l is independent of u. The existence of such a constant C_l must be shown in any case to verify the boundedness of l in assumption (i) of Theorem 1. Then, instead of (2) the inequality

$$|l(u)| \leq C_{BH} C_l |u|_{k,p,\Omega} \quad \text{for each } u \in W_p^k(\Omega) \tag{2'}$$

holds. ■

2.2. A modification of the Bramble-Hilbert Lemma.

If the functional l vanishes for polynomials q_{k-2} only, then the following modification of Theorem 1 can be proved.

Theorem 2 (modification of Theorem 1)

(i) Let the assumption (i) of Theorem 1 be fulfilled, with $k \geq 2$.

(ii) Let the functional l satisfy

$$l(q) = 0 \text{ for each } q \in P_{k-2} := \left\{ a : a(x) := \sum_{\gamma} a_\gamma x^\gamma,\ |\gamma| \leq k-2 \right\}. \tag{4}$$

Then, instead of (2) the inequality

$$|l(u)| \leq \tilde{C}_{BH} \|l\|^*_{k,p,\Omega} \{|u|^2_{k-1,p,\Omega} + |u|^2_{k,p,\Omega}\}^{1/2} \tag{5}$$

holds for each $u \in W^k_p(\Omega)$. ■

<u>Proof:</u> For any $u \in W^k_p(\Omega)$ and $q \in P_{k-2}$, the function $\tilde{u} := u - q$ belongs to $W^k_p(\Omega)$, and the inequality (3) holds for this $\tilde{u}$, too, i.e.: $|l(\tilde{u})| \leq \|l\|^*_{k,p,\Omega} \|\tilde{u}\|_{k,p,\Omega}$. Taking into account (4), the estimate

$$|l(u)| = |l(\tilde{u})| \leq \|l\|^*_{k,p,\Omega} \|\tilde{u}\|_{k,p,\Omega} = \|l\|^*_{k,p,\Omega} \{\|u-q\|^2_{k-2,p,\Omega} + |u|^2_{k-1,p,\Omega} + |u|^2_{k,p,\Omega}\}^{1/2} \tag{6}$$

can be derived. For the bounded domain Ω satisfying the cone condition and for any $u \in W^k_p(\Omega)$, there exists a polynomial $q^* \in P_{k-2}$ such that

$$\|u - q^*\|_{k-2,p,\Omega} \leq C|u|_{k-1,p,\Omega} \tag{7}$$

holds, with a positive constant C depending on the diameter of Ω but not on u, cf. Theorem 3.13 in ODEN/REDDY [1, p. 85]. The inequality (5) is a simple consequence of (6), (7), and this completes the proof.

<u>Remark 2.</u> Note that the constants in the inequalities presented in Appendix ES can be estimated, e.g. C_1 from (3), by means of the constants occurring in the inequalities of the imbedding theorems, cf. Appendix IM. The estimation of the constant C_{BH} in (7) is implicitly performed in ODEN/REDDY [1], ib. cf. Lemma 6.3 (p. 277), Theorem 3.13 (see the proof on pp. 85, 86, 87) and Theorem 3.14 (pp. 87, 88). ■

3. Appendix EX: Extension of functions

3.1. Calderon's extension

Theorem 1 (Calderon). Let Ω be a bounded domain, with $\Omega \subset R^N$ and $\partial\Omega \in C^{0,1}$, and $\tilde{\Omega}$ a domain containing $\bar{\Omega}$, i.e. $\tilde{\Omega} \supset \bar{\Omega}$. Let k, p be integers, with $k \geq 0$ and $1 < p < \infty$. Then there exists an extension of the class of functions $W_p^k(\Omega)$ defined on Ω to functions on R^N. If Cu denotes the extension of u, then the following relations hold

$$\begin{aligned} &(i) \quad Cu \in W_p^k(\tilde{\Omega}),\ Cu(x) = u(x) \text{ for } x \in \Omega, \\ &(ii) \quad \|Cu\|_{k,p,\tilde{\Omega}} \leq C_E \|u\|_{k,p,\Omega} \text{ for each } u \in W_p^k(\Omega), \end{aligned} \tag{1}$$

where the constant C_E does not depend on u. The symbol C will be mostly omitted. ■

3.2. Geymonat's extension. If seminorms defined for u on $\tilde{\Omega}$ are to be bounded by seminorms of u on Ω, the following assertion seems to be useful.

Theorem 2 (Geymonat). Let E be the operator which extends a polynomial defined on $\Omega \subset R^N$ to the same polynomial defined on R^N, and let Π_{k-1} denote the projection of $W_2^k(\Omega)$ $(k \geq 1)$ onto the polynomials of degree less than k. Moreover, let C be the extension operator from Theorem 1. Then there exists an extension of the class of functions $W_2^k(\Omega)$ defined on Ω to functions on R^N, which is described by the extension operator

$$C_k := C(I - \Pi_{k-1}) + E\,\Pi_{k-1},\ \text{I: identity operator,} \tag{2}$$

and which is bounded with respect to the seminorm with the kth derivative. That is, for the extension $C_k u$ with $C_k u(x) = u(x)$, $x \in \Omega$, the inequality $|C_k u|_{k,R^N} \leq C_E' |u|_{k,\Omega}$ holds, where C_E' does not depend on u. Therefore, instead of (1) the inequality

$$|C_k u|_{k,\tilde{\Omega}} \leq C_E' |u|_{k,\Omega} \tag{3}$$

holds. The symbol C_k is often omitted. ■

For Theorem 1 and the constant C_E see MICHLIN [1, pp. 21], [2]; for Theorem 2 cf. STRANG [1, p. 95].

4. Appendix GE: Some relations of geometry

4.1. Asymptotic behaviour of an arc $\mathfrak{b}$ over a chord $\mathfrak{h}$

Let the assumptions $\mathfrak{b} \subset \partial_t\Omega$, $\partial_t\Omega \in C^2$ and mes $\mathfrak{h}$ = O(h) be fulfilled, where $\partial_t\Omega$ and $\mathfrak{b} := \mathfrak{b}(x,\xi)$ denote arcs on the boundary $\partial\Omega \in C^{0,1} \cap PC^2$ of a domain $\Omega \subset R^2$, and $\mathfrak{h} := \mathfrak{h}(x,\xi)$ is the straight segment between the boundary points of the arc $\mathfrak{b}$, h: the mesh size parameter. Then the following asymptotic relations can be proved:

1. $\delta(\eta) \leq \max\limits_{\eta \in \mathfrak{h}} \delta(\eta) \leq K_1 h^2$, cf. STRANG/FIX [1], (1)

2. $0 < \text{mes}\, \mathfrak{b} - \text{mes}\, \mathfrak{h} \leq K_2 h^3$, (2)

3. $0 < \text{mes}\, \widehat{R_i M'} - \frac{1}{2}\,\text{mes}\, \mathfrak{h} \leq K_3 h^3$, i = 1,2. (3)

Here, M is the midpoint of the segment $\mathfrak{h}$, and, at the same time, the orthogonal projection of some point $M' \in \mathfrak{b}$ onto the chord $\mathfrak{h}$ (cf. figure); R_i (i=1,2) are the two points of intersection of the arc $\mathfrak{b}$ and the segment $\mathfrak{h}$, i.e.

$\mathfrak{b} := \mathfrak{b}(R_1,R_2) := \widehat{R_1 R_2}$

$\mathfrak{h} := \mathfrak{h}(R_1,R_2) := [R_1,R_2]$

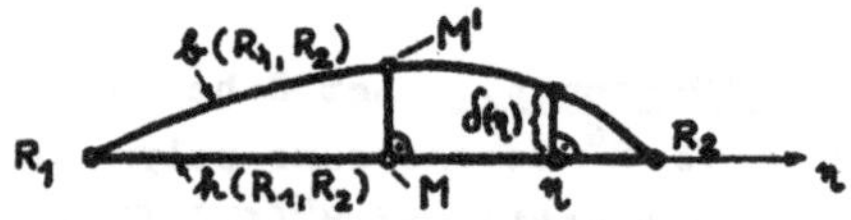

In this monograph, the constants K_i (i=1,2,3) can be taken as universal constants for all mesh segments $\mathfrak{h}(x,\xi)$ on the polygonal boundary $\partial\Omega_h (=\Gamma_h)$ approximating $\partial\Omega$.

4.2. Relations between the angles of triangles and segments of the perpendicular bisectors.

Let $U^{\pm}(r^{\pm}, M^{\pm})$ denote the circumcircles of the triangles $\Delta^{\pm}$, with the radii $r^{\pm}$ and the circumcentres $M^{\pm}$. Then, the following relations involving $r^{\pm}$, $h(x,\xi) :=$ mes $\mathfrak{h}(x,\xi)$, $s(x,\xi) :=$ =mes $\mathfrak{S}(x,\xi)$, $s^{\pm}(x,\xi) :=$ mes $(\mathfrak{S}(x,\xi) \cap \Delta^{\pm})$ and $\theta_{\pm}(x,\xi)$ hold and can be proved by elementary trigonometric considerations:

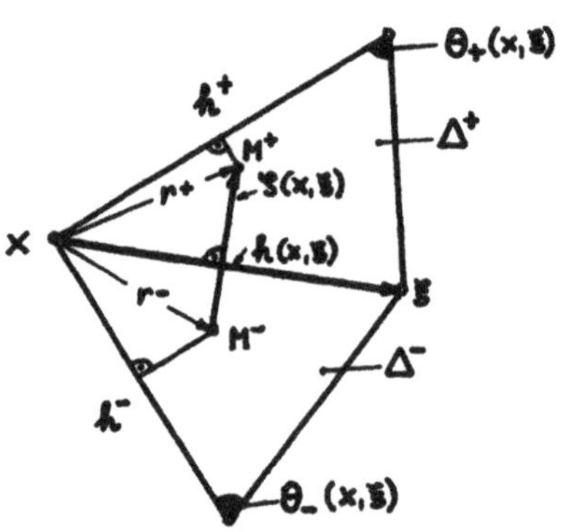

$\mathfrak{S}(x,\xi) := \overline{M^- M^+}$

1. $r^{\pm} = \dfrac{h(x,\xi)}{2} \dfrac{1}{\sin \theta_{\pm}(x,\xi)}$ (4)

   for $0 < \theta_{\pm} < \pi$,

2. $s^{\pm}(x,\xi) = \dfrac{h(x,\xi)}{2} \cot \theta_{\pm}(x,\xi) \geq 0$

   for $0 < \theta_{\pm} \leq \frac{\pi}{2}$, (5)

3. $s(x,\xi) = \dfrac{h(x,\xi)}{2} (\cot \theta_+ + \cot \theta_-) \geq 0$

   for $0 < \theta_- + \theta_+ \leq \pi$.

5. Appendix IM: Imbedding and trace theorems

5.1. An imbedding theorem and assumptions on $\partial\Omega$. For further statements involving Ω and $\partial\Omega$, let the following assumption $V(\partial\Omega)$ be fulfilled:

(i) Ω is an open bounded subset in R^N, i.e. $\Omega \subset R^N$, $\Omega \neq \emptyset$,

(ii) $\partial\Omega \in C^{0,1}$, i.e., $\partial\Omega$ is a Lipschitz-continuous boundary or regular in the sense of Calderon. For an exact definition of "$C^{0,1}$", cf. ADAMS [1, pp. 65-70], LADYSHENSKAYA/URALT'SEVA [1, pp. 30-32] or GAJEWSKI et al. [1], NEČAS [1], WLOKA [1]. Furthermore, let the following condition be satisfied:

(iii) $\partial\Omega \in PC^k$ $(k \geq 1)$; unless otherwise stated $k = 2$ is assumed, i.e. $\partial\Omega$ is "piecewise continuously curved", with at most a finite number of corners on $\partial\Omega$. If a corner at $x_0 \in \partial\Omega$ is directed outwards with respect to Ω, then there is a cone $\mathcal{K}$ which can be put in such a position that x_0 is the vertex of the cone and $\mathcal{K} \setminus \{x_0\} \subset \Omega$ holds.

Note that some of the partial assumptions in (iii) are consequences of (ii), which, however, will not be discussed here, cf. ADAMS [1].

Remark 1. Bounded domains Ω with a $C^{0,1}$-boundary $\partial\Omega$ satisfy the cone condition, cf. ADAMS [1]. Moreover, $\bar{\Omega}$ can be represented by $\bar{\Omega} = \bigcup_{i=1}^{n_0} \bar{\Omega}_i$ where n_0 is a finite number, $\Omega_i \cap \Omega_j = \emptyset$ for $i \neq j$, and each Ω_i is star-shaped with respect to a ball lying in Ω_i. ■

Theorem 1. For Ω, with $V(\partial\Omega$, i,ii), and for $0 \leq j < k$, $1 \leq p < \infty$, $0 < \alpha < 1$ (j,k,p integer), the imbedding

$$W_p^k(\Omega) \subset C^{j,\alpha}(\bar{\Omega}) \subset C^j(\bar{\Omega}) \text{ for } N < p(k-j-\alpha) \tag{1}$$

is completely continuous, and the imbedding

$$W_p^k(\Omega) \subset W_q^j(\Omega),\ \mathring{W}_p^k(\Omega) \subset \mathring{W}_q^j(\Omega) \text{ for } 1 \leq q < \infty, \quad p^{-1} - N^{-1}(k-j) \leq q^{-1}, \tag{2}$$

is continuous. ■

For this, cf. ADAMS [1], LADYSHENSKAYA/URAL'TSEVA [1], NEČAS [1], SMIRNOW [1, pp. 305, pp. 317], WLOKA [1], ZEIDLER [1, p. 231].

Corollary 1. From Theorem 1 and with some constant C_I which does not depend on u, we get the inequalities

$$\|u\|_{C(\bar{\Omega})} \leq C_I \|u\|_{k,p,\Omega} \quad \text{for } N < pk, \tag{3}$$

especially

$$\|u\|_{C(\bar{\Omega})} \leq C_I \|u\|_{2+j,2,\Omega} \quad \text{for } N = 2, 3 \text{ and } j = 0, 1, 2, \dots \ ■ \tag{4}$$

Remark 2. The constants C_I or bounds of C_I depend, in general, on Ω, N, j, k, p, q, α. But as to Ω, the dependence os often traced back to the cone $\mathcal{K}$, which verifies the cone property of Ω, or to mes Ω; cf. ADAMS [1], MICHLIN [2].

5.2. Two trace theorems

Theorem 2 (case $M_s \subset \Omega$). For $\Omega \subset R^N$, with $V(\partial\Omega, i, ii)$, let $1 < p < \infty$, $0 \leq N - pk < s \leq N$, $q < \frac{ps}{N-pk}$ be satisfied, and let M_s be a sufficiently smooth s-dimensional manifold, with $M_s \subset \Omega$. Then the imbedding of $W_p^k(\Omega)$ in $L_q(M_s)$ is completely continuous. ■

Corollary 2. For segments M_s of hyperplanes, with int $M_s \subset \Omega$ and $s := N - 1$, the inequality

$$\|\gamma_0 u\|_{L_2(M_s)} \leq C_I \|u\|_{1,2,\Omega} \qquad \text{for } N = 2, 3 \quad (\gamma_0\text{: trace operator}) \tag{5}$$

holds, with C_I depending on M_s and Ω but not on u. ■

Theorem 3 (case $\partial\Omega$). For $\Omega \subset R^N$, with $V(\partial\Omega, i, ii)$, and $q \leq \frac{p(N-1)}{N-p}$ for $N > p$, $q < \infty$ for $N = p$, $q = \infty$ for $N < p$, there exists a continuous linear operator (trace operator) $\gamma_0 : W_p^1(\Omega) \longrightarrow L_q(\partial\Omega)$, and the inequality

$$\|\gamma_0 u\|_{L_q(\partial\Omega)} \leq C_I \|u\|_{1,p,\Omega} \tag{6}$$

holds, with a constant C_I independent of u. The symbol γ_0 will be mostly omitted. ■

For Theorem 2 and 3, see GAJEWSKI et al. [1], KUFNER et al. [1], LADYSHENSKAYA/URAL'TSEVA [1], NEČAS [1], SMIRNOW [1], ZEIDLER [1].

Corollary 3. Theorem 3 implies, especially for N = 2, N = 3, and $\partial_t\Omega \subset \partial\Omega$, the inequalities

$$\|u\|_{L_2(\partial_t\Omega)} \leq \|u\|_{L_2(\partial\Omega)} \leq C_I \|u\|_{1,2,\Omega} \text{ for } u \in W_2^1(\Omega),$$

$$\||\nabla u|\|_{L_2(\partial_t\Omega)} \leq C_I \||\nabla u|\|_{1,2,\Omega} \leq C_I \{|u|_{1,\Omega}^2 + |u|_{2,\Omega}^2\}^{1/2} \tag{7}$$

$$\text{for } u \in W_2^2(\Omega). \quad ■$$

5.3. Imbedding theorem for convergent boundary strips

Theorem 4. For $\Omega \subset R^2$, with $V(\partial\Omega, i, ii, iii)$ and thus $\partial\Omega \in C^{0,1} \cap PC^2$, let ω_h denote a "convergent boundary strip", which has a boundary $\partial\omega_h$ "parallel to $\partial\Omega$" and $\partial\Omega$ as the "centre line" of ω_h. Let the boundary strip have the width $C_0 h$ ($C_0 > 0$) such that mes $\omega_h = O(h)$ holds. Then there exists a constant C_1 independent of h, $h \leq h_0$, and u such that the inequality

$$\|u\|_{L_2(\omega_h)} \leq C_1\, h^{1/2}\, \|u\|_{W_2^1(\Omega)} \quad \text{for } u \in W_2^1(\Omega) \tag{9}$$

holds. Here, the function u was extended by Calderon to $u \in W_2^1(\Omega \cup \omega_h)$. ∎
The proof of Theorem 4 for the case $\partial\Omega \in C^2$ and an interior boundary strip $\omega_h \subset \Omega$ was given in OGANESYAN/RUKHOVETS [2, p. 24] or in OGANESYAN et al. [1] (part 1), cf. also the literature cited in ASTRAKHANTSEV [1]. If $\partial\Omega \in PC^2$ is assumed instead of $\partial\Omega \in C^2$, then $\partial\Omega$ consists of a finite number of C^2-arcs $\partial_t\Omega$, and the proof from OGANESYAN/RUKHOVETS [2] can be performed in sections for the strip ω_h, i.e. (9) can be proved, too. If additionally $\omega_h \not\subset \Omega$ is admitted, cf. Theorem 4, then the technique of the proof used in OGANESYAN/RUKHOVETS [2, pp. 24, 25] will be applied separately to the interior and exterior parts (with respect to Ω) of the boundary strip ω_h. In the part of the proof taking into account the exterior portion of ω_h, the local coordinate systems considered in OGANESYAN et al. [1], OGANESYAN/RUKHOVETS [2], is also to be taken on the arcs $\partial_t\Omega$ of the fixed boundary $\partial\Omega$. Here, the norm $\|u\|_{1,2,\tilde{\Omega}}$ for $\tilde{\Omega} \supset \omega_h$ and $h \leq h_0$ occurs temporarily on the right-hand side of the inequality (9). Applying Theorem 1 from Appendix EX, especially $\|u\|_{1,2,\tilde{\Omega}} \leq C_E \|u\|_{1,2,\Omega}$, the W_2^1-norm is reduced to that with Ω as actually noted in (9).

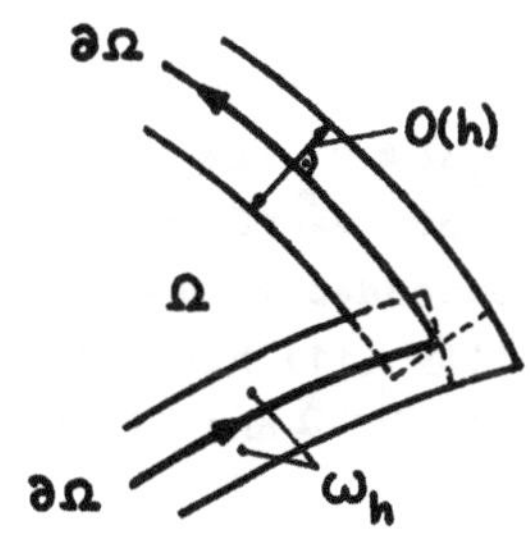

<u>Corollary 4.</u> Especially for $u \in W_2^{1+j}(\Omega)$, the inequalities

$$\|u\|_{j,2,\omega_h} \leq C_2\, h^{1/2}\, \|u\|_{1+j,2,\Omega} \quad \text{for } j = 0,1,2 \text{ and } \omega_h \text{ from (9)} \tag{10}$$

hold.

<u>5.4. Trace theorems with Sobolev spaces on Ω, $\partial\Omega$.</u> In KUFNER et al. [1] and in some other monographs cited in this appendix, boundaries $\partial\Omega \in C^{k-1,1}$ Sobolev spaces $W_p^k(\partial\Omega)$ on the boundary $\partial\Omega$ and trace theorems for $u \in W_p^{k+1}(\Omega)$ are considered, with $k \geq 1$, k integer.

<u>Theorem 5 (case $\partial\Omega$).</u> For $\Omega \subset R^N$ with $\partial\Omega \in C^{k-1,1}$, $k \geq 1$, $p \geq 1$, $q = p(N-1)(N-p)^{-1}$ for $p < N$ or $q \geq 1$ for $p \geq N$, there exists a continuous linear operator (trace operator) $\gamma_0: W_p^{k+1}(\Omega) \to W_q^k(\partial\Omega)$. ∎
For Theorem 5, see KUFNER et al. [1] (Theorem 6.7.8 and Remark 6.7.9), cf. GAJEWSKI et al. [1]. For N = p = 2 and q = 2, Theorem 5 implies the inequality (γ_0 is omitted)

$$\|u\|_{W_2^k(\partial\Omega)} \leq C_I\, \|u\|_{W_2^{k+1}(\Omega)} \quad \text{for } k \in \{1,2\},\ \partial\Omega \in C^{k-1,1}. \tag{11}$$

Remark 3. The relation $\partial\Omega\in C^{k-1,1}$ means that the (k-1)th derivative (concerning $\partial\Omega$) is Lipschitz-continuous. Thus, for k = 2 corners are not admitted. But for k = 2 and under the assumption $V(\partial\Omega,i,ii,iii)$, an inequality of the type (11) holds for parts $\partial_t\Omega\in C^2$ of the boundary $\partial\Omega\in C^{0,1}\cap PC^2$, i.e.

$$\|u\|_{W_2^2(\partial_t\Omega)} \leq C_I\|u\|_{W_2^3(\Omega)} \quad \text{for } u\in W_2^3(\Omega). \blacksquare \tag{12}$$

The proof of (12) is based on the following considerations: The arc $\partial_t\Omega\in C^2$ can be extended to a closed boundary $\partial\hat{\Omega}$ of a domain $\hat{\Omega}$ such that $\partial\hat{\Omega}\in C^2$ and $\hat{\Omega}\cap\Omega\neq\emptyset$ hold. Then the function $u\in W_2^{k+1}(\Omega)$ will be extended to $u\in W_2^{k+1}(\tilde{\Omega})$ (see Appendix EX), with $\tilde{\Omega}\supset\hat{\Omega}\cup\Omega$. Now inequality (11) holds for $\partial\hat{\Omega},\hat{\Omega}$ instead of $\partial\Omega,\Omega$. Finally, (11) (with $\partial\hat{\Omega}$, $\hat{\Omega}$), $\hat{\Omega}\subset\tilde{\Omega}$ and (1) from Appendix EX imply (12).

6. Appendix TR: Affine transformation of coordinates and functionals

6.1. Affine transformation of domains and coordinates. Let the two domains Ω and $\hat{\Omega}$ of R^N be affine-equivalent, i.e., there exists an invertible affine mapping F such that $\Omega = F(\hat{\Omega})$, with

$$F:\hat{x}\in R^N \longrightarrow x = F(\hat{x}) = B\hat{x} + b\in R^N,$$

$$B = (b_{ij})_{N\times N}\in L(R^N,R^N),\ b\in R^N,\ \det B\neq 0, \tag{1}$$

$$F^{-1}:x\in R^N \longrightarrow \hat{x} = F^{-1}(x) = B^{-1}x - B^{-1}b.$$

6.2. Transformation of derivatives. The affine mapping F from (1), with $x = F(\hat{x})$, transforms the derivatives of a sufficiently smooth function $\hat{u}(\hat{x})$ defined over $\hat{\Omega}$ into the derivatives of the function $u(x)$ over Ω. The corresponding transformation rules for the kth derivative ($0\leq k\leq 2$) and the "inverse rules" originating from $\hat{x} = F^{-1}(x)$ are presented in (2), where the symbol $B^{T-1}:=(B^T)^{-1}$ is employed.

functions $u,\hat{u}$: $\quad u(x) = \hat{u}(\hat{x})$,

first derivatives: $\nabla u(x) = B^{T-1}\hat{\nabla}\hat{u}(\hat{x})$, $\hat{\nabla}\hat{u}(\hat{x}) = B^T\nabla u(x)$, (2)

second derivatives: $(D^2u\xi,\eta) = (B^{T-1}\hat{D}^2\hat{u}B^{-1}\xi,\eta)$,

$\quad (\hat{D}^2\hat{u}\xi,\eta) = (B^TD^2uB\xi,\eta)$,

for all $\xi,\eta\in R^N$. Especially, if $\xi = e_i$ and $\eta = e_j$ (e_i: the unit vector of the x_i-axis) are inserted in the transformation formulas for the second derivatives, then $\partial_{ij}u$ or $\hat{\partial}_{ij}\hat{u}$ will be obtained. The symbols D^2 and $\hat{D}^2$ are defined by

$$D^2u(x) := \left(\frac{\partial^2 u(x)}{\partial x_i \partial x_j}\right)_{N\times N}, \qquad \hat{D}^2\hat{u}(\hat{x}) := \left(\frac{\partial^2 \hat{u}(\hat{x})}{\partial \hat{x}_i \partial \hat{x}_j}\right)_{N\times N}.$$

Formulas of the type (2) can be found in the literature on analysis.

<u>6.3. Transformation of multiple and line integrals.</u> The affine transformation $\Omega = F(\hat{\Omega})$ from (1) implies the following integral identities:

a) $\int_\Omega u(x)dx = \det B \int_{\hat{\Omega}} \hat{u}(\hat{x})d\hat{x}$, $\Omega = F(\hat{\Omega})$, $dx = \det B\, d\hat{x}$,

b) moreover, for $N = 2$ and the affine-equivalent straight segments $\mathcal{S}$ and $\hat{\mathcal{S}}$: (3)

$$\int_{\mathcal{S}} u\,ds = \frac{|B\hat{\mathcal{S}}|}{|\hat{\mathcal{S}}|}\int_{\hat{\mathcal{S}}} \hat{u}\,d\hat{s}, \quad \mathcal{S} = B\hat{\mathcal{S}}, \quad \mathcal{S} \subset \Omega \subset R^2:$$

$$ds = \frac{|B\hat{\mathcal{S}}|}{|\hat{\mathcal{S}}|}\, d\hat{s}, \quad ds \le \|B\|d\hat{s},$$

cf. standard works on analysis.

<u>6.4. Estimates for transformed derivatives</u>

a) first derivatives: $|\nabla u| \le \|B^{-1}\|\,|\hat{\nabla}\hat{u}|$, $|\hat{\nabla}\hat{u}| \le \|B\|\,|\nabla u|$,

b) second derivatives: $\|D^2u\|_F \le \|B^{-1}\|^2\|\hat{D}^2\hat{u}\|_F$, $\|\hat{D}^2\hat{u}\|_F \le \|B\|^2\,\|D^2u\|_F$,

with $\|A\|_F := \{\text{trace}\,(A^TA)\}^{\frac{1}{2}}$ for $A \in L(R^N, R^N)$, $|\cdot|$ taken for the Euclidian norm and $\|\cdot\|$ for the matrix norm induced by $|\cdot|$. Estimates for derivatives of higher order can be found in ODEN/REDDY [1, pp. 267, 268], CIARLET [2, pp. 118, 119].

<u>6.5. Estimates of the seminorms $|\cdot|_{k,\Omega}$ in Sobolev spaces $W_2^k(\Omega)$</u>

Assume that $u \in W_2^k(\Omega)$, $k \ge 0$, $\hat{\Omega} = F^{-1}(\Omega)$, with F from (1), hold. Then $\hat{u} = u \circ F$ ($\hat{u}(\hat{x}) = u(x)$) belongs to the space $W_2^k(\hat{\Omega})$, and the following relations can be verified:

a) $\|u\|^2_{0,\Omega} := \int_\Omega u^2(x)dx = \int_{\hat{\Omega}} \hat{u}^2(\hat{x}) \det B\, d\hat{x} = |\det B|\,\|\hat{u}\|^2_{0,\hat{\Omega}}$,

b) $|u|^2_{1,\Omega} := \int_\Omega |\nabla u|^2\, dx = \int_{\hat{\Omega}} |B^{T-1}\,\hat{\nabla}\hat{u}|^2 \det B\, d\hat{x} \le |\det B\,|\,\|B^{-1}\|^2|\hat{u}|^2_{1,\hat{\Omega}}$, (4)

conversely: $|\hat{u}|^2_{1,\hat{\Omega}} \le |\det B|^{-1}\,\|B\|^2\,|u|^2_{1,\Omega}$,

c) $|u|^2_{2,\Omega} := \int_\Omega \sum_{i,j=1}^N \left(\frac{\partial^2 u}{\partial x_i \partial x_j}\right)^2 dx = \int_\Omega \|D^2u\|^2_F\, dx \le |\det B|\,\|B^{-1}\|^4|\hat{u}|^2_{2,\hat{\Omega}}$,

conversely: $|\hat{u}|^2_{2,\hat{\Omega}} \le |\det B|^{-1}\|B\|^4|u|^2_{2,\Omega}$.

Generally, for $k \ge 0$ the following relations containing (4) as a special case hold:

$$|u|_{k,\Omega} \leq \|B^{-1}\|^k |\det B|^{1/2} |\hat{u}|_{k,\hat{\Omega}} \qquad \hat{\Omega} = F^{-1}(\Omega),$$
$$\text{for} \qquad\qquad\qquad\qquad (5)$$
$$|\hat{u}|_{k,\hat{\Omega}} \leq \|B\|^k |\det B|^{-1/2} |u|_{k,\Omega} \qquad \Omega = F(\hat{\Omega}),$$

with F from (1). For (5), see ODEN/REDDY [1, p. 279].

6.6. The estimation of $|\det B|$, $\|B\|$, $\|B^{-1}\|$

a) For det B the relation

$$|\det B| = \frac{\text{mes}\,\Omega}{\text{mes}\,\hat{\Omega}}\ , \text{ with } \Omega = F(\hat{\Omega}),\ F \text{ and } B \text{ from (1)}, \qquad (6)$$

holds. Therefore, lower and upper bounds of $|\det B|$ can be derived by suitable use of lower and upper bounds of mes Ω and mes $\hat{\Omega}$, cf. (11).

b) Using the definitions $B^{T-1} := (B^T)^{-1}$,

$$|x|^2 := \sum_{i=1}^{N} x_i^2,\ x := (x_1, x_2, \ldots, x_N)^T \in R^N,\ \|B\| := \sup_{|x|=1} |Bx|, \qquad (7)$$

the norms of B, B^T, B^{-1}, B^{T-1} can be estimated in the following way:

$$\left.\begin{array}{l} \|B\| = \|B^T\| \leq \dfrac{d}{\hat{\varrho}} \\ \|B^{-1}\| = \|B^{T-1}\| \leq \dfrac{\hat{d}}{\varrho} \end{array}\right., \text{ with } \left\{\begin{array}{l} d := d(\Omega) := \text{diam}\,(\Omega),\ \hat{d} := d(\hat{\Omega}), \\ \varrho := \varrho(\Omega) := \sup\{d(S) : S \text{ is a ball contained in } \Omega\}, \\ \hat{\varrho} := \varrho(\hat{\Omega}). \end{array}\right. \qquad (8)$$

For (6) and (8), cf. CIARLET [2] and ODEN/REDDY [1].

c) In the special case $\Omega := \Delta$ (triangle) and under the assumptions

$$0 < \theta_0 \leq \theta \leq \pi - \theta_0,\quad 0 < \varepsilon_0 h \leq h_\Delta \leq \varepsilon_0^{-1} h,\quad 0 < \varepsilon_0 \leq 1,\quad 0 < h \leq h_0, \qquad (9)$$

where θ and h_Δ denote the angle at a vertex of Δ and the length of one side of Δ, respectively, the following estimates hold:

$$d \leq \varepsilon_0^{-1} h,\quad \varrho^{-1} \leq 2\,(\varepsilon_0 h \sin\theta_0)^{-1},\quad \tfrac{1}{2}\,\varepsilon_0 h^2 \sin\theta_0 \leq \text{mes}\,\Delta \leq \tfrac{1}{2}\,\varepsilon_0^{-2} h^2. \qquad (10)$$

For instance, if $\hat{\Omega} := \hat{\Delta}$ is an equilateral triangle with the vertices $(1/2, 0)$, $(-1/2, 0)$ and $(0, \sqrt{3}/2)$ (length of the sides: $\hat{h}_\Delta = 1$), then the relations $\hat{d} = 1$, $\hat{\varrho} = \sqrt{3}/3$ and mes $\hat{\Delta} = \sqrt{3}/4$ hold, and, moreover,

$$\frac{2\sqrt{3}}{3}\,\varepsilon_0^2 h^2 \sin\theta_0 \leq |\det B| \leq \frac{2\sqrt{3}}{3}\,\varepsilon_0^{-2} h^2,\ \|B\| \leq \sqrt{3}\,\varepsilon_0^{-1} h, \qquad (11)$$
$$\|B^{-1}\| \leq 2(\varepsilon_0 h \sin\theta_0)^{-1}.$$

6.7. The transformation of finite difference quotients. The finite difference quotients are transformed in the same manner as the differential quotients. For instance, the following relations hold, if the mappings F and F^{-1} from (1) as well as $\Omega = F(\hat{\Omega})$ and $\hat{\Omega} = F^{-1}(\Omega)$ are taken into account:

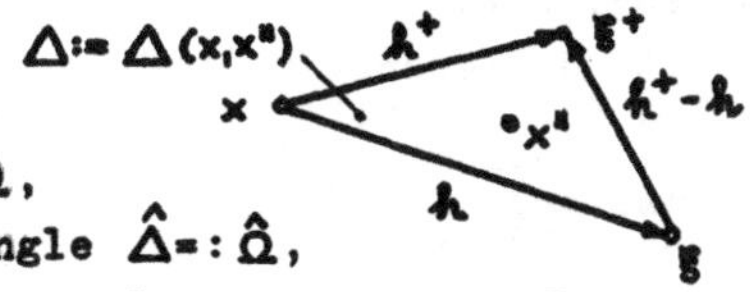

a) $h = B\hat{h}$, $h^+ = B\hat{h}^+$,
h, h^+: sides of the triangle $\Delta =: \Omega$,
$\hat{h}$, $\hat{h}^+$: sides of the reference triangle $\hat{\Delta} =: \hat{\Omega}$, (12)

b) $h^\circ = (B\hat{h})^\circ = \det B\{B^{T-1}\hat{h}^\circ\}$, $h := (h_1, h_2)^T$, $h^\circ := (h_2, -h_1)^T$,
similarly: $h^\bullet = \det B\{B^{T-1}\hat{h}^\bullet\}$; (13)

c) the value of the function u at $x \in \overline{\omega}$: $u(x) = \hat{u}(\hat{x})$,

d) finite difference quotient of first order on the triangle $\Delta(x,x")$:

$$\nabla_h u(x,x") = B^{T-1}\hat{\nabla}_{\hat{h}}\hat{u}(\hat{x},\hat{x}"), \text{ with}$$

$$\nabla_h u(x,x") := \frac{[u(\xi)-u(x)]\,h^\bullet - [u(\xi^+)-u(x)]\,h^\circ}{(h^\bullet, h)}, \tag{14}$$

$\hat{\nabla}_{\hat{h}}\hat{u}$ is similarly defined, with $\hat{h}^\bullet$ and $\hat{h}^\circ$ (after some transformations $\hat{\nabla}_{\hat{h}}\hat{u}$ can be noted explicitly);

e) finite difference quotient of first order on the segment $h(x,\xi)$ connecting two points x, ξ :

$$u_h(x,\xi) = \frac{|\hat{h}|}{|h|}\,\hat{u}_{\hat{h}}(\hat{x},\hat{\xi}), \text{ with } h = B\hat{h}, \ \frac{|\hat{h}|}{|h|} \le \|B^{-1}\|, \ h := h(x,\xi)$$

$$u_h(x,\xi) := \frac{u(\xi)-u(x)}{h(x,\xi)}, \quad \hat{u}_{\hat{h}}(\hat{x},\hat{\xi}) := \frac{\hat{u}(\hat{\xi})-\hat{u}(\hat{x})}{\hat{h}(\hat{x},\hat{\xi})}. \tag{15}$$

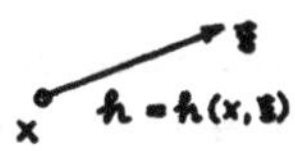

REFERENCES

Adams, R.A.

[1] Sobolev spaces, Acad. Press, New York 1975.

Astrakhantsev, G.P.

[1] A finite difference method for solving the third boundary value problem for elliptic and parabolic equations in an arbitrary domain. Iterative solution of the finite difference equations, parts I and II (Russian), Zh. Vychisl. Mat. Mat. Fiz. 11 (1971), No. 1, 105-120 (part I), and No. 3, 677-687 (part II).

Axelsson, O.

[1] Stability and error estimates of Galerkin finite element approximations for convection-diffusion equations, IMA J. Numer. Anal. 1 (1981), No. 3, 329-346.

Axelsson, O., and V.A. Barker

[1] Finite element solution of boundary value problems (Theory and Computation), series 'Computer science and applied mathematics', Acad. Press, Inc. London, 1984.

Aziz, A.K. (ed.)

[1] The mathematical foundations of the finite element method with applications to partial differential equations, Acad. Press, New York and London 1972.

Azzam, A., and E. Kreyszig

[1] On solutions of elliptic equations satisfying mixed boundary conditions, SIAM J. Math. Anal. 13 (1982), No. 2, 254-262.

Bank, R.E.

[1] PLTMG users' guide - June 1981 version, Dep. Math., Univ. of California, San Diego, Tech. Rep., Aug. 1982.

Bank, R.E., D.J. Rose and W. Fichtner

[1] Numerical methods for semiconductor device simulation, IEEE Trans. Electron Devices 30 (1983), No. 9, 1031-1041.

Beck, J.V., B. Blackwell, and Ch. R.S. Clair (Jr.)

[1] Inverse heat conduction, Wiley-Interscience Publ., New York 1985.

Book, L. (ed.)

[1] Finite-difference techniques for vectorized fluid dynamics calculations, Springer-Verlag, New York, Heidelberg, Berlin 1981.

Bramble, J.H., and S.R. Hilbert

[1] Estimation of linear functionals on Sobolev spaces with application to Fourier transforms and spline interpolation, SIAM J. Numer. Anal. 7 (1970), 112-124.

Bramble, J.H., and B.E. Hubbard

[1] Approximation of solutions of mixed boundary value problems for Poisson's equation by finite differences, J. Ass. Comp. Mach. 12 (1965), 114-123.

Ciarlet, Ph.G.

[1] Discrete maximum principle for finite-difference operators, Aequat. Math. 4 (1970), 338-352.

[2] The finite element method for elliptic problems, North-Holland Publ. Comp., Amsterdam, New York, Oxford 1978.

Ciarlet, Ph.G., and P.A. Raviart

[1] Maximum principle and uniform convergence for the finite element method, Comp. Meth. Appl. Mech. Engin. 2 (1973), 17-31.

Collatz, L.

[1] Funktionalanalysis und numerische Mathematik, Springer-Verlag, Berlin, Göttingen, Heidelberg 1964.

[2] The numerical treatment of differential equations, (third ed.) Springer-Verlag, Berlin, Göttingen, Heidelberg 1960.

Conous, P.

[1] Numerical solution of the nonlinear magnetostatic-field equation in two dimensions, J. Comp. Physics 1 (1967), 330-342.

Costabel, M.

[1] Starke Elliptizität von Randintegraloperatoren erster Art, Habilitationsschrift, Preprint-Nr. 868, TH Darmstadt 1984.

Costabel, M., and E. Stephan

[1] Boundary integral equations for mixed boundary value problems in polygonal domains and Galerkin approximation, Preprint-Nr. 593, TH Darmstadt 1981.

[2] Curvature terms in the asymptotic expansions for solutions of boundary integral equations on curved polygons, Preprint-Nr. 673, TH Darmstadt 1982.

Dobrowolski, M.

[1] Numerical approximation of elliptic interface and corner problems, Habilitationsschrift, Universität Bonn 1981.

Dupont, T., and R. Scott

[1] Polynomial approximation of functions in Sobolev spaces, Math. Comp. 34 (1980), No. 150, 441-463.

Frehse, J.

[1] Existenz und Konvergenz von Lösungen nichtlinearer elliptischer Differenzengleichungen unter Dirichletrandbedingungen, Math. Zeitschr. 109 (1969), 311-343.

Fryasinov, I.V., and L.A. Maslyankina

[1] On a finite difference approximation of elliptic and parabolic equations on an irregular network, (Russian) Preprint No. 49 (1977), Inst. Prikl. Mat. im. Keldyša, Akad. Nauk SSSR, Moskva.

Fryasinov, I.V.

[1] On a finite difference approximation of Poisson's equation, (Russian) Differ. Uravn. 12 (1976), No. 3, 540-548.

[2] On a finite difference approximation of problems with elliptic equation, (Russian) Zh. Vychisl. Mat. Mat. Fiz. 16 (1976), No. 1, 102-118.

[3] Balance method and variational-difference schemes. Schemes of alternating directions on irregular networks, (Russian) Preprint No. 53 (1979), Inst. Prikl. Mat. im. Keldyša, Akad. Nauk SSSR, Moskva.

[4] Balance method and variational-difference schemes, (Russian) Differ. Uravn. 16 (1980), No. 7, 1332-1343.

Gajewski, H., K. Gröger and K. Zacharias

[1] Nichtlineare Operatorgleichungen und Operatordifferentialgleichungen, Akademie-Verlag, Berlin 1974.

Gilbarg, D., and N.S. Trudinger

[1] Elliptic partial differential equations of second order, Springer-Verlag, Berlin, Heidelberg, New York 1977.

Girault, V.

[1] Theory of a finite difference method on irregular networks, SIAM J. Numer. Anal. 11(1974), 260-282.

[2] Nonelliptic approximations of a class of partial differential equations with Neumann boundary conditions, Math. Comp. 30 (1976), No. 133, 68-91.

Goering, H., A. Felgenhauer, G. Lube, H.-G. Roos and L. Tobiska

[1] Singularly perturbed differential equations, Math. Research, vol. 13, Akademie-Verlag, Berlin 1983.

Gosman, A.D., W.M. Pun, A.K. Runchal, D.B. Spalding and M. Wolfshtein

[1] Heat and mass transfer in recirculating flows, Acad. Press, London, New York 1969.

Greenfield, J.A., and R.W. Dutton

[1] Nonplanar VLSI device analysis using the solution of Poisson's equation, IEEE Trans. Electron Devices 27 (1980), No. 8, 1520-1532.

Grisvard, P.

[1] Behaviour of the solutions of an elliptic boundary value problem in a polygonal or polyhedral domain, in 'Num. sol. of part. diff. equat.' - III - Synspade 1975 (B. Hubbard ed.), pp. 207-274.

Hageman, L.A., and D.M. Young

[1] Applied iterative methods, Acad. Press, New York 1981.

Hartwig, K.-H., und W. Weinelt

[1] Methode der wechselnden Dreieckszerlegungen zur iterativen Auflösung elliptischer Differenzengleichungen auf unregelmäßigen Netzen, Wiss. Informationen der Sektion Mathematik Nr. 42 (1983), TH Karl-Marx-Stadt.

[2] Numerische Lösung linearer elliptischer Variationsungleichungen mittels FDM, Wiss. Schriftenreihe der TH Karl-Marx-Stadt, Nr. 8, 1985.

Heimeier, H.

[1] Zweidimensionale numerische Lösung eines nichtlinearen Randwertproblems am Beispiel des Transistors im stationären Zustand, Dissert., TH Aachen 1973.

Heinrich, B.

[1] Beiträge zum Differenzenverfahren für elliptische Differentialgleichungen mit gemischten Randbedingungen, Dissert. A, TH Karl-Marx-Stadt 1976.

[2] Monotone Differenzenapproximationen für lineare elliptische Differentialgleichungen mit gemischten Randbedingungen, Beiträge Numer. Math. 7 (1979), 49-64.

[3] Zur Fehlerabschätzung beim Differenzenverfahren für elliptische Differentialgleichungen nach Gerschgorin-Batschelet, Beiträge Numer. Math. 9 (1981), 73-85.

[4] The finite-difference method for a mildly nonlinear elliptic problem, Abhandl. der AdW DDR, Abt. Math.-Naturw.-Technik, Jahrg. 1981, Nr. 2 N, 119-128.

[5] Integralbilanzmethode für elliptische Probleme I: Konstruktion von Differenzenapproximationen, Beiträge Numer. Math. 11 (1983), 41-53.

[6] Integralbilanzmethode für elliptische Probleme II: Fehlerabschätzung und Konvergenz, Beiträge Numer. Math. 12 (1984), 75-94.

[7] Verallgemeinerte Differenzenverfahren für Klassen elliptischer Differentialgleichungen auf unregelmäßigen Netzen, Forschungsbericht (322 S.), Sektion Mathematik der TH Karl-Marx-Stadt 1983.

[8] Verallgemeinerte Differenzenverfahren für elliptische Differentialgleichungen auf unregelmäßigen Netzen, Dissert. B, TH Karl-Marx-Stadt 1984.

[9] Diskrete Analoga für Ungleichungen vom Friedrichs-Poincaré-Typ, Wiss. Informationen der Sektion Mathematik Nr. 45 (1984), TH Karl-Marx-Stadt.

[10] Differenzenverfahren für ein elliptisches Problem mit schiefer Richtungsableitung in der Randbedingung, Wiss. Zeitschr. der TH Karl-Marx-Stadt 26 (1984), Nr. 6, 893-901.

Heinrich, B. und J. Förster

[1] Differenzenverfahren für elliptische Probleme: Konstruktion unregelmäßiger Netze, Wiss. Zeitschr. der TH Karl-Marx-Stadt 23 (1981) Nr. 4, 395-403.

Höhn, W. und W. Törnig

[1] Ein Maximum-Minimum-Prinzip für Lösungen von finite-Element-Approximationen für quasilineare elliptische Randwertprobleme, Computing 18 (1977), No. 3, 267-270.

Ikeda, T.

[1] Maximum principle in finite element models for convection-diffusion phenomena, Lecture Notes in Numerical and Applied Analysis, vol. 4, North-Holland Publ. Comp., Amsterdam, New York, Oxford 1983.

Kondrat'ev, V.A., and O.A. Oleinik

[1] Boundary value problems for partial differential equations in non-smooth domains, (Russian) Usp. Mat. Nauk 38 (1983), No. 2 (230), 3-76.

Korneev, V.G.

[1] Finite element schemes with accuracy of higher order, (Russian) Izdatel'stvo Leningradskogo Univ., Leningrad 1977.

Kufner, A., O. John and S. Fučík

[1] Function spaces, Academia, Prague 1977.

Ladyshenskaya, O.A.

[1] Boundary value problems of mathematical physics, (Russian) Izdatel'stvo Nauka, Moskva 1973.

Ladyshenskaya, O.A., and N.N. Ural'tseva

[1] Linear and quasilinear equations of elliptic type, (Russian) Izdatel'stvo Nauka, Moskva 1973.

Langer, U., und M. Jung

[1] Zur numerischen Bestimmung der Konstanten in den Ungleichungen von Friedrichs und Poincaré, Wiss. Zeitschr. der TH Karl-Marx-Stadt 25 (1983), Nr. 3, 368-375.

Lazarov, R.D.

[1] On the question of convergence of finite difference schemes for generalized solutions of Poisson's equation, (Russian) Differ. Uravn. 17 (1981), No. 7, 1285-1294.

[2] On the convergence of finite difference solutions to generalized solutions of the biharmonic equation in a rectangle, (Russian) Differ. Uravn. 17 (1981), No. 7, 1295-1303.

Lazarov, R.D., V.L. Makarov and W. Weinelt

[1] On the convergence of difference schemes for the approximation of solutions $u \in W_2^m$ $(m > 0.5)$ of elliptic equations with mixed derivatives, Numer. Math. 44 (1984), 223-232.

Lazarov, R.D., V.L. Makarov and A.A. Samarskiĭ

[1] Finite difference methods for differential equations with generalized solutions, (Russian) Izdatel'stvo Nauka, Moskva 1986 (to appear).

Lorenz, J.

[1] Zur Inversmonotonie diskreter Probleme, Numer. Math. 27 (1977), Fasc. 2, 227-238.

Mc Neal, R.H.

[1] An asymmetrical finite difference network, Quart. Appl. Math. 11 (1953), 295-310.

Meyer, A.G.

[1] Schranken für die Lösung von Randwertaufgaben mit elliptischer Differentialgleichung, Arch. Rat. Mech. Anal. 6 (1960), 277-298.

Michlin, S.G., und Ch. L. Smolizki

[1] Näherungsmethoden zur Lösung von Differential- und Integralgleichungen, B.G. Teubner-Verlagsgesellschaft, Leipzig 1969.

Michlin, S.G.

[1] Approximation auf dem kubischen Gitter, Akademie-Verlag, Berlin 1976.

[2] Konstanten in einigen Ungleichungen der Analysis, Teubner-Texte zur Mathematik, Bd. 35, B.G. Teubner-Verlagsgesellschaft, Leipzig 1981 .

Mitchell, A.R., and D.F. Griffiths

[1] The finite difference method in partial differential equations, J. Wiley & Sons, Chichester, New York, Brisbane, Toronto 1980.

[2] Upwinding by Petrov-Galerkin methods in convection-diffusion problems, J. Comp. Appl. Math. 6 (1980), No. 3, 219-227.

Nävert, U.

[1] A finite element method for convection-diffusion problems, Dissert., Dep. of comp. sc., Chalmers Univ. of Techn., Göteborg 1982.

Nečas, J.

[1] Les méthodes directes en théorie des équations elliptiques, Academia, Prague 1967.

Oden, J.T., and J.N. Reddy

[1] An introduction to the mathematical theory of finite elements, J. Wiley & Sons, New York, London, Sydney, Toronto 1976.

Oganesyan, L.A., and L.A. Rukhovets

[1] On variational-difference schemes for linear elliptic equations of second order in a two-dimensional domain with piecewise smooth boundary, (Russian) Zh. Vychisl. Mat. Mat. Fiz. 8 (1968), No. 1, 97-114.

[2] Variational-difference methods for solving elliptic equations, (Russian) Izdatel'stvo Akad. Nauk Arm. SSR, Jerevan 1979.

Oganesyan, L.A., V.J. Rivkind and L.A. Rukhovets

[1] Variational-difference methods for solving elliptic equations, (Russian) Trud. Sem. 'Differ. Uravn. i ikh Primen.' 1973, part I, No. 5, pp. 1-394, and 1974, part II, No. 8, pp. 1-319, Akad. Nauk Lit. SSR, Vil'njus 1973, 1974.

Ortega, J.M., and W.C. Rheinboldt

[1] Iterative solution of nonlinear equations in several variables, Acad. Press, New York, London 1970.

Pohl, A.

[1] Beitrag zur numerischen Berechnung elektromagnetischer Energiewandler, Dissert., TH Karl-Marx-Stadt 1983.

Reichert, K.

[1] Über ein Verfahren zur numerischen Berechnung von Magnetfeldern und Wirbelströmen in elektrischen Maschinen, Habilitationsschrift, TH Stuttgart, 1968.

Reid, J.K.

[1] On the construction and convergence of a finite element solution of Laplace's equation, J. Inst. Math. Appl. 9 (1972), 1-13.

Reissmann, Ch.

[1] Das Bilanzierungsverfahren für finite Elementarbereiche (BFE); Teil I: Die Anwendung des BFE zur numerischen Lösung von Potential- und Bipotentialproblemen, Schiffbauforschung 14 (1975), No. 1/2, 1-12; Teil II: Die Anwendung des BFE zur numerischen Integration der Laméschen Verschiebungsgleichungen bei ebenen Spannungs- und Verformungszuständen, Schiffbauforschung 14 (1975), No. 3/4, 139-148, Univ. Rostock.

[2] Zum Stand und zur Weiterentwicklung des Bilanzierungsverfahrens für finite Elementarbereiche, Wiss. Zeitschr. Univ. Rostock 24 (1975), Math.-nat. Reihe, Heft 9, 1135-1144.

[3] Über die näherungsweise Berechnung von Feldproblemen durch elementweise Bilanzierung physikalischer Mengen, Tagungsband A der Tagung Festkörpermechanik, Festigkeitslehre und Materialverhalten, TU Dresden, 25. - 28. Mai 1976 (26 Seiten).

[4] Berechnung der Kerbspannungen in tordierten Wellen, Maschinenbautechnik 26 (1977), No. 4, 160-163.

Reissmann, Ch., und W. Haug

[1] Die finiten Bilanzgleichungen der Platte mit beliebigen Randbedingungen und ihre praktische Anwendung, Wiss. Zeitschr. Univ. Rostock 29 (1980), Math.-nat. Reihe, Heft 7, 51-57.

Rice, J.R., and R.F. Boisvert

[1] Solving elliptic problems using ELLPACK, Springer Series in Comp. Math. 2, Springer-Verlag, New York, Berlin, Heidelberg, Tokyo 1985.

Ruas Santos, V.

[1] On the strong maximum principle for some piecewise linear finite element approximate problems of non-positive type, J. Fac. Sci. Univ. Tokyo 29 (1982), No. 2, 473-491.

Samarskij, A.A.

[1] Theorie der Differenzenverfahren, Akad. Verlagsgesellschaft Geest & Portig, Leipzig 1984.

Samarskiĭ, A.A., and V.B. Andreev

[1] Finite difference methods for elliptic equations, (Russian) Izdatel'stvo Nauka, Moskva 1976.

Samarskiĭ, A.A., and E.S. Nikolaev

[1] Methods for solving net equations, (Russian) Izdatel'stvo Nauka, Moskva 1978.

Samarskiĭ, A.A., V.F. Tishkin, A.P. Favorskiĭ, J.M. Shashkov

[1] Operator finite difference schemes, (Russian) Preprint No. 9 (1981), Inst. Prikl. Mat. im. Keldyša, Akad. Nauk SSSR, Moskva.

Schatz, A.H., and L.B. Wahlbin

[1] Maximum norm estimates in the finite element method on plane polygonal domains, part 1, Math. Comp. 32 (1978), No. 141, 73-109; part 2, Math. Comp. 33 (1979), No. 146, 465-492.

Schwarz, H.R., H. Rutishauser und E.Stiefel

[1] Numerik symmetrischer Matrizen, BSB B.G. Teubner-Verlagsgesellschaft, Leipzig 1968, Lizenzausgabe von B.G.Teubner, Stuttgart 1968.

Selberherr, S.

[1] Analysis and simulation of semiconductor devices, Springer-Verlag Wien 1984.

Selberherr, S., A. Schütz und H.W. Pötzl

[1] MINIMOS - a two-dimensional MOS transistor analyzer, IEEE Trans. Electron Devices 27 (1980), No. 8, 1540-1549.

Slotboom, J.W.

[1] Computer-aided two-dimensional analysis of bipolar transistors, IEEE Trans. Electron Devices 20 (1973), No. 8, 669-679.

Smirnow, W.I.

[1] Lehrgang der höheren Mathematik, Teil V, VEB Deutscher Verlag der Wissenschaften, Berlin 1976.

Stepleman, R.S.

[1] Finite-dimensional analogues of variational problems in the plane, SIAM J. Numer. Anal. 8 (1971), 11-23.

Strang, G.

[1] Approximation in the finite element method, Numer. Math. 19 (1972), Heft 1, 81-98.

Strang, G., and G.J. Fix

[1] An analysis of the finite element method, Prentice-Hall Inc., Englewood Cliffs, New Jersey, 1973.

Thomasset, F.

[1] Implementation of finite element methods for Navier-Stokes equations, Springer-Verlag, New York, Heidelberg, Berlin 1981.

Temam, R.

[1] Numerical Analysis, D. Reidel Publ. Comp., Dordrecht, Boston 1973.

[2] Navier-Stokes Equations, Theory and Numerical Analysis, North-Holland Publ. Comp., Amsterdam, New York, Oxford 1979.

Törnig, W.

[1] Monotonie- und Randmaximumsätze bei Diskretisierungen des Dirichletproblems allgemeiner nichtlinearer elliptischer Differentialgleichungen, Computing 11 (1973), 391-401.

Vainikko, G.

[1] Projektionsmethoden, Spezialvorlesung, (72 S.), Sektion Mathematik TH Karl-Marx-Stadt 1973.

Varga, R.S.

[1] Matrix iterative analysis, Prentice-Hall Inc., Englewood Cliffs, New Jersey 1962.

Weiland, Th.

[1] Zur Berechnung der Wirbelströme in beliebig geformten, lamellierten dreidimensionalen Eisenkörpern, Teil I: Die Methode, Archiv für Elektrotechnik 60 (1978), 345-351.

Weinelt, W.

[1] Untersuchungen zur Konvergenzgeschwindigkeit bei Differenzenverfahren, Wiss. Zeitschr. TH Karl-Marx-Stadt 20 (1978), Nr. 6, 763-769.

[2] Untersuchungen zum Differenzenverfahren für lineare elliptische Variationsungleichungen, Dissert. B, TH Karl-Marx-Stadt 1982.

Wigley, N.M.

[1] Mixed boundary value problems in plane domains with corners, Math. Zeitschr. 115 (1970), 33-52.

Winslow, A.M.

[1] Numerical solution of the quasilinear Poisson equation in a nonuniform triangle mesh, Journ. of Comput. Physics 2 (1967), 149-172.

Wloka, J.

[1] Partielle Differentialgleichungen, BSB B.G. Teubner-Verlagsgesellschaft, Leipzig 1982, Lizenzausgabe von B.G.Teubner, Stuttgart 1982.

Wolff, W., und W. Müller

[1] Allgemeine Lösung der magnetostatischen Gleichungen, Wiss. Berichte AEG-Telefunken 49 (1976), Nr. 3, 77-86.

Zeidler, E.

[1] Vorlesungen über nichtlineare Funktionalanalysis, Teil II, BSB B.G.Teubner-Verlagsgesellschaft, Leipzig 1977.

Ženišek, A.

[1] Discrete forms of Friedrichs' inequalities in the finite element method, R.A.I.R.O. Numerical analysis 15 (1981), No. 3, 265-286.

Zlámal, M.

[1] Curved elements in the finite element method, I, SIAM J. Numer. Anal. 10 (1973), Nr. 1, 229-240.

[2] Lectures on the finite element method, Vorlesungsskripte (52 S.), Sektion Mathematik TH Karl-Marx-Stadt, 1974.

SUBJECT INDEX

LIST OF FIGURES

ABBREVIATIONS

BVP	boundary value problem
FDM	finite difference method
FEM	finite element method
FDS	finite difference scheme
PB, PB- ...	perpendicular bisector ... (in compounds)
MD, MD- ...	median ... (in compounds)
V ..., V(...)	assumption
a.e.	almost everywhere

NOTATIONS

$Au = F$	boundary value problem
$Lu = f$	differential equation
$lu = g$	boundary condition
Ω	bounded domain, $\bar{\Omega}$: closure of Ω
$\Gamma := \partial\Omega$	boundary of Ω
Γ_i $(i=1,2,3)$	part of the boundary Γ, $\Gamma = \bigcup_{i=1}^{3} \Gamma_i$, with boundary conditions of i-th kind
h	mesh size parameter
$A_h y = F_h$	finite difference scheme approximating $Au = F$
$L_h y = f_h$, $l_h y = g_h$	finite difference analogues approximating $Lu = f$, $lu = g$
Ω_h	polygonal domain approximating Ω
$\Gamma_h := \partial\Omega_h$	boundary of Ω_h
$\bar{\omega} := \omega + \gamma$	$\bar{\omega}$: grid associated with $\bar{\Omega}$, ω: set of grid points in Ω and Ω_h, γ: set of grid points on Γ and Γ_h
$\hbar(x, \xi)$	straight mesh segment of the primary network
$\mathcal{S}(x, \xi)$	mesh segment of the secondary network
$x' := x'_\xi$	midpoint of the mesh segment $\hbar(x, \xi)$
$\bar{\omega}'$	set of the midpoints of all mesh segments $\hbar(x, \xi)$
$\varkappa(x)$	box associated with the grid point x
$H(x)$	measure of the box $\varkappa(x)$
$e = \Delta, \mathit{\Delta}, \Box$	finite elements e (straight and curved triangles, rectangles)
x''	centre of gravity of e or an index
$\bar{\omega}''$	set of the centres of gravity of all straight elements
$\mathcal{T}_{\mathit{\Delta}}, \mathcal{T}_\Delta$	primary triangulations (primary networks)
$\mathcal{T}_\varkappa, \mathcal{T}_{\tilde{\varkappa}}$	secondary triangulations (secondary networks)
$u_\hbar(x'_\xi)$, $\nabla_\hbar u(x'')$, $\frac{\delta u}{\delta x_i}$	difference analogues of $\frac{\partial u}{\partial \hbar}$ on $\hbar(x,\xi)$, ∇u on $e(x'')$ and $\frac{\partial u}{\partial x_i}$, respectively
$z := y - u$	local approximation error of the approximate solution y
ψ	local approximation error of A_h and F_h
$\varkappa, \gamma_H$	error functionals (principal part of A_h)
γ_N	error functional which corresponds to the secondary part of A_h and to F_h
C^k, PC^k	k times continuously or piecewise continuously differentiable functions or boundaries

For further notations, especially for Sobolev spaces, cf. APPENDICES.

Band 1

Inverse and Improperly Posed Problems in Differential Equations

Proceedings of the Conference on Mathematical and Numerical Methods, held in Halle Saale (GDR) from May 29 to June 2, 1979

Edited by Gottfried Anger
296 Seiten, 1979 (vergriffen)

Band 2

Fundamentals of Computation Theory FCT '79

Proceedings of the Conference on Algebraic, Arithmetic and Categorial Methods in Computation Theory, held in Berlin/Wendisch-Rietz (GDR) September 17–21, 1979

Edited by Lothar Budach
575 Seiten, 1979

Band 3

Decidability and Generalized Quantifiers

By Andreas Baudisch, Detlef Seese, Hans-Peter Tuschik, and Martin Weese
XII, 235 Seiten, 1980

Band 4

Zur Elastizitäts- und Thermoelastizitätstheorie

Potentialtheoretische Methoden in Anwendung auf verschiedene Modelle

Von Lothar Jentsch und Johannes Maul
157 Seiten, 1980

Band 5

Systems Analysis and Simulation 1980

Proceedings of the International Symposium held at Berlin (GDR) September 1–5, 1980

Edited by Achim Sydow
496 Seiten, 1980 (vergriffen)

Band 6

Nichtlineare diskrete Optimierung

von Eberhard Girlich und Michail M. Kowaljow
218 Seiten, 1981

Band 7

Completeness Problems in the Structural Theory of Automata

By Jürgen Dassow
148 Seiten, 1981

Band 8

Darstellung von Lösungen linearer elliptischer Differentialgleichungen

Von Günther Wildenhain
92 Seiten, 1981

Band 9

Introduction to the Spectral Theory of Operators in Spaces with an Indefinite Metric

By I. S. Iohvidov, M. G. Krein, and H. Langer
120 Seiten, 1982 (vergriffen)

Band 10

Nonlinear Time-discrete Systems

A General Approach by Nonlinear Superposition
By Michael Gössel
115 Seiten, 1982 (vergriffen)

Band 11

Preprimal Algebras

By Klaus Denecke
162 Seiten, 1982

Band 12

Ergodic Theory and Related Topics

Proceedings of the Conference held in Vitte/Hiddensee (GDR) October 19–23, 1981
Edited by Horst Michel
232 Seiten, 1982

Band 13

Singularly Perturbed Differential Equations

By H. Goering, A. Felgenhauer, G. Lube, H.-G. Roos, and L. Tobiska
176 Seiten, 1983

Band 14

Leben wir in einer Volterra-Welt?

Ein ökologischer Zugang zur angewandten Systemanalyse
Von Manfred Peschel und Werner Mende
233 Seiten, 1983

Band 15

Markov Decision Problems with Countable State Spaces

Optimality Criteria – Algorithms – Clustering

By Hans Michael Dietz and Volker Nollau
174 Seiten, 1983

Band 16

Stochastische Verfahren zur Suche nach dem Minimum einer Funktion

Von Ryszard Zieliński und Peter Neumann
133 Seiten, 1983

Band 17

Invertibility and Asymptotics of Toeplitz Matrices

By Albrecht Böttcher and Bernd Silbermann
200 Seiten, 1983

Band 18

Structural Induction on Partial Algebras

Introduction to Theory and Application of Partial Algebras – Part II
By Horst Reichel
205 pages, 1984

Band 19

Algebraic Methods for Toeplitz-like Matrices and Operators

By Georg Heinig and Karla Rost
212 pages, 1984

Band 20

Frege Conference 1984

Proceedings of the International Conference held at Schwerin (GDR), September 10–14, 1984

Edited by Gerd Wechsung

408 pages, 1984

Band 21

Numerical Analysis of Selected Semilinear Differential Equations

Edited by Christian Großmann, Thomas Riedrich, and Hartmut Schönheinz

203 pages, 1984

Band 22

Singular Ordinary Differential Operators and Pseudodifferential Equations

By Johannes Elschner

200 pages, 1985

Band 23

Lotka-Volterra-Approach to Cooperation and Competition in Dynamic Systems

Proceedings of the 5th Meeting of UNESCO's Working Group on Systems Theory held on the Wartburg, Eisenach (GDR) at March 5–9, 1984

Edited by Werner Ebeling and Manfred Peschel

271 pages, 1985

Band 24

Convergence Structures 1984

Proceedings of the Conference on Convergence held in Bechyně (Czechoslovakia); September 24–28, 1984

Edited by Josef Novák, Werner Gähler, Horst Herrlich, and Jan Mikusiński

254 pages, 1985

Band 25

Parcella '84

Proceedings of the II. International Workshop on Parallel Processing by Cellular Automata and Arrays held in Berlin (GDR), September 25–27, 1984

Edited by Wolfgang Händler, Tamás Legendi, and Gottfried Wolf

239 pages, 1985

Band 26

Multiobjective and Stochastic Optimization Based on Parametric Optimization

By Jürgen Guddat, Francisco Guerra Vasquez, Klaus Tammer, and Klaus Wendler

175 pages, 1985

Band 27/28

Systems Analysis and Simulation 1985
I: Theory and Foundations
II: Applications

Proceedings of the International Symposium held in Berlin (GDR), August 26–31, 1985

Edited by Achim Sydow, Manfred Thoma, and Robert Vichnevetsky

per volume 495 pages, 1985

Band 29

Parcella '86

Proceedings of the III. International Workshop on Parallel Processing by Cellular Automata and Arrays held in Berlin (GDR), September 9–11, 1986

Edited by Tamás Legendi, Dennis Parkinson, Roland Vollmar, and Gottfried Wolf

about 272 pages, 1986

Band 30

Fuzzy Sets Applications, Methodological Approaches, and Results

Proceedings of the International Workshop on Fuzzy Sets Applications cosponsored by IIASA and convened in cooperation with IAEA held on the Wartburg, Eisenach (GDR), at March 3–8, 1985

Edited by Steffen Bocklisch, Sergei Orlovski, Manfred Peschel, and Yasushi Nishiwaki

233 pages, 1986

Band 31

Mathematical Methods of Specification and Synthesis of Software Systems '85

Proceedings of the International Spring School, held in Wendisch-Rietz (GDR), April 22–26, 1985

Edited by Wolfgang Bibel and Klaus P. Jantke

245 pages, 1986

Band 32

A Model Theoretic Oriented Approach to Partial Algebras

Introduction to Theory and Application of Partial Algebras – Part I

By Peter Burmeister

319 pages, 1986

Band 33

Finite-difference Methods on Irregular Networks

A Generalized Approach to Elliptic Problems of Second Order

By Bernd Heinrich

about 200 pages, 1986